AF411339

Untargeted Alternative Routes of Arbovirus Transmission

Untargeted Alternative Routes of Arbovirus Transmission

Editor

Julien Pompon

MDPI • Basel • Beijing • Wuhan • Barcelona • Belgrade • Manchester • Tokyo • Cluj • Tianjin

Editor
Julien Pompon
MIVEGEC
France

Editorial Office
MDPI
St. Alban-Anlage 66
4052 Basel, Switzerland

This is a reprint of articles from the Special Issue published online in the open access journal *Pathogens* (ISSN 2076-0817) (available at: https://www.mdpi.com/journal/pathogens/special_issues/Untargeted_Alternative_Routes_Arbovirus_Transmission).

For citation purposes, cite each article independently as indicated on the article page online and as indicated below:

LastName, A.A.; LastName, B.B.; LastName, C.C. Article Title. *Journal Name* **Year**, *Volume Number*, Page Range.

ISBN 978-3-03943-767-2 (Hbk)
ISBN 978-3-03943-768-9 (PDF)

Contents

About the Editor

Julien Pompon completed his Ph.D. at the University of New Brunswick (Canada) on insect physiology in 2010. Dr Pompon then conducted a four-year postdoctoral appointment where he researched mosquito immunity at the Institute for Molecular and Cell Biology (IBMC) in Strasbourg (France), followed by a two-year postdoctoral position where he examined arbovirus transmission at Duke-NUS medical school (Singapore). Dr Pompon was appointed as research track Assistant Professor at Duke-NUS and co-led a team on arbovirus transmission until 2019. He then relocated to IRD in Montpellier (France), where he leads a group studying interhost arbovirus transmission. Dr Pompon's current research focuses on the molecular mechanisms of arbovirus transmission.

Preface to "Untargeted Alternative Routes of Arbovirus Transmission"

Arboviruses have emerged as global pathogens of significant importance in the past 50 years. Globalization, unplanned urbanization, and climate change have all contributed to their developments. The main culprits that transmit these viruses are mosquitoes, mostly the species Aedes aegypti and Aedes albopictus. These mosquitoes thrive in man-made environments and have been dispersed throughout the world by global transport. Their ecological acquaintance with and proclivity to bite humans together with their susceptibility for arbovirus replication make them ideal vectors. Consequently, Ae. aegypti is considered the primary vector for flaviviruses such as dengue, Zika, and yellow fever viruses, and alphaviruses such as chikungunya and Mayaro viruses. All these viruses cause flu-like symptoms that can develop into life-threatening (hemorrhage for flaviviruses) or life-debilitating (arthralgia for alphaviruses) conditions. Because of the vectors' worldwide distribution, close to half of the world population is at risk of being infected. Dengue virus (DENV) is the most widespread arbovirus and infect an estimated 400 million people per year, while Zika virus (ZIKV) became famous after spreading like wildfire in South America, where close to 70% of the population was infected in just 4 years. Based on this recent history, it is expected that other arboviruses will emerge as global threats.

This alarming situation has stimulated research to develop control methods. The difficulties in developing an effective vaccine (for example, the last DENV vaccine developed by Sanofi Pasteur) and therapeutic treatments favor interventions that target the vector. Tools using transgenic, radiation, or microbiome modifications of vector populations have been developed. The most promising tools make use of the bacterium Wolbachia to sterilize progeny and suppress populations or to reduce virus replication and block transmission. Both strategies are being tested in field trials with interesting results. The technologies can be deployed and restrict transmission at least in a limited area. However, these vector-based interventions mainly target the main vector, Ae. aegypti, making room for so-called "secondary" or alternative routes of transmission to become preponderant.

With regards to the potential for these new tools to significantly reduce the global burden of arboviruses, we thought it was time to gather information about the untargeted alternative routes of arbovirus transmission. These untargeted routes may become significant in arbovirus epidemiology. This Special Issue aims to present knowledge about potential substitutive transmission routes. This information will help in devising long-term control measures to sustain low arbovirus transmission.

Julien Pompon
Editor

 *pathogens*

Opinion

From Anonymous to Public Enemy: How Does a Mosquito Become a Feared Arbovirus Vector?

Didier Fontenille [1,*] and Jeffrey R. Powell [2]

[1] MIVEGEC unit, Université de Montpellier, Institut de Recherche pour le Développement (IRD), CNRS, BP 64501, 34394 Montpellier, France

[2] Department of Ecology and Evolutionary Biology, Yale University, 21 Sachem Street, New Haven, CT 06511-8934, USA; jeffrey.powell@yale.edu

* Correspondence: didier.fontenille@ird.fr

Received: 6 March 2020; Accepted: 2 April 2020; Published: 5 April 2020

Abstract: The past few decades have seen the emergence of several worldwide arbovirus epidemics (chikungunya, Zika), the expansion or recrudescence of historical arboviruses (dengue, yellow fever), and the modification of the distribution area of major vector mosquitoes such as *Aedes aegypti* and *Ae. albopictus*, raising questions about the risk of appearance of new vectors and new epidemics. In this opinion piece, we review the factors that led to the emergence of yellow fever in the Americas, define the conditions for a mosquito to become a vector, analyse the recent example of the new status of *Aedes albopictus* from neglected mosquito to major vector, and propose some scenarios for the future.

Keywords: mosquito; culicidae; *Aedes aegypti*; *Aedes albopictus*; emergence; arbovirus

1. Introduction

In a time of major social, climatic and environmental changes, several old concepts are back in fashion: "health is one" (one health approach), "the microbe is nothing, the context is everything" (Antoine Béchamp, Louis Pasteur), "diseases will always continue to emerge" [1]. All these old, but still very relevant views require a holistic approach, taking into account the complexity of interactions between diseases, microbes, hosts, vectors, environment, and their evolution as described by Mirko Grmek [2], under the term pathocenosis. Emergence of SARS-Cov2 viruses (Covid-19 disease) in 2019, dramatically confirms these predictions.

It is generally accepted that, other than our own species, the "most dangerous animal in the world" for humankind is the mosquito. This figure of speech, however forceful it may be, is misleading. Yes, there are mosquitoes responsible for the transmission of major human pathologies (malaria, filariasis, yellow fever, dengue, Zika, chikungunya, haemorrhagic fevers, encephalitis, etc.), and it is likely that most blood-sucking mosquitoes transmit some pathogens to some vertebrates. However, given that only a small minority of the thousands of mosquito species account for the vast majority of human diseases, "the mosquito" should be afforded the principal of *"habeas corpus"*, innocent until proven guilty.

In this opinion piece, we discuss the factors that lead to the emergence of a vector in the human environment causing the transmission of viruses pathogenic to humans. A related, but distinct phenomenon which we do not discuss, is the establishment of newly introduced pathogenic viruses in sylvatic vectors, leading to new cycles of enzootic transmission.

2. Vector or not yet Vector: Guilty or Presumed Innocent

Two species of mosquitoes of particular relevance to human public health, *Aedes aegypti* and *Aedes albopictus*, are good models for understanding, estimating and, if possible, anticipating and limiting the risk of emergence, establishment and transmission of pathogens to humans by vectors [3,4]. *Aedes aegypti* was suspected to be a vector of the yellow fever virus by Beauperthuy and later Finlay since the middle of the 19th century [5,6], and experimentally demonstrated as a vector in 1900 by Reed et al. [7], and therefore an enemy to be killed [8,9]. In contrast, *Ae. albopictus* became widely known only over the last 40 years when it was gradually detected on all continents (except Antarctica) outside its native region of Southeast Asia and implicated in the recent chikungunya pandemic [10,11]. These two species (*aegypti* and *albopictus*) of genus *Aedes* and subgenus *Stegomyia* are responsible for virtually all major outbreaks of diseases caused by the major primate arboviruses: dengue, Zika, chikungunya, and yellow fever. Both mosquitoes are invasive species, spreading thanks to man-made means of transport, and are extremely well adapted to human environments.

Transmission of viruses to humans by mosquitoes seems to be the rule for many public health officers. Actually, it is an exception. The vast majority of mosquitoes do not transmit human diseases. Of the 3585 species of mosquitoes currently formally described, less than 30 transmit the yellow fever virus, either naturally or in the laboratory. In other words, more than 3500 species of mosquitoes do not transmit the yellow fever virus. They are not guilty. There are many reasons for this: (1) the virus is not present in the geographical area of the vector mosquito species; either by "luck" because it was never introduced, or due to the absence of vertebrate amplifying hosts allowing the maintenance of cycles (primates in the case of yellow fever); (2) the mosquito is not competent to transmit (absence of required receptors or innate immunity controlling the pathogen); (3) the biology of the mosquito is not compatible with transmission (life span too short, does not take blood meals from primates, etc.).

However, our present understanding should not be taken as dogma. We must keep an open mind and explore what factors can make a non-vector mosquito become a vector, or what can cause a so-called "secondary" vector to become a major vector for humans. To continue our metaphor, what turns an "innocent" mosquito into a "guilty" menace; then into an effective vector to humans? In theory, it is very simple. Only two conditions are necessary: (1) the mosquito is physiologically capable of sustaining a virus infection that spreads to saliva and takes multiple blood meals from primates, i.e., was always a potential vector, (2) an event or several events occur in its environment that allow it to express this vector potential. Subsequently, adaptive mechanisms, under selection pressure, may cause this mosquito to become responsible for severe outbreaks. However, while this seems simple, in reality, a large number of factors are necessary for the conditions to be met that allow the emergence of a new vector.

3. Short History of Yellow Fever: Once Upon a Time in America

The yellow fever virus circulates in African forests from monkey to monkey, transmitted by forest *Aedes* (*Ae. africanus*, *Ae. luteocephalus*, *Ae. furcifer*, *Ae. simpsoni* s.l., *Ae. opok*, etc.), with incursions into villages where domestic anthropophilic *Ae. aegypti* are established, and able to replicate and transmit YF virus [12]. These domestic *Ae. aegypti* are themselves the result of a slow adaptation of forest ancestors to human habitats and blood [13]. The virus, once in humans, following cross species transmission, can be introduced through the movement of humans into African cities, where it finds *Ae. aegypti* populations adapted to urbanisation and able to generate outbreaks.

This village or urban yellow fever virus, already adapted to human and *Ae. aegypti*, was introduced into tropical America, probably via viremic people and/or *Ae. aegypti* transported on ships, during the triangular slave trade, which started in the 16th century [14,15]. Upon arrival to the Central and South American rainforest, not only did the virus find susceptible New World monkeys but also new competent endemic mosquitoes from two genera, *Sabethes* sp. and especially *Haemagogus* sp. Yellow fever virus thus had a second phase of cross species transmission (African monkeys to American monkeys, African Aedes mosquitoes to American mosquitoes).

The evolutionary history of *Haemagogus* genus mosquitoes, which are only found in the New World, is not well known. Phylogenetically, they seem to be quite close to the *Aedes* (Figure 1), which may explain their ability to transmit, more or less, the same viruses as *Aedes aegypti*: yellow fever, dengue [16], Zika [17], and chikungunya [18], all introduced into South America. New cycles of jungle yellow fever have thus developed, involving South American primates and tree-dwelling *Haemagogus* mosquitoes. These cycles have spread throughout South and Central America, where they are still active, as shown by the 2016–2018 yellow fever epidemic in Brazil, with more than 700 deaths [19]. This combination of both a sylvan and an urban cycle for yellow fever in the Americas regularly fuelled epidemics that greatly affected the history of colonization of the New World [20]. As *Ae. aegypti* adapted to American cities, even non-tropical locations were subject to yellow fever epidemics, such as the well-known Philadelphia epidemic in 1793 (5000 deaths). A largely successful eradication program against *Ae. aegypti* began initially in the 1930s and was started by the Rockefeller Foundation [21], and this continued with DDT after World War II. At the same time, wide vaccination coverage was undertaken. This led to almost no outbreaks of yellow fever due to *Ae. aegypti* in the Americas, since about 1970, although *Ae. aegypti* is again abundant [22], and able to transmit dengue fever virus and the recently introduced Zika virus. It is thought that the Brazil 2016-18 yellow fever outbreak was almost entirely due to spill over from the sylvan cycle and involved *Haemagogus* and *Sabethes* mosquitoes [23,24].

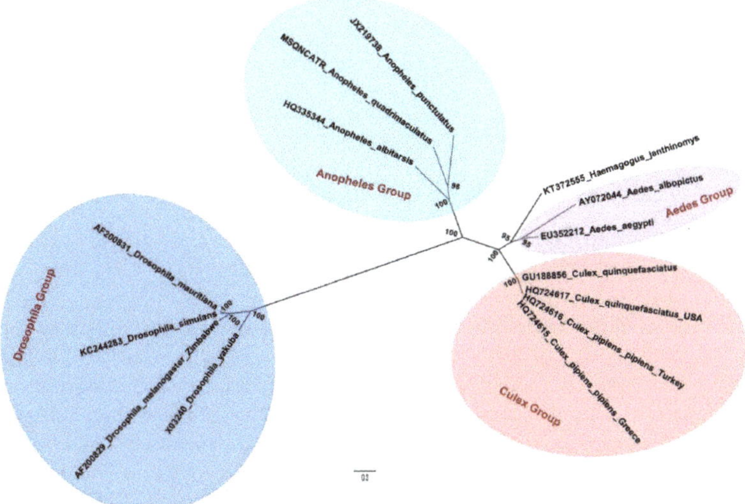

Figure 1. Phylogenetic tree built by the method of maximumlLikelihood, from sequences of mitochondrial genes COX1, COX2, NAD4, NAD5 and CYOB of *Haemagogus janthinomys* and other species of Diptera, from da Silva Lemos et al. 2017 [25]. The bootstrap values are represented in each node.

Interestingly, endemic yellow fever has never been reported in Asia, Madagascar or other Indian Ocean islands, for reasons that are still poorly understood. Primates (monkeys and humans) from these regions are susceptible, and *Ae. aegypti* and other experimentally competent *Aedes* mosquitoes are present there [26–28]. Importantly, *Haemagogus* sp. and *Sabethes* sp. are absent from Africa, the Indian Ocean and Asia.

4. The Necessary Conditions

The history of yellow fever leads us to the consideration of the factors necessary, but not always sufficient, for a mosquito species that was initially of little interest to humans, to become a public health problem and an enemy to fight.

First of all, the mosquito, or to be precise, a given population of a given species of mosquito, must be biologically able to transmit the virus. This ability is called vectorial capacity, which includes vector competence (i.e., the ability of a mosquito to become infected after ingestion of an infected blood meal and later transmit the virus via its saliva). Both terms, vectorial capacity and vector competence, have been formalized since MacDonald [29], summarized by Cohuet and coll. [30], and include the following parameters:

- The vector–host ratio (i.e. the vector density in relation to vertebrate host): m (the mosquito abundance);
- The human feeding rate: the number of human bites per mosquito, per day: a (mosquito - human contact);
- The daily survival rate (i.e. the probability of a mosquito surviving each day): p (mosquito longevity);
- The extrinsic development time, the time necessary for viruses to complete development from ingestion in midgut to the saliva: n;
- The infectiousness of the mosquito to the vertebrate host: b (largely dependent on virus titre in saliva)
- The susceptibility of the vertebrate host to the virus (e.g., immune state, age, health, etc): c;
- The vertebrate host infectious period: 1/r (how long the virus titre in the vertebrate remains at a level needed to infect a mosquito);

Knowing the values of these parameters makes it possible to calculate the basic reproductive rate of the virus, R_0.

$$R_0 = (ma^2 \times p^n / - \ln p) \times bc \times 1/r \tag{1}$$

R_0 is the total number of cases derived from one infective case that the mosquito population would distribute to vertebrate hosts. R_0 must equal at least 1 for the disease to persist or spread. For values less than 1, the disease will go extinct.

However, while we can write the simple equation above, in reality the parameters (m, a, p, n, b, c, r) are themselves complex, being dependent on many other factors. For just the mosquito, a non-exhaustive list includes mosquito genotype, mosquito microbiome including viruses, integrated RNA virus sequences in mosquito genomes, predation, competition at larval and adult stage, previous exposure to related viruses, etc. [31,32]. Moreover, it is not yet known which receptors in the cells of the stomach and salivary glands of mosquitoes are involved in the entry and exit of flaviviruses, like yellow fever virus [33].

Just this intrinsic biology of the mosquito is not sufficient to make it a vector. Its environment may, or may not, allow this vector potential to manifest itself. According to Euzet and Combes [33], the specificity of the vector–pathogen interaction passes through four stages, which they named encounter and compatibility filters (Figure 2). These four stages are: (1) to co-occur in space and time, (2) to meet each other (behavior), (3) to recognize each other (receptors), and (4) to accept each other (immunity).

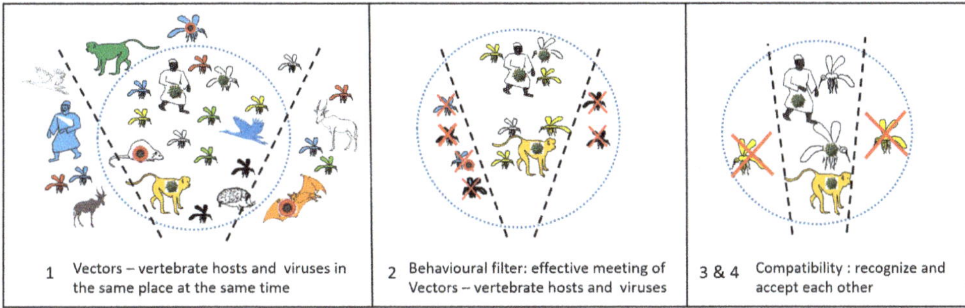

Figure 2. Encounter and compatibility filters, from Euzet and Combes [34].

All these conditions are rarely met, and it is therefore understandable why being a vector is an exception. However, considering the very large number of mosquito species and large number of viruses, while we are presently facing only a limited number of dangerous cycles, this must be a very small fraction of thousands, if not millions, of other potential cycles that have failed. The understanding of the current efficient cycles allows us to conceive possible future cycles.

We can hypothesize that only a few species in the genera *Aedes*, *Haemagogus*, and *Sabethes* have the capacity to replicate and transmit yellow fever virus to primates, i.e. (1) possess the receptors for cellular penetration of the virus, (2) have insufficient innate immunity to suppress virus replication, (3) live in the same environments as viremic vertebrates (monkeys or humans), (4) take blood from these vertebrates, and (5) survive long enough to retransmit the virus.

5. *Aedes albopictus*: From Local to Global Concern

Aedes albopictus became widely known only over the last 50 years, when it was gradually discovered on all continents outside its native Asia and caused a chikungunya pandemic [10,11,35,36]. A total of more than 40 viruses can be transmitted naturally or experimentally by *Ae. albopictus* [36]. Experimentally, it will take blood from many vertebrate species [37]. It also lives longer than *Aedes aegypti* [38]. *Aedes albopictus* therefore perfectly fulfils the necessary conditions in terms of biology and events to be an excellent vector of virus to humans.

Aedes albopictus is considered to be phylogeographically native to Southeast Asia [10]. This assumption is based on the fact that this mosquito is present everywhere in the forested areas of this region and that many species close to the Albopictus subgroup, member of the Scutellaris group, are present in Southeast Asia. However, the notoriety of *Ae. albopictus*, compared to relatives, comes from the fact that it has moved out of its area of origin, becoming worldwide in 50 years, adapting perfectly to urbanization, temperate climates, and international transport, as well as being involved in several dengue and chikungunya epidemics.

In tropical Southeast Asia, the five morphologically closely related species of the Albopictus subgroup live in forests (*Aedes novalbopictus*, *Aedes patriciae*, *Aedes seatoi*, *Aedes subalbopictus* and *Aedes pseudalbopictus*) and lay their eggs in tree holes and bamboo stumps [39,40]. These species are likely to be vectors of arboviruses to vertebrates from which they take blood. These forest-confined viruses are largely unknown, but may emerge through increased contact with humans or domestic animals (see below). They could then be transmitted by domestic vectors, such as *Ae. aegypti* and *Ae. albopictus*, in a pattern similar to the emergence of the yellow fever virus in Africa from forests to villages [12].

In Laos, *Ae. albopictus* is reported from deep natural forest [41], rubber forest and secondary forest [42], as well as in towns and villages. In contrast, in neighbouring countries, such as in Cambodia, this mosquito was found along hundreds of meters of forest edge, whereas *Ae. pseudalbopictus*, is found in deeper natural forests. *Aedes albopictus* is frequently found in cities such as the capital, Phnom Penh (*Boyer pers. com.*). Similarly, in Malaysia, *Ae. albopictus* is rare in forests [43], and in Yunnan Province,

China, *Ae. albopictus* is sometimes more abundant than *Ae. pseudalbopictus* in bamboo forests, but is often absent from deep forests [40].

Outside Asia, *Ae. albopictus* populations display highly variable success in invading the deep forest. In Brazil, where *Ae. albopictus* was observed for the first time in the 1980s, Pereira dos Santos et al. found that *Ae. albopictus* was able to enter degraded forest in the Manaus region, up to 750 metres from the edge [44]. In Gabon, where *Ae. albopictus* was first reported in 2007, it was captured 12 years later in the Lopé natural forest, several kilometres from villages or clearings (*Paupy pers. com.*).

In both its native range Asia and recently invaded South America and Africa, *Ae. albopictus* has retained its ancestral capacity to colonize forest environments, laying eggs in natural pools of water and taking blood from nondomestic vertebrates. It is logical to think that under these conditions, *Ae. albopictus* populations would take blood from vertebrates carrying as yet unknown viruses, such as from monkeys, terrestrial mammals, birds or reptiles. If these forest-breeding *Ae. albopictus* are able to replicate these viruses, and then retransmit them to humans, they would then be excellent bridge vectors, allowing the emergence of forest viruses hitherto confined to sylvatic *Aedes*-vertebrates cycles. We can speculate that this is most likely to occur in Asia. In Africa, *Ae. aegypti* has already brought many, perhaps most, human adapted viruses from the forest to human habitats, such as yellow fever, dengue, chikungunya, and Zika viruses [13].

Mogi et al. [40], suggest that the spreading of *Ae. albopictus* from its original tropical forest region was possible following evolution from an ancestral wild species, due to adaptation to man-made habitats and then migration with humans to temperate climate regions, where *Ae. albopictus* developed a winter diapause. In most introduced localities, it probably encountered only limited competition from native mosquitoes and when it did, *Ae. albopictus* proved to be a robust competitor, e.g., with *Ae. aegypti* [45]. It is likely that the ancestral wild species was already a vector of forest vertebrate arboviruses, for example, dengue-like flaviviruses, or chikungunya-like alphaviruses, which are monkey viruses [46].

From the foregoing, it appears that *Ae. albopictus*, while being ancestrally a forest-breeding mosquito like its close relatives, among these relatives it is the most closely adapted to forest margins (ecotone), the transition from forest to degraded or secondary forests, open grasslands or scrub. *Aedes aegypti* in Africa is similar [47]. This may have pre-adapted these two *Aedes,* among all their congeners in Asia and Africa, to come into contact with human settlements and thus to adapt to this new niche, human settlements.

From regions recently colonized by *Ae. albopictus*, "modern" populations, highly adapted to the urban environment and able to transmit viruses such as chikungunya, Zika and dengue, spread to all continents. It is likely that invasive populations re-invaded regions where ancestral populations already existed, such as in Asia and the Indian Ocean. These ancestral populations then found themselves in unfavourable competition and modern *Ae. albopictus* replaced them [48].

It is not too difficult to predict the epidemiological future of *Ae. albopictus*. Its geographical distribution will increase, especially in temperate regions, its control by insecticides will be more difficult due to the emergence of resistance, and human pathogenic viruses will adapt to this new vector, increasing its efficiency to transmit. *Ae. albopictus*-vectored epidemics are almost certain to increase. It could be responsible for the epidemics of yellow fever, or of currently unknown viruses transmitted by forest edge *Aedes* mosquitoes of South American, South East Asia or Central Africa. A major unknown for *Ae. albopictus'* impact in temperate regions is whether arbovirus replication at lower temperatures is selected to be rapid enough to reach saliva before the female dies.

6. Scenarios for the Future: The Worst Doesn't Always Happen

To quote Charles Nicolle [1], it is certain that new epidemics will appear. Mosquitoes and viruses are poised to ambush humans following changes in vector ecosystems [49]. The process of establishment may be abrupt and rapid, as was the recent Zika pandemic with *Ae. aegypti* as vector, or more gradual, going through adapting mechanisms, as was the case with the emergence of the West Nile virus in several temperate cities, particularly in the USA [50].

The optimistic view, but not very realistic, is that the worst is already behind us. Urban environments, where the majority of the humans now live, are already colonized by *Ae. aegypti*, *Ae albopictus*, and/or *Culex pipiens*, which transmit or can transmit known viruses of tropical or equatorial origin: dengue, chikungunya, Zika, yellow fever, West Nile, as well as, at least potentially, many other tropical viruses like Japanese encephalitis and Rift Valley fever, with variable success depending on species and environments [13]. However, this optimistic view needs to be tempered. With global changes (environment, climate, demography, movements), the distribution areas of some of these human adapted vectors and viruses will continue to expand and we will see a steady increase of epidemics of viruses, including in temperate regions. While we cannot predict details of time and place, we know it will happen, and we can prepare for it.

A more pessimistic view is that the reservoir of unknown forest viruses, not presently affecting humans, but potentially transmissible to domestic animals and humans, is immense in Asia, Central and South America, and Africa. More than 500 arboviruses have so far been identified and described, but this is only the tip of the iceberg. From about 1930–1980, the Rockefeller Foundation sponsored a repository for viruses from field-collected arthropods, primarily mosquitoes, that reached more than 4000 isolates of insect-specific viruses or arboviruses, the vast majority of which are undescribed [51], and even this collection is far from complete. Consequently, several non-exclusive scenarios for the emergence of new human viral epidemics are possible:

(1) Via primate-biting bridge vectors such as *Ae. albopictus* from forest edges in South America, Africa and Asia, forest *Aedes* of the Albopictus group in South East Asia, *Haemagogus* in South America, *Stegomyia* from forest galleries in Africa. In South America, *Haemagogus* and *Sabethes* are likely to transmit any new potential human viruses among primates and, as displayed by yellow fever, if these viruses are capable of transitioning to transmission in human habitats, it is likely to have already occurred.

(2) By misfortune, as happened with the establishment of a sylvan cycle of yellow fever in Central and South America 500 years ago, new viruses may appear in transmission cycles that are heretofore unknown. The increase in trade and travel, and the establishment of invasive species (such as *Ae. albopictus*, *Ae. koreicus*, and *Ae. japonicus*, and even *Ae. aegypti* in Europe) suggest that this risk should not be overlooked.

(3) Via increased human contacts with wild cycles due to deforestation and irrational forest exploitation. These ecological modifications favour the emergence of viruses from forest edges, and then to the human environment.

(4) Via zoonotic cycles. For example, many viruses (West Nile, Japanese encephalitis, St Louis encephalitis virus, Murray valley, usutu) which are bird viruses, can be transmitted from birds (or mammals such as pigs) to humans, via vectors taking blood meals from both birds and humans, such as *Cx. tritaeniorhynchus* or *Cx. pipiens*, *Cx. quinquefasciatus*. It is very likely that some of these viruses, presently confined to wild cycles, will emerge in the coming years somewhere in the world, as a result of socio-ecological changes. Their shift from endemic to epidemic will be facilitated by close contact between humans and vertebrate hosts (urban commensal birds or rodents, farm animals).

(5) By opening an ecological niche in urbanized areas. It is conceivable that local effective vector control by way of elimination of *Ae. aegypti* or *Ae. albopictus* would open their ecological niche in some areas, allowing a new species, such as *Ae. malayensis* in South-East Asia, already recognized as a vector of dengue and chikungunya viruses, to begin to colonize even closer to human populations. This scenario has not happened yet. *Aedes albopictus* may have replaced *Ae. aegypti* or vice versa, but for the moment no third species, with high vectorial capacity, has occupied their niches.

(6) Via the evolution of viruses already known, but which have not yet found the conditions for emerging and spreading. Genetic changes in virus strains could lead to better adaptation to new vectors and a better transmissibility, as happened with the chikungunya virus and *Ae. albopictus* [52]. Viruses may also evolve resistance to drugs, when any are used, or human immune defences. Given the short generation time, large population size, and high mutation rate of RNA viruses (like yellow fever, dengue, and most pathogenic arboviruses), virus adaptation to efficient transmission by a human-preferring mosquito is rapid. That is, the virus more readily adapts to the mosquito (and vertebrate host), not the mosquito to the virus [53].

Among the hundreds of wild animal viruses, several viruses are regularly cited as worthy of surveillance. This is the case of the Mayaro alphavirus, closely related to the chickungunya virus. It is a South American virus of mammals, including monkeys, mainly transmitted by *Haemagogus* mosquitoes, and which could adapt to urban *Ae. aegypti* or *Ae. albopictus*, or to some other anthrophilic Culicidae. However, these other mosquitoes are considered poor vectors [54,55]. In Africa, the Spondweni virus is genetically close to the Zika virus, which places it on the list of viruses to keep an eye on. While it is indeed pathogenic to humans, very few human cases have been detected; it has no known domestic cycle. It is poorly transmitted experimentally by *Aedes (Stegomyia)*, being efficiently transmitted by zoophilic mosquitoes (*Aedes circumluteolus*, *Mansonia africana*), but this situation could change [56]. Indeed, this virus has already been found in Haiti [57] in *Cx. quinquefasciatus*, a human-biting mosquito recorded in all tropical regions.

7. Conclusions

On a final note of optimism, it is expected that new outbreaks can be detected more quickly, thanks to surveillance and warning systems that have improved greatly in recent decades and thanks to the development of efficient, rapid and less costly diagnostic techniques. Moreover, not all emergences will find favourable environments, including encounter and compatibility filters, and are likely to quickly die out. For public health authorities to succeed in limiting new outbreaks, The International Health Regulations, published in 2005 by WHO [58], provide a framework and obligations for member states, and give hope that new epidemics could be brought under control through early detection, efficient vector control, vaccination when available, and other public health measures. While worst case scenarios should be planned for, they seldom actually arise. However, they happen regularly and the recent Covid-19 pandemic, which was not caused by an arbovirus, reminds us that, although we are aware that this can happen, we are not always properly prepared.

Author Contributions: Both authors, D.F. and J.R.P. wrote the paper. All authors have read and agree to the published version of the manuscript.

Funding: Publication fees were covered by Institut de Recherche pour le Développement (IRD).

Acknowledgments: We thank Poliana da Silva Lemos, for permission to use Figure 1 on phylogeny of *Haemagogus janthinomys*, Christophe Paupy for useful discussion and non-published data, and Sébastien Boyer for non-published data from Cambodia.

Conflicts of Interest: The authors declare no conflict of interest.

References

1. Nicolle, C. *Destin des Maladies Infectieuses*; Librairie Félix Alcan: Paris, France, 1933; p. 216.
2. Grmek, M. *Les Maladies à L'aube de la Civilisation Occidentale*; PAYOT: Paris, France, 1983; p. 527.
3. Weaver, S.C. Prediction and prevention of urban arbovirus epidemics: A challenge for the global virology community. *Antivir. Res.* **2018**, *156*, 80–84. [CrossRef]
4. Brady, O.J.; Hay, S.I. The Global Expansion of Dengue: How *Aedes aegypti* Mosquitoes Enabled the First Pandemic Arbovirus. *Ann. Rev. Entomol.* **2020**, *65*, 191–208. [CrossRef]
5. Agramonte, A. An account of Dr. Louis-Daniel Beauperthuy, a pioneer in yellow fever research. *Boston Med. Surg. J.* **1908**, *158*, 927–930. [CrossRef]

6. Finlay, C.J. The mosquito hypothetically considered as the agent of transmission of yellow fever. Presented at the Real Academia de Ciencias Medicas, Fısicas y Naturales de La Habana, Havana, Cuba, 14 August 1881.

7. Reed, W.; Carroll, J.; Agramonte, A.; Lazear, J.W. The etiology of yellow fever: A preliminary note. *Philad. Med. J.* **1900**, *6*, 790–796.

8. Christophers, S.R. Aedes aegypti (L.), the Yellow Fever Mosquito. In *Its Life History, Bionomics, and Structure*; Cambridge University Press: New York, NY, USA, 1960.

9. Souza-Neto, J.A.; Powell, J.R.; Bonizzoni, M. *Aedes aegypti* vector competence studies: A review. *Infect. Genet. Evol.* **2019**, *67*, 191–209. [CrossRef]

10. Hawley, A.H. The biology of *Aedes albopictus*. *J. Am. Mosq. Control Assoc.* **1988**, *1*, 1–39.

11. Kotsakiozi, P.; Richardson, J.B.; Pichler, V.; Favia, G.; Martins, A.J.; Urbanelli, S.; Armbruster, P.A.; Caccone, A. Population genomics of the Asian tiger mosquito, *Aedes albopictus*: Insights into the recent worldwide invasion. *Ecol. Evol.* **2017**, *7*, 10143–10157. [CrossRef]

12. WHO. *Prevention and Control of Yellow Fever in Africa*; World Health Organization: Geneva, Switzerland, 1986; p. 96.

13. Powell, J.R. Mosquito-Borne Human Viral Diseases: Why *Aedes aegypti*? *Am. J. Trop. Med. Hyg.* **2018**, *98*, 1563–1565. [CrossRef]

14. Powell, J.R.; Gloria-Soria, A.; Kotsakiozi, P. Recent history of *Aedes aegypti*: Vector genomics and epidemiology records. *Bioscience* **2018**, *68*, 854–860. [CrossRef]

15. Chippaux, J.; Chippaux, A. Yellow fever in Africa and the Americas: A historical and epidemiological perspective. *J. Venom. Anim. Toxins Incl. Trop. Dis.* **2018**, *24*, 20. [CrossRef]

16. de Figueiredo, M.L.; de C Gomes, A.; Amarilla, A.A.; de S Leandro, A.; de S Orrico, A.; de Araujo, R.F.; do S M Castro, J.; Durigon, E.L.; Aquino, V.H.; Figueiredo, L.T. Mosquitoes infected with dengue viruses in Brazil. *Virol. J.* **2010**, *7*, 152. [CrossRef]

17. Fernandes, R.S.; Bersot, M.I.; Castro, M.G.; Telleria, E.L.; Ferreira-de-Brito, A.; Raphael, L.M.; Bonaldo, M.C.; Lourenço-de-Oliveira, R. Low vector competence in sylvatic mosquitoes limits Zika virus to initiate an enzootic cycle in South America. *Sci. Rep.* **2019**, *9*, 20151. [CrossRef]

18. Figueiredo, L.T.M. Human Urban Arboviruses Can Infect Wild Animals and Jump to Sylvatic Maintenance Cycles in South America. *Front. Cell. Infect. Microbiol.* **2019**, *9*, 259. [CrossRef]

19. Jácome, R.; Carrasco-Hernández, R.; Campillo-Balderas, J.A.; López-Vidal, Y.; Lazcano, A.; Wenzel, R.P.; Ponce de León, S. A yellow flag on the horizon: The looming threat of yellow fever to North America. *Int. J. Infect. Dis.* **2019**, *87*, 143–150. [CrossRef]

20. McNeill, J.R. Mosquito Empires. In *Ecology and War in the Greater Caribbean, 1620–1914*; Cambridge University Press: New York, NY, USA, 2010.

21. Soper, F.L. The elimination of urban yellow fever in the Americas through eradication of *Aedes aegypti*. *Am. J. Publ. Health* **1963**, *53*, 7–16. [CrossRef]

22. Webb, J.L., Jr. *Aedes aegypti* suppression in the Americas: Historical perspectives. *Lancet* **2016**, *388*, 556–557. [CrossRef]

23. Couto-Lima, D.; Madec, Y.; Bersot, M.I.; Compos, S.S.; de Albuquerque Motta, M.; Barreto dos Santos, F.; Vazeille, M.; Vasconcelos, P.F.; Lourenco-de-Oliveira, R.; Failloux, A.B. Potential risk of re-emergence of urban transmission of yellow fever virus in Brazil facilitated by competent *Aedes* populations. *Sci. Rep.* **2017**, *7*, 4848. [CrossRef]

24. Moreira-Soto, A.; Torres, M.C.; Lima de Mendonça, M.C.; Mares-Guia, M.A.; Dos Santos Rodrigues, C.D.; Fabri, A.A.; Dos Santos, C.C.; Machado Araújo, E.S.; Fischer, C.; Ribeiro Nogueira, R.M.; et al. Evidence for multiple sylvatic transmission cycles during the 2016–2017 yellow fever virus outbreak, Brazil. *Clin. Microbiol. Infect.* **2018**, *24*, 1019. [CrossRef]

25. Lemos, P.S.; Monteiro, H.A.O.; Castro, F.C.; Lima, C.P.S.; Silva, D.E.A.; Vasconcelos, J.M.; Oliveira, L.F.; Silva, S.P.; Cardoso, J.F.; Vianez-Júnior, J.L.S.G.; et al. Caracterização do genoma mitocondrial de *Haemagogus janthinomys* (Diptera: Culicidae). *DNA Mitocondrial* **2017**, *28*, 50–51. [CrossRef]

26. Tabachnick, W.J.; Wallis, G.P.; Aitken, T.H.G.; Miller, B.R.; Amato, G.D.; Lorenz, L.; Powell, J.R.; Beaty, B.R. Oral infection of *Aedes aegypti* with yellow fever virus: Geographical variation and genetic considerations. *Am. J. Trop. Med. Hyg.* **1985**, *34*, 1219–1224. [CrossRef]

27. Wasserman, S.; Tambyah, P.A.; Lim, P.L. Yellow fever cases in Asia: Primed for an epidemic. *Int. J. Infect. Dis.* **2016**, *48*, 98–103. [CrossRef]

28. Brey, P.T.; Fontenille, D.; Tang, H. Re-evaluate yellow fever risk in Asia-Pacific region. *Nature* **2018**, *554*, 31. [CrossRef]

29. Macdonald, G. *The Epidemiology and Control of Malaria*; Oxford University Press: Oxford, UK, 1957.

30. Cohuet, A.; Harris, C.; Robert, V.; Fontenille, D. Evolutionary forces on Anopheles: What makes a malaria vector? *Trends Parasitol.* **2010**, *26*, 130–136. [CrossRef]

31. Azar, S.R.; Weaver, S.C. Vector competence: What has Zika virus taught us. *Viruses* **2019**, *11*, 867. [CrossRef]

32. Houé, V.; Gabiane, G.; Dauga, C.; Suez, M.; Madec, Y.; Mousson, L.; Marconcini, M.; Yen, P.S.; de Lamballerie, X.; Bonizzoni, M.; et al. Evolution and biological significance of flaviviral elements in the genome of the arboviral vector *Aedes albopictus*. *Emerg. Microbes Infect.* **2019**, *8*, 1265–1279. [CrossRef]

33. Smith, D.R. An update on mosquito cell expressed dengue virus receptor proteins. *Insect Mol. Biol.* **2012**, *21*, 1–7. [CrossRef]

34. Combes, C. *Parasitism: The Ecology and Evolution of Intimate Interactions*; University of Chicago Press: Chicago, IL, USA, 2001; p. 728.

35. Paupy, C.; Delatte, H.; Bagny, L.; Corbel, V.; Fontenille, D. *Aedes albopictus*, an arbovirus vector: From the darkness to the light. *Microbes Infect.* **2009**, *11*, 1177–1185. [CrossRef]

36. Reiter, P.; Fontenille, D.; Paupy, C. *Aedes albopictus* as an epidemic vector of Chikungunya virus: Another emerging problem? *Lancet Inf. Dis.* **2006**, *6*, 463–464. [CrossRef]

37. Delatte, H.; Desvars, A.; Bouetard, A.; Bord, S.; Gimonneau, G.; Vourc'h, G.; Fontenille, D. Blood-feeding behavior of *Aedes albopictus*, vector of chikungunya on La Reunion. *Vector Borne Zoonotic Dis.* **2008**, *8*, 25–34. [CrossRef]

38. Brady, O.J.; Johansson, M.A.; Guerra, C.A.; Bhatt, S.; Golding, N.; Pigott, D.M.; Delatte, H.; Grech, M.G.; Leisnham, P.T.; Maciel-de-Freitas, R.; et al. Modelling adult *Aedes aegypti* and *Aedes albopictus* survival at different temperatures in laboratory and field settings. *Parasites Vectors* **2013**, *6*, 351. [CrossRef]

39. Huang, Y.M. Contributions to the mosquito fauna of Southeast Asia. XIV. The subgenus *Stegomyia* of *Aedes* in Southeast Asia. I. The *scuterallis* group of species. Contrib. *Am. Entomol. Inst.* **1972**, *9*, 1–109.

40. Mogi, M.; Armbruster, P.A.; Tuno, N.; Aranda, C.; Yong, H.S. The climate range expansion of *Aedes albopictus* (Diptera: Culicidae) in Asia inferred from the distribution of Albopictus subgroup species of *Aedes* (*Stegomyia*). *J. Med. Entomol.* **2017**, *54*, 1615–1625. [CrossRef] [PubMed]

41. Miot, E. Potential of the mosquito *Aedes malayensis* as an arbovirus vector in South East Asia. Ph.D. Thesis, Sorbonne University, Paris, France, 20 December 2019.

42. Tangena, J.A.; Thammavong, P.; Malaithong, N.; Inthavong, T.; Ouanesamon, P.; Brey, P.T.; Lindsay, S.W. Diversity of mosquitoes (Diptera: Culicidae) attracted to human subjects in rubber plantations, secondary forests, and villages in Luang Prabang Province, northern Lao PDR. *J. Med. Entomol.* **2017**, *54*, 1589–1604. [CrossRef] [PubMed]

43. Lee, J.M.; Wasserman, R.J.; Gan, J.Y.; Wilson, R.F.; Rahman, F.; Yek, S.H. Human activities attract harmful mosquitoes in a tropical urban landscape. *EcoHealth* **2020**, *17*, 52–63. [CrossRef]

44. Pereira Dos Santos, T.; Roiz, D.; Santos de Abreu, F.V.; Luz, S.L.B.; Santalucia, M.; Jiolle, D.; Santos Neves, M.S.A.; Simard, F.; Lourenço-de-Oliveira, R.; Paupy, C. Potential of *Aedes albopictus* as a bridge vector for enzootic pathogens at the urban-forest interface in Brazil. *Emerg. Microbes Infect.* **2018**, *7*, 191. [CrossRef]

45. Lounibos, L.P.H. Invasions by Insect Vectors of Human Disease. *Ann. Rev. Entomol.* **2002**, *47*, 233–266. [CrossRef]

46. Pandit, P.S.; Doyle, M.M.; Smart, K.M.; Young, C.C.W.; Drape, G.W.; Johnson, C.K. Predicting wildlife reservoirs and global vulnerability to zoonotic *Flaviviruses*. *Nat. Commun.* **2018**, *9*, 5425. [CrossRef]

47. Lounibos, L.P. Habitat segregation among African treehole mosquitoes. *Ecol. Entomol.* **1981**, *6*, 129–154. [CrossRef]

48. Delatte, H.; Bagny, L.; Brengue, C.; Bouetard, A.; Paupy, C.; Fontenille, D. The invaders: Phylogeography of dengue and chikungunya viruses *Aedes* vectors, on the South West islands of the Indian Ocean. *Infect. Genet. Evol.* **2011**, *11*, 1769–1781. [CrossRef]

49. Gould, E.; Pettersson, J.; Higgs, S.; Charrel, R.; de Lamballerie, X. Emerging arboviruses: Why today? *One Health* **2017**, *4*, 1–13. [CrossRef]

50. Hadfield, J.; Brito, A.F.; Swetnam, D.M.; Vogels, C.B.F.; Tokarz, R.E.; Andersen, K.G.; Smith, R.C.; Bedford, T.; Grubaugh, N.D. Twenty years of West Nile virus spread and evolution in the Americas visualized by Nextstrain. *PLoS Pathog.* **2019**, *15*, e1008042. [CrossRef] [PubMed]

51. Murphy, F.A.; Calisher, C.H.; Tesh, R.B.; Walker, D.H. In memoriam: Robert Ellis Shope (1929–2004). *Emerg. Infect. Dis.* **2004**, *10*, 762–765. [CrossRef]
52. Schuffenecker, I.; Iteman, I.; Michault, A.; Murri, S.; Frangeul, L.; Vaney, M.C.; Lavenir, R.; Pardigon, N.; Reynes, J.M.; Pettinelli, F.; et al. Genome Microevolution of Chikungunya Viruses Causing the Indian Ocean Outbreak. *PLoS Med.* **2006**, *3*, e263. [CrossRef]
53. Powell, J.R. An evolutionary perspective on vector-borne diseases. *Front. Gent.* **2019**, *10*, 1266. [CrossRef] [PubMed]
54. Abad-Franch, F.; Grimmer, G.H.; de Paula, V.S.; Figueiredo, L.T.; Braga, W.S.M.; Luz, S.L.B. Mayaro Virus Infection in Amazonia: A Multimodel Inference Approach to Risk Factor Assessment. *PLoS Negl. Trop. Dis.* **2012**, *6*, e1846. [CrossRef] [PubMed]
55. Pezzi, L.; A Diallo, M.; Rosa-Freitas, M.G.; Vega-Rua, A.; Ng, L.F.P.; Boyer, S.; Drexler, J.F.; Vasilakis, N.; Lourenço-de-Oliveira, R.; Weaver, S.C.; et al. GloPID-R report on chikungunya, o'nyong-nyong and Mayaro virus, part 5: Entomological aspects. *Antivir. Res.* **2020**, *174*, 104670. [CrossRef] [PubMed]
56. Haddow, A.D.; Nasar, F.; Guzman, H.; Ponlawat, A.; Jarman, R.G.; Tesh, R.B.; Weaver, S.C. Genetic Characterization of Spondweni and Zika Viruses and Susceptibility of Geographically Distinct Strains of *Aedes aegypti*, *Aedes albopictus* and *Culex quinquefasciatus* (Diptera: Culicidae) to Spondweni Virus. *PLoS Negl. Trop. Dis.* **2016**, *10*, e0005083. [CrossRef]
57. White, S.K.; Lednicky, J.A.; Okech, B.A.; Morris, J.G., Jr.; Dunford, J.C. Spondweni Virus in Field-Caught *Culex quinquefasciatus* Mosquitoes, Haiti, 2016. *Emerg. Infect. Dis.* **2018**, *24*, 1765–1767. [CrossRef]
58. WHO. *International Health Regulations*, 3rd ed.; World Health Organization: Geneva, Switzerland, 2005; p. 74.

Review

A Review: *Wolbachia*-Based Population Replacement for Mosquito Control Shares Common Points with Genetically Modified Control Approaches

Pei-Shi Yen * and Anna-Bella Failloux *

Unit Arboviruses and Insect Vectors, Department of Virology, Institut Pasteur, F-75724 Paris, France
* Correspondence: pei-shi.yen@pasteur.fr (P.-S.Y.); anna-bella.failloux@pasteur.fr (A.-B.F.);
Tel.: +33-01-40613617 (A.-B.F.)

Received: 15 April 2020; Accepted: 20 May 2020; Published: 22 May 2020

Abstract: The growing expansion of mosquito vectors has made mosquito-borne arboviral diseases a global threat to public health, and the lack of licensed vaccines and treatments highlight the urgent need for efficient mosquito vector control. Compared to genetically modified control strategies, the intracellular bacterium *Wolbachia*, endowing a pathogen-blocking phenotype, is considered an environmentally friendly strategy to replace the target population for controlling arboviral diseases. However, the incomplete knowledge regarding the pathogen-blocking mechanism weakens the reliability of a *Wolbachia*-based population replacement strategy. *Wolbachia* infections are also vulnerable to environmental factors, temperature, and host diet, affecting their densities in mosquitoes and thus the virus-blocking phenotype. Here, we review the properties of the *Wolbachia* strategy as an approach to control mosquito populations in comparison with genetically modified control methods. Both strategies tend to limit arbovirus infections but increase the risk of selecting arbovirus escape mutants, rendering these strategies less reliable.

Keywords: mosquito control; replacement strategy; *Wolbachia*; environmental factors; arbovirus; viral adaptation

---

1. Half of the World's Population Exposed to Arboviral Diseases

Mosquito-borne arboviral diseases such as chikungunya, dengue, yellow fever, and Zika have been one of the major public health issues over the last few decades, threatening more than half of the world's population [1]. These arboviruses have dispensed with the need for amplification in wild animals to cause outbreaks in the human population. Human hosts serve simultaneously as the reservoir, amplifier, and disseminator, and the major vectors are the anthropophilic mosquitoes *Aedes aegypti* and *Aedes albopictus* [2]. Due to their complex transmission mechanisms with interactions between viruses, mosquito vectors, and vertebrate hosts, all three evolving in changing environments, the control of arboviral diseases is still extremely difficult. The frustrating lack of development of broad-spectrum vaccines against arboviruses has pointed out the importance of new alternatives for arboviral diseases control [3,4].

Chikungunya is a rapidly reemerging arboviral disease. In recent years, as the Indian Ocean lineage (IOL) evolved, this *Ae. aegypti* adapted chikungunya virus (CHIKV) has caused several outbreaks in tropical countries [5]. The newly emerged CHIKV IOL contains a mutation from Alanine to Valine at position 226 in the E1 protein, which influences the pH threshold for fusion, facilitating the virus entry [6]. Moreover, the E1-A226V mutation also increases the vector competence for *Aedes* mosquitoes, particularly for *Aedes albopictus* [7]. Therefore, the epidemic areas have extended beyond tropical regions, reaching the temperate countries in Europe where *Ae. albopictus* mosquitoes have been established since 1979 [8]; this species is now present in 20 European countries with the highest infestation levels in France [9,10] and Italy [11]. Dengue virus (DENV) alone contributes to approximately 390 million infections and tens of thousands fatal cases annually [12]. There are four DENV serotypes, namely dengue-1, -2, -3, and -4, which trigger distinct immune responses in humans [13]. Although a life-long immunity against a given DENV serotype could be acquired after a first infection, a secondary infection with another DENV serotype could facilitate dengue severe syndrome or dengue hemorrhagic fever (DHF) that increases the burden of this disease even if a secondary infection does not always necessarily lead to DHF [14,15]. Moreover, the nearly 80% of dengue asymptomatic cases enhance the risk of DENV transmission, thus complexifying the disease control. Yellow fever virus (YFV) has not yet become a global threat but heavily affects Sub-Saharan Africa and South America (mainly in Brazil), causing between 94,336 and 118,500 infection incidents annually [16]. In Africa, the urban cycle results from sporadic spillover transmission from the jungle cycle, while in America, human infections are only acquired in forests [17]. Although the YFV live-attenuated vaccine has been available since the 1930s, the global stockpile of YFV vaccines is still an issue due to the production process that requires specific pathogens-free (SPF) eggs; the limited supply of SPF eggs makes it difficult to launch an urgent vaccination campaign [18]. The recent outbreaks in Brazil and the imported cases from Angola to China have raised many concerns on the potential risk of major outbreaks, especially for the immune-naïve populations in Asia [19–21]. Zika virus (ZIKV) was first isolated from an infected monkey in Uganda in 1947. Starting from Yap Island in 2007 [22], a much larger outbreak in French Polynesian islands followed [23], and reached Brazil at the end of 2013 [24]. Although ZIKV infection formerly caused only mild illness and was self-limiting [22], more severe outcomes were reported, including microcephaly in newborns and the neurological affections [25,26] giving it the status of a Public Health Emergency of International Concern in 2016 by the World Health Organization (WHO). It should be noted that other emerging arboviruses have not yet caused large outbreaks; they could be the next arboviral threats owing to current global changes, including climate warming and increasing international exchanges [27]. *Aedes* mosquitoes are also transmitting Mayaro (MAYV; genus: *Alphavirus*), and Usutu (USUV; genus: *Flavivirus*) viruses that originated from South America and Sub-Saharan Africa, respectively [28,29]. Since 2000, MAYV has caused several outbreaks in South America, although the number of reported infections remains low. The spread of MAYV to non-endemic areas was reported in Europe after 2008, expanding the risk area of this disease [30]. Similar to MAYV, the USUV was first identified in 1959 [29], but the first human infection was reported at the beginning of 1981 in Africa [31]. Since then, USUV has been introduced into Europe and was repeatedly reported in mosquitoes, birds, and horses in 12 European countries [32].

2. Virus Overcomes Mosquito Immune Barriers to Be Transmitted by Generating Viral Quasispecies

Aedes mosquitoes are the major vectors in transmitting arboviruses. They have expanded their geographic distribution as a consequence of population growth, human activities, and climate change, creating conditions favorable for their proliferation and introducing the means of passive transportation [33]. Their distribution is no longer restricted to tropical regions, spreading to new geographic regions over long distances and stepping up the global impact of arboviral diseases [33].

A mosquito vector acquires an arbovirus by taking a blood meal from a viremic host, after which the virus enters into the midgut epithelial cell and replicates. After infection of the midgut, the virus needs to escape through the midgut basal lamina and disseminate into the hemocele, infecting different tissues or organs including salivary glands. Finally, in a subsequent blood meal, the virus is excreted from the salivary glands and transmitted by the mosquito bite [34].

From the midgut to the salivary glands, the virus encounters different mosquito immune responses, such as RNAi (RNA interference), the Toll, IMD (immune deficiency), and JAK-STAT (Janus kinase signal transducer and activator of transcription protein) pathways, limiting the virus infection, dissemination, and transmission [35]. Viral infections in mosquitoes result from a subtle balance between virus replication and cellular immunity, manifested by a persistent asymptomatic infection in vectors [36]. Considering the error-prone nature of viral RNA-dependent RNA polymerase, RNA viruses generate a collection of variants (namely quasispecies) during replication that facilitates viral adaptation and transmission by mosquitoes [37,38]. Thus, after passing all these filters, only a fraction of the viral population is transmitted to the vertebrate host [39]. This remains dependent on each pairing of virus-mosquito that conditions the mosquito vector competence [34].

3. Population Replacement Strategy to Control Mosquito-Borne Diseases

As the pivot for arboviruses transmission, the mosquito vector is considered the target for efficient arboviral diseases control. Depending on the outcomes, the mosquito population control strategies could be roughly divided into two categories, population reduction and population replacement, which by reducing the target population size or introducing an anti-pathogen phenotype into the target population, respectively, minimizes the contact between arboviruses-carrying mosquitoes and human hosts. The current strategy is to reduce the mosquito vector population with insecticides; however, in the absence of a vector surveillance program, the impact of insecticides on local biodiversity is unpredictable, in addition there is the risk in generating insecticide-resistant mosquitoes that decreases the control efficacy, calling for more species-specific alternative methods [40]. Instead of reducing the target population from the field, which might cause ecological disruption [41,42] and risk of secondary pest emergence, population replacement has been proposed as an alternative [43]. Replacing the target population with a pathogen-refractory strain could specifically reduce the pathogens' transmission while maintaining the population in its original ecological niche, limiting the risk of secondary pest emergence [43].

A modified genetic-based population replacement approach is composed of an anti-pathogen gene and a gene-drive system, in order to suppress the pathogen replication and to spread the phenotype within the target population [44]. Different mosquito antiviral factors such as siRNA [45,46], miRNA [47–49], ribozymes [50,51], immune factors [52], and neutralizing antibodies [53], can act as effectors to reduce the virus infection and transmission in genetically modified mosquitoes. Combined with a proper gene-drive system, the genetically modified mosquitoes expressing a virus-refractory phenotype can replace the wild population in a few generations, that is to say in a few months for mosquitoes. Although the biosafety concerns about using genetically modified insects is still debatable [54], the highly-specific synthetic antiviral immunity used as effectors (RNAi, antiviral ribozymes, overexpressed immune genes, and neutralizing antibodies) has raised issues regarding the selection of escape virus mutants. Thus, strategies generating weaker selection in virus populations might be more sustainable [55].

4. *Wolbachia*-Based Mosquito Control

The microbiota of mosquito vectors has a strong impact on arbovirus infections [56,57]. The endosymbiont bacteria *Wolbachia* has been in the spotlight with the discovery of its properties in suppressing the replication of vector-borne human arboviruses such as DENV, YFV, and ZIKV [58,59]. The *Wolbachia*-based insect control approach is more acceptable than the genetic modification-based approach for the public as it is a naturally existing microbe. *Wolbachia* is a Gram-negative bacterium, a member of the Alphaproteobacteria (Rickettsiales order). In arthropods, the genome of *Wolbachia* ranges from 1.2 to 1.6 Mb and contains WO prophages (named after *Wolbachia*) [60]. It was first discovered in the *Culex pipiens* mosquito in 1924 by Hertig and Wolbach [61], opening a new avenue of research owing to its high diversity and wide distribution in arthropods [62]. To date, 18 supergroups of *Wolbachia* have been identified, most of them present in arthropods [63], and more than 65% of insect species harbor *Wolbachia* [64]. Horizontal transfers of *Wolbachia* have been demonstrated between species within the same supergroup and, conversely, the same host species can host different *Wolbachia* strains. *Wolbachia* bacteria are involved in different symbiotic interactions ranging from parasitic to mutualistic [62]. Mainly transmitted vertically, it intervenes in manipulating the host reproduction in order to optimize its maternal transmission through the eggs. *Wolbachia* can induce different sex ratio distortion phenotypes in the progeny to favor females: parthenogenesis, feminization, male-killing, and cytoplasmic incompatibility (CI) [65]. In the CI phenotype, *Wolbachia*-infected females are favored over the non-infected females and males (Figure 1). The molecular mechanisms controlling the CI are today better understood: CI and its rescue are driven by toxin-antidote interactions, whose affinity between partners determines the success of the rescue [65]. For example, in *Cx. pipiens* during the first embryonic mitosis, cidA and cidB of *w*Pip are the key elements of CI traits [66], where the B factor expresses its toxic effect inducing the CI effect and the A factor acts as an antidote for the rescue. In addition to colonizing reproductive organs, *Wolbachia* bacteria are also present in somatic tissues; *Wolbachia* can be acquired from infected embryonic lineages or by passing from cell to cell [67].

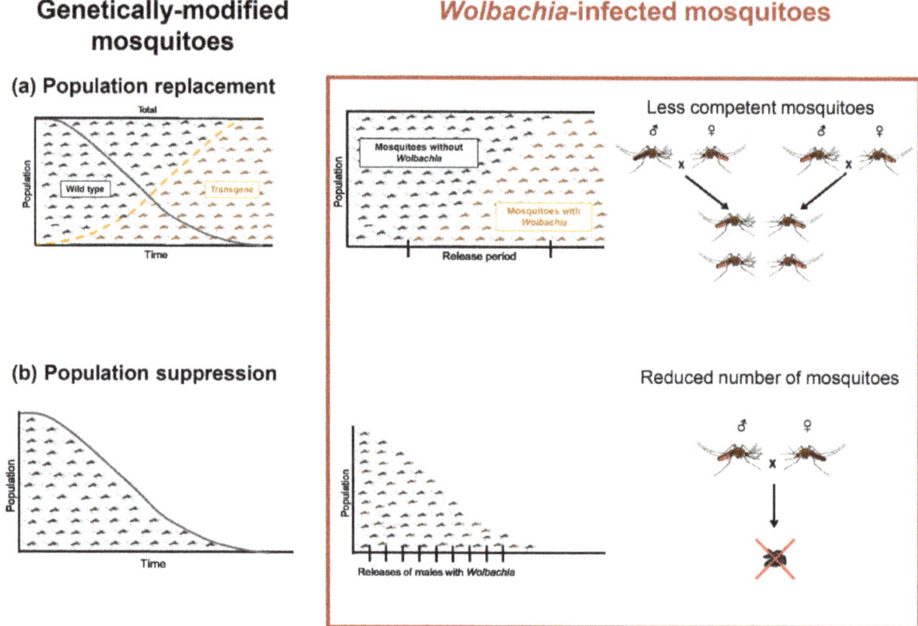

Figure 1. *Wolbachia*-based mosquito control strategy can be viewed as a genetically modified control approach, population replacement, or population suppression. (**a**) The population replacement strategy is to release *Wolbachia*-infected female mosquitoes that, after mating with males (*Wolbachia*-infected or not), produce viable offspring, allowing a wide spread of *Wolbachia* in the field population that harbors less competent individuals, even as the total number of mosquitoes remains unchanged. (**b**) Population suppression aims to release *Wolbachia*-infected male mosquitoes that, after mating with wild females, do not produce viable offspring, thus reducing the total number of mosquitoes.

5. *Wolbachia* in Limiting Arbovirus Transmission

Wolbachia can provide fitness advantages for host fertility and/or survival and can alter responses to infections to reduce arbovirus transmission. As an example, the *w*Mel *Wolbachia* strain artificially introduced in *Drosophila melanogaster* inhibited *Drosophila* C virus (DCV) infection, and *Wolbachia*-infected flies were much more resistant to DCV than the uninfected flies [68,69].

Combining their ability to invade the host population by inducing CI and to interfere negatively with the transmission of disease pathogens, *Wolbachia*-based control methods have been deployed to prevent the transmission of mosquito-borne diseases [70,71]. The bacterium *Wolbachia* has been introduced in natural populations of the mosquito *Ae. aegypti*, the urban vector of dengue, Zika, chikungunya, and yellow fever [72]. *Wolbachia* bacteria are stably maintained in field mosquito populations and attributed to the pathogen-blocking phenotype [73]. The critical question is whether this effect will persist and if adaptive changes in the mosquito vector, the bacteria, or virus may occur, hampering the success of this strategy.

Wolbachia can share the same niche with the virus, colonizing ovaries, gut, and salivary glands, organs that are essential for the replication and transmission of arboviruses [74]. Usually, high antiviral resistance is associated with high densities of *Wolbachia*, which could reach several hundred bacteria per cell [75] and cause a significant fitness cost (e.g., reduced fecundity, fertility, and survival) [76]. Mosquito adaptive changes may occur, leading to the evolution towards lower *Wolbachia* densities and, therefore, a reduction or loss of the antiviral phenotype. Another concern is the evolution of the virus itself; since *Wolbachia* block the replication of the virus, viral populations could be shaped to overcome such inhibition. Variants able to replicate despite the presence of *Wolbachia* could be advantaged and spread, weakening the sustainability of the *Wolbachia* control strategy. As an example, in *D. melanogaster*, *Wolbachia* did not present antiviral effects for the *Wolbachia*-adaptive DCV, which were genetically different from viral populations in *Wolbachia*-free controls [77]. These findings were obtained from cell cultures, which are far from replicating the conditions in a natural host, with a succession of reduction and restoration of viral diversities after crossing anatomical barriers (midgut and salivary glands) in the mosquito vector [39].

6. The Hypothesized Mechanisms of *Wolbachia*-Mediated Pathogen Blocking Activity

Several mechanisms have been proposed to explain the molecular basis of the pathogen-blocking phenotype: regulation of immune genes, indirect host gene regulation through other cellular machinery (RNAi, sfRNA), production of reactive oxygen species, or competition for a limited resource such as cholesterol. Growing evidence has shown that the mosquito transcriptome profiles were altered after *Wolbachia* infection [52,70,78,79]. Insect immune pathways such as Toll, IMD, and JAK/STAT pathways were activated in *Wolbachia*-infected *Ae. aegypti*, and led to an efficient reduction in replication of CHIKV, DENV, and *Plasmodium* [52,70,78,79]. In addition to the immune factors, many genes were reported to be upregulated and subsequently may be able to suppress virus replication; the genes regulating reactive oxygen species production and the upregulation of *Wolbachia*-mediated methyltransferase are also reported to suppress the replication of DENV in *Ae. aegypti* [52,80,81]. Mosquito non-coding RNA expression is also influenced upon *Wolbachia* infection, and might regulate virus replication in infected cells [82]. Although the direct interaction between virus replication and *Wolbachia*-induced miRNA is not fully understood, an extensively-expressed aae-miRNA-2940 in *Wolbachia*-infected mosquito cells was proven to upregulate methyltransferase expression and, subsequently, reduce DENV replication [81]. Moreover, this methyltransferase upregulation negatively controls the metalloprotease expression leading to a reduction in West Nile virus replication [83]. Recent studies have suggested that *Wolbachia* bacteria suppress virus replication through cellular resources allocation (e.g., intracellular space in *Wolbachia*-infected cells [70,84]), vascular trafficking, and lipid metabolism caused by *Wolbachia*-mediated cholesterol perturbation [85]. In fact, on one hand, *Wolbachia* do not have any functional lipopolysaccharide synthase and need cholesterol for membrane formation, while on the other hand, viruses rely on host cholesterol for replication. Therefore, both behave as competitors for access to cholesterol ingested by the mosquito, which is auxotrophic for this element [85]. Moreover, the limited intracellular space in *Wolbachia* infected cells is also reported to restrict DENV replication [70,84].

7. Viruses Evolve towards an Adaption to *Wolbachia*-Blocking Activity

Recent studies have raised concerns regarding *Wolbachia*-mediated selection pressure for arboviruses that might facilitate viral adaptive evolution and escape from pathogen-blocking effects [77]. Evidence has been brought that *Wolbachia* bacteria do not suppress the replication of viral RNA genomes in mosquito cells and, consequently, are not able to accumulate adaptive viral genomes [86,87]. The same alteration can be observed in *Wolbachia*-infected *D. melanogaster* for DCV, where *Wolbachia*-mediated selective DCV strains could be found after 10 passages in *Wolbachia*-infected flies. Despite the fact that the adapted DCV strains did not exhibit the same resistance phenotype to *Wolbachia* in newly *Wolbachia*-infected flies, evidence points to the risk of *Wolbachia*-mediated selection for virus adaptive mutations. The resulting adapted viruses might have a better chance of escaping the *Wolbachia*-mediated virus-blocking effects in drosophila [77]. Moreover, many studies have indicated the *Wolbachia* naturally infecting *Ae. albopictus*, *w*AlbA, and *w*AlbB do not show significant pathogen-blocking activity against virus infection in this mosquito [75,88–91]. Even if *w*AlbB is transinfected in other mosquito species, it cannot trigger the pathogen-blocking phenotype, as observed with other *Wolbachia* strains [92,93]. These results raise the question of the outcome of co-evolution between *Wolbachia* and viruses, highlighting the risk for viruses to escape from *Wolbachia*-mediated immunity (Figure 2a).

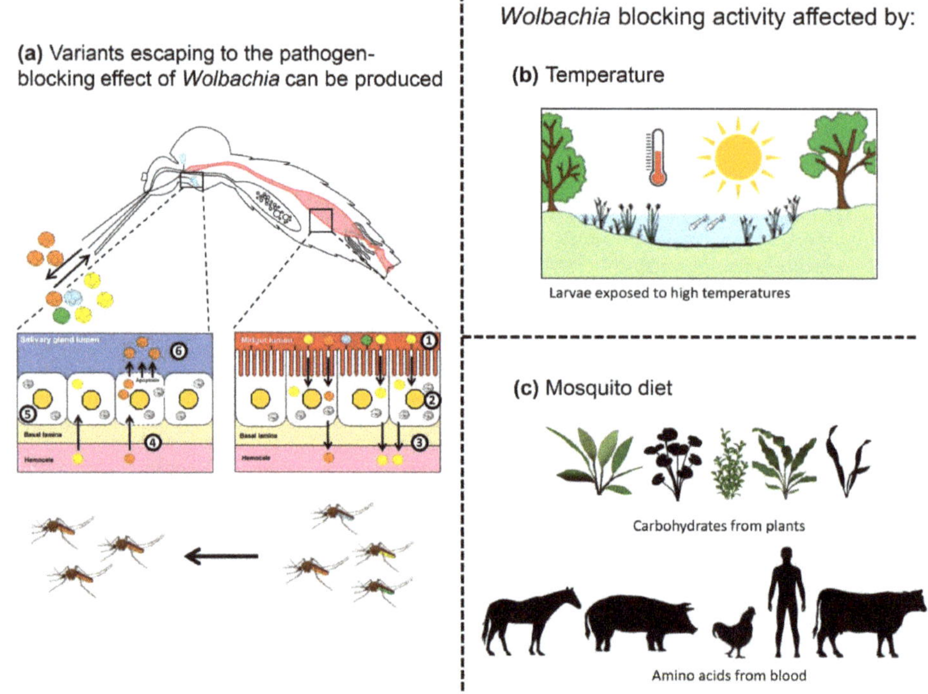

Figure 2. Viruses escaping from *Wolbachia* blocking activity. (**a**) After ingestion of an infectious blood meal (1), viruses enter into the midgut epithelial cells and replicate (2). Newly produced viruses are released into the hemocoel, infecting internal organs/tissues (3). After reaching the salivary glands (4), viruses replicate (5), and new viruses are excreted with saliva expectorated (6) by females when they bite. In the presence of *Wolbachia* in the midgut and salivary glands, escaping variants can be produced, leading mosquitoes to transmit viruses less sensitive to the pathogen-blocking effect. (**b**) *Wolbachia* blocking activity can be altered by environmental factors such as temperature. Mosquito larvae submitted to high temperatures produce adults with a lower density of *Wolbachia*. (**c**) Mosquito adults feeding on enriched diets (carbohydrates and amino acids) show reduced *Wolbachia* loads. Both (**b**) and (**c**) lead to a diminished blocking activity of *Wolbachia* as their bacterial densities are much lower.

8. Environmental Factors That May Affect *Wolbachia*-Based Control Programs

8.1. An Abiotic Factor, the Temperature

In the field, *Wolbachia*-infected mosquitoes are exposed (naturally or not) to fluctuating and even extreme temperatures (Figure 2b). Larvae reared under cycling temperatures between 26 °C and 37 °C resulted in adults with a lower density of *Wolbachia*, an effect persisting across generations [94], and a diminished CI effect that resulted in the inability of *Wolbachia* to be transmitted vertically [95]. More interesting is the possibility that these effects could be strain-dependent. Temperature significantly affects *w*Mel and *w*MelPop-CLA in *Ae. aegypti*, whereas no effect was observed with *w*AlbB, which naturally colonizes the mosquito *Ae. albopictus*, suggesting that the *Wolbachia* adaptation might reduce the pathogen-blocking activity. Infections with *w*AlbB seem to be more applicable in a population reduction strategy than pathogen suppression in field conditions. The duration of heat stress also alters *Wolbachia* density [96]. A long duration of heat stress of *Ae. aegypti* immature stages leads to lower *w*Mel densities in adults. Complementary studies using local mosquitoes and field temperatures are required in *Wolbachia*-released sites [97].

Temperature might influence *Wolbachia* density in *Ae. aegypti* through bacteriophage WO infection [98]. High temperatures may reduce *Wolbachia* densities in *Ae. aegypti* through interactions with WO, which infects *w*Mel [99] and *w*AlbB [100]. This phage undergoes cycles of lysogenic and lytic phases; heat shock triggers the lytic phase during which the phage replicates and causes *Wolbachia* lysis, reducing its densities [101]. However, the temperature has a positive effect on virus replication with significantly higher titers of DENV at temperature > 28 °C in *Ae. aegypti* mosquitoes at day 10 post-infection in salivary glands [102]. More globally, high temperature is likely to weaken the *Wolbachia*-mediated pathogen-blocking activity by shortening the extrinsic incubation period (the time necessary for mosquitoes to become infectious after an infectious blood meal) [103–105], and reducing *Wolbachia* density through bacteriophage WO.

8.2. A Biotic Factor, the Host Diet for Mosquito Vectors

Host diet notably influences *Wolbachia* load (Figure 2c) [106]. Flies reared with yeast-enriched diets show reduced *Wolbachia* loads in the host female germline, and flies with a high sucrose diet show an increase of *Wolbachia* titer in oocytes. Thus, *Wolbachia* bacteria rely upon host uptake of amino acids and carbohydrates. Exposure to yeast-enriched food alters *Wolbachia* nucleoid morphology in oogenesis [106]. The yeast-induced *Wolbachia* depletion is mediated by the somatic target of rapamycin (TOR) and insulin signaling pathways. These findings are critical in programs combating arboviral diseases by releasing *Wolbachia*-infected vectors [73,107]. Thus, the natural environment including host diet (i.e., nature of blood and sugar absorbed by released mosquitoes) should be considered when evaluating the efficiency of *Wolbachia* strategies in the field.

9. Obstacles for Population Replacement Program

9.1. Unclear Wolbachia Distribution in Wild Mosquito Populations

Ae. aegypti populations were thought to be free of naturally-harboring *Wolbachia*; as a result, many *Wolbachia*-based mosquito control strategies were deployed worldwide to reduce or replace the target mosquito populations via the CI nature of *Wolbachia*. However, accumulating evidence shows the existence of naturally-occurring *Wolbachia* in wild *Ae. aegypti* populations in India [108], Malaysia [109], Panama [110], Thailand [111], Philippines [112], and the United States [113–115]. Even though only small proportions of *Ae. aegypti* were reported in naturally-harboring *Wolbachia* in field-collected populations, the presence of naturally-infecting *Wolbachia* tends to increase the instability for replacing the target populations, thus weakening the efficacy of mosquito control programs.

9.2. The Robust CI Effect Limits the Solutions for Wolbachia Re-Replacement

On the one hand, the naturally existing *Wolbachia* in the target population might be an obstacle for the mosquito population replacement program. On the other hand, the robust CI effect also increases the difficulty to re-replace the target population that has been already replaced by *Wolbachia*-harboring mosquitoes. Because of this, the *Wolbachia*-mediated pathogen-blocking strategy implemented can deviate from the objectives initially planned. Thus, a more flexible replacement program is required to adapt to this scenario. The results of computational modeling have suggested that unless a correct *Wolbachia* strain is used, the selection could favor the strain with a lower fitness cost (commensal infection), thus replacing the existing *Wolbachia*-harboring population. Furthermore, the second replacement can be less efficient than the first one [116].

10. Conclusions

Mosquito control, as an essential step for mosquito-borne diseases transmission management, either by reducing the target population size or replacing the target population with a pathogen-refractory strain, could efficiently reduce the contact between mosquitoes and hosts, thereby interrupting the disease transmission. Compared to a genetically modified control strategy, a *Wolbachia*-based control strategy is considered a promising alternative to control mosquito populations due to its natural and environmentally friendly features: It can carry a broad pathogen-blocking activity and robust CI effect, which together ensure the efficacy in reducing mosquito-borne diseases transmission. However, complex interactions and resulting co-evolution processes among mosquito, virus, host, and the environment, make difficult any mosquito-borne disease control strategies. Thus, monitoring and prevention programs to avoid escape mutants in viral populations must be attentively planned if the targeted objective is to reach a sustainable control strategy.

Author Contributions: P.-S.Y. and A.-B.F. wrote the paper. Both authors have read and agreed to the published version of the manuscript.

Funding: This study was funded by the French Government's Investissement d'Avenir program, Laboratoire d'Excellence "Integrative Biology of Emerging Infectious Diseases, IBEID" (grant n_ANR-10-LABX-62-IBEID) (A.-B.F.) and the National Health Research Institutes (NHRI-PP-108-0324-01-17-07) (P.-S.Y.).

Acknowledgments: We thank Marie Vazeille, Adrien Blisnick, and Rachel Bellone for discussions. We also thank Laura Manuellan for help in making figures and Peter Sahlins for editing.

Conflicts of Interest: The authors declare no conflict of interest.

References

1. Gould, E.; Pettersson, J.; Higgs, S.; Charrel, R.; De Lamballerie, X. Emerging arboviruses: Why today? *One Health* **2017**, *4*, 1–13. [CrossRef] [PubMed]
2. Weaver, S.C.; Reisen, W.K. Present and future arboviral threats. *Antiviral Res.* **2010**, *85*, 328–345. [CrossRef] [PubMed]
3. Flasche, S.; Wilder-Smith, A.; Hombach, J.; Smith, P.G. Estimating the proportion of vaccine-induced hospitalized dengue cases among Dengvaxia vaccinees in the Philippines. *Wellcome Open Res.* **2019**, *4*, 165. [CrossRef] [PubMed]
4. Wilder-Smith, A.; Flasche, S.; Smith, P.G. Vaccine-attributable severe dengue in the Philippines. *Lancet* **2019**, *394*, 2151–2152. [CrossRef]
5. Powers, A.M.; Logue, C.H. Changing patterns of chikungunya virus: Re-emergence of a zoonotic arbovirus. *J. Gen. Virol.* **2007**, *88*, 2363–2377. [CrossRef]
6. Tsetsarkin, K.A.; Vanlandingham, D.L.; McGee, C.E.; Higgs, S. A single mutation in chikungunya virus affects vector specificity and epidemic potential. *PLoS Pathog.* **2007**, *3*, e201. [CrossRef]
7. Vazeille, M.; Moutailler, S.; Coudrier, D.; Rousseaux, C.; Khun, H.; Huerre, M.; Thiria, J.; Dehecq, J.S.; Fontenille, D.; Schuffenecker, I.; et al. Two Chikungunya isolates from the outbreak of La Reunion (Indian Ocean) exhibit different patterns of infection in the mosquito, Aedes albopictus. *PLoS ONE* **2007**, *2*, e1168. [CrossRef]

8. Adhami, J.; Reiter, P. Introduction and establishment of Aedes (Stegomyia) albopictus skuse (Diptera: Culicidae) in Albania. *J. Am. Mosq. Control Assoc.* **1998**, *14*, 340–343.

9. Delisle, E.; Rousseau, C.; Broche, B.; Leparc-Goffart, I.; L'Ambert, G.; Cochet, A.; Prat, C.; Foulongne, V.; Ferre, J.B.; Catelinois, O.; et al. Chikungunya outbreak in Montpellier, France, September to October 2014. *Euro Surveill.* **2015**, *20*. [CrossRef]

10. Grandadam, M.; Caro, V.; Plumet, S.; Thiberge, J.M.; Souares, Y.; Failloux, A.B.; Tolou, H.J.; Budelot, M.; Cosserat, D.; Leparc-Goffart, I.; et al. Chikungunya virus, southeastern France. *Emerg. Infect. Dis.* **2011**, *17*, 910–913. [CrossRef]

11. Angelini, R.; Finarelli, A.C.; Angelini, P.; Po, C.; Petropulacos, K.; Macini, P.; Fiorentini, C.; Fortuna, C.; Venturi, G.; Romi, R.; et al. An outbreak of chikungunya fever in the province of Ravenna, Italy. *Euro Surveill.* **2007**, *12*. [CrossRef] [PubMed]

12. Bhatt, S.; Gething, P.W.; Brady, O.J.; Messina, J.P.; Farlow, A.W.; Moyes, C.L.; Drake, J.M.; Brownstein, J.S.; Hoen, A.G.; Sankoh, O.; et al. The global distribution and burden of dengue. *Nature* **2013**, *496*, 504–507. [CrossRef]

13. St John, A.L.; Rathore, A.P.S. Adaptive immune responses to primary and secondary dengue virus infections. *Nat. Rev. Immunol.* **2019**, *19*, 218–230. [CrossRef] [PubMed]

14. Guzman, M.G.; Alvarez, M.; Halstead, S.B. Secondary infection as a risk factor for dengue hemorrhagic fever/dengue shock syndrome: An historical perspective and role of antibody-dependent enhancement of infection. *Arch. Virol.* **2013**, *158*, 1445–1459. [CrossRef] [PubMed]

15. Kliks, S.C.; Nisalak, A.; Brandt, W.E.; Wahl, L.; Burke, D.S. Antibody-dependent enhancement of dengue virus growth in human monocytes as a risk factor for dengue hemorrhagic fever. *Am. J. Trop Med. Hyg.* **1989**, *40*, 444–451. [CrossRef]

16. Shearer, F.M.; Longbottom, J.; Browne, A.J.; Pigott, D.M.; Brady, O.J.; Kraemer, M.U.G.; Marinho, F.; Yactayo, S.; De Araujo, V.E.M.; Da Nobrega, A.A.; et al. Existing and potential infection risk zones of yellow fever worldwide: A modelling analysis. *Lancet Glob. Health* **2018**, *6*, e270–e278. [CrossRef]

17. Silva, N.I.O.; Sacchetto, L.; De Rezende, I.M.; Trindade, G.S.; LaBeaud, A.D.; De Thoisy, B.; Drumond, B.P. Recent sylvatic yellow fever virus transmission in Brazil: The news from an old disease. *Virol. J.* **2020**, *17*, 9. [CrossRef]

18. WHO Expert Committee on Biological Standardization. Forty-fifth report. *World Health Organ. Tech. Rep. Ser.* **1995**, *858*, 1–101.

19. Schlagenhauf, P.; Chen, L.H. Yellow Fever importation to China - a failure of pre- and post-travel control systems? *Int. J. Infect. Dis.* **2017**, *60*, 91–92. [CrossRef]

20. Wilder-Smith, A.; Lee, V.; Gubler, D.J. Yellow fever: Is Asia prepared for an epidemic? *Lancet Infect. Dis.* **2019**, *19*, 241–242. [CrossRef]

21. Wasserman, S.; Tambyah, P.A.; Lim, P.L. Yellow fever cases in Asia: Primed for an epidemic. *Int. J. Infect. Dis.* **2016**, *48*, 98–103. [CrossRef] [PubMed]

22. Duffy, M.R.; Chen, T.H.; Hancock, W.T.; Powers, A.M.; Kool, J.L.; Lanciotti, R.S.; Pretrick, M.; Marfel, M.; Holzbauer, S.; Dubray, C.; et al. Zika virus outbreak on Yap Island, Federated States of Micronesia. *N. Engl. J. Med.* **2009**, *360*, 2536–2543. [CrossRef] [PubMed]

23. Musso, D.; Bossin, H.; Mallet, H.P.; Besnard, M.; Broult, J.; Baudouin, L.; Levi, J.E.; Sabino, E.C.; Ghawche, F.; Lanteri, M.C.; et al. Zika virus in French Polynesia 2013-14: Anatomy of a completed outbreak. *Lancet Infect. Dis.* **2018**, *18*, e172–e182. [CrossRef]

24. Faria, N.R.; Azevedo, R.d.S.d.S.; Kraemer, M.U.G.; Souza, R.; Cunha, M.S.; Hill, S.C.; Thézé, J.; Bonsall, M.B.; Bowden, T.A.; Rissanen, I.; et al. Zika virus in the Americas: Early epidemiological and genetic findings. *Science* **2016**, *352*, 345–349. [CrossRef]

25. Paixao, E.S.; Leong, W.Y.; Rodrigues, L.C.; Wilder-Smith, A. Asymptomatic Prenatal Zika Virus Infection and Congenital Zika Syndrome. *Open Forum. Infect. Dis.* **2018**, *5*, 073. [CrossRef]

26. Moghadas, S.M.; Shoukat, A.; Espindola, A.L.; Pereira, R.S.; Abdirizak, F.; Laskowski, M.; Viboud, C.; Chowell, G. Asymptomatic Transmission and the Dynamics of Zika Infection. *Sci. Rep.* **2017**, *7*, 5829. [CrossRef]

27. Kraemer, M.U.G.; Reiner, R.C., Jr.; Brady, O.J.; Messina, J.P.; Gilbert, M.; Pigott, D.M.; Yi, D.; Johnson, K.; Earl, L.; Marczak, L.B.; et al. Past and future spread of the arbovirus vectors Aedes aegypti and Aedes albopictus. *Nat. Microbiol.* **2019**, *4*, 854–863. [CrossRef]

28. Anderson, C.R.; Downs, W.G.; Wattley, G.H.; Ahin, N.W.; Reese, A.A. Mayaro virus: A new human disease agent. II. Isolation from blood of patients in Trinidad, B.W.I. *Am. J. Trop. Med. Hyg.* **1957**, *6*, 1012–1016. [CrossRef]

29. Engel, D.; Jost, H.; Wink, M.; Borstler, J.; Bosch, S.; Garigliany, M.M.; Jost, A.; Czajka, C.; Luhken, R.; Ziegler, U.; et al. Reconstruction of the Evolutionary History and Dispersal of Usutu Virus, a Neglected Emerging Arbovirus in Europe and Africa. *mBio* **2016**, *7*. [CrossRef]

30. Acosta-Ampudia, Y.; Monsalve, D.M.; Rodriguez, Y.; Pacheco, Y.; Anaya, J.M.; Ramirez-Santana, C. Mayaro: An emerging viral threat? *Emerg. Microbes Infect.* **2018**, *7*, 163. [CrossRef]

31. Nikolay, B.; Diallo, M.; Boye, C.S.; Sall, A.A. Usutu virus in Africa. *Vector Borne Zoonotic Dis.* **2011**, *11*, 1417–1423. [CrossRef] [PubMed]

32. Ashraf, U.; Ye, J.; Ruan, X.; Wan, S.; Zhu, B.; Cao, S. Usutu virus: An emerging flavivirus in Europe. *Viruses* **2015**, *7*, 219–238. [CrossRef] [PubMed]

33. Kraemer, M.U.; Sinka, M.E.; Duda, K.A.; Mylne, A.Q.; Shearer, F.M.; Barker, C.M.; Moore, C.G.; Carvalho, R.G.; Coelho, G.E.; Van Bortel, W. The global distribution of the arbovirus vectors Aedes aegypti and Ae. albopictus. *Elife* **2015**, *4*, e08347. [CrossRef] [PubMed]

34. Kramer, L.D.; Ebel, G.D. Dynamics of flavivirus infection in mosquitoes. *Adv. Virus Res.* **2003**, *60*, 187–232. [PubMed]

35. Sim, S.; Jupatanakul, N.; Dimopoulos, G. Mosquito immunity against arboviruses. *Viruses* **2014**, *6*, 4479–4504. [CrossRef]

36. Blair, C.D. Mosquito RNAi is the major innate immune pathway controlling arbovirus infection and transmission. *Future Microbiol.* **2011**, *6*, 265–277. [CrossRef]

37. Patterson, E.I.; Khanipov, K.; Rojas, M.M.; Kautz, T.F.; Rockx-Brouwer, D.; Golovko, G.; Albayrak, L.; Fofanov, Y.; Forrester, N.L. Mosquito bottlenecks alter viral mutant swarm in a tissue and time-dependent manner with contraction and expansion of variant positions and diversity. *Virus Evol.* **2018**, *4*, 001. [CrossRef]

38. Grubaugh, N.D.; Weger-Lucarelli, J.; Murrieta, R.A.; Fauver, J.R.; Garcia-Luna, S.M.; Prasad, A.N.; Black, W.C.t.; Ebel, G.D. Genetic Drift during Systemic Arbovirus Infection of Mosquito Vectors Leads to Decreased Relative Fitness during Host Switching. *Cell Host Microbe* **2016**, *19*, 481–492. [CrossRef]

39. Coffey, L.L.; Forrester, N.; Tsetsarkin, K.; Vasilakis, N.; Weaver, S.C. Factors shaping the adaptive landscape for arboviruses: Implications for the emergence of disease. *Future Microbiol.* **2013**, *8*, 155–176. [CrossRef]

40. Flores, H.A.; O'Neill, S.L. Controlling vector-borne diseases by releasing modified mosquitoes. *Nat. Rev. Microbiol.* **2018**, *16*, 508–518. [CrossRef]

41. Fang, J. Ecology: A world without mosquitoes. *Nature* **2010**, *466*, 432–434. [CrossRef] [PubMed]

42. Ostera, G.R.; Gostin, L.O. Biosafety concerns involving genetically modified mosquitoes to combat malaria and dengue in developing countries. *JAMA* **2011**, *305*, 930–931. [CrossRef]

43. Carballar-Lejarazu, R.; James, A.A. Population modification of Anopheline species to control malaria transmission. *Pathog. Glob. Health* **2017**, *111*, 424–435. [CrossRef] [PubMed]

44. Champer, J.; Buchman, A.; Akbari, O.S. Cheating evolution: Engineering gene drives to manipulate the fate of wild populations. *Nat. Rev. Genet.* **2016**, *17*, 146–159. [CrossRef] [PubMed]

45. Franz, A.W.; Sanchez-Vargas, I.; Adelman, Z.N.; Blair, C.D.; Beaty, B.J.; James, A.A.; Olson, K.E. Engineering RNA interference-based resistance to dengue virus type 2 in genetically modified Aedes aegypti. *Proc. Natl. Acad. Sci. USA* **2006**, *103*, 4198–4203. [CrossRef]

46. Mathur, G.; Sanchez-Vargas, I.; Alvarez, D.; Olson, K.E.; Marinotti, O.; James, A.A. Transgene-mediated suppression of dengue viruses in the salivary glands of the yellow fever mosquito, Aedes aegypti. *Insect Mol. Biol.* **2010**, *19*, 753–763. [CrossRef]

47. Yen, P.S.; James, A.; Li, J.C.; Chen, C.H.; Failloux, A.B. Synthetic miRNAs induce dual arboviral-resistance phenotypes in the vector mosquito Aedes aegypti. *Commun. Biol.* **2018**, *1*, 11. [CrossRef]

48. Ramyasoma, H.; Dassanayake, R.S.; Hapugoda, M.; Capurro, M.L.; Silva Gunawardene, Y.I.N. Multiple dengue virus serotypes resistant transgenic Aedes aegypti fitness evaluated under laboratory conditions. *RNA Biol.* **2020**, 1–21. [CrossRef]

49. Buchman, A.; Gamez, S.; Li, M.; Antoshechkin, I.; Li, H.H.; Wang, H.W.; Chen, C.H.; Klein, M.J.; Duchemin, J.B.; Paradkar, P.N.; et al. Engineered resistance to Zika virus in transgenic Aedes aegypti expressing a polycistronic cluster of synthetic small RNAs. *Proc. Natl. Acad. Sci. USA* **2019**, *116*, 3656–3661. [CrossRef]

50. Mishra, P.; Furey, C.; Balaraman, V.; Fraser, M.J. Antiviral Hammerhead Ribozymes Are Effective for Developing Transgenic Suppression of Chikungunya Virus in Aedes aegypti Mosquitoes. *Viruses* **2016**, *8*, 163. [CrossRef]

51. Carter, J.R.; Keith, J.H.; Fraser, T.S.; Dawson, J.L.; Kucharski, C.A.; Horne, K.M.; Higgs, S.; Fraser, M.J., Jr. Effective suppression of dengue virus using a novel group-I intron that induces apoptotic cell death upon infection through conditional expression of the Bax C-terminal domain. *Virol. J.* **2014**, *11*, 111. [CrossRef] [PubMed]

52. Pan, X.; Zhou, G.; Wu, J.; Bian, G.; Lu, P.; Raikhel, A.S.; Xi, Z. Wolbachia induces reactive oxygen species (ROS)-dependent activation of the Toll pathway to control dengue virus in the mosquito Aedes aegypti. *Proc. Natl. Acad. Sci. USA* **2012**, *109*, E23–E31. [CrossRef]

53. Buchman, A.; Gamez, S.; Li, M.; Antoshechkin, I.; Li, H.H.; Wang, H.W.; Chen, C.H.; Klein, M.J.; Duchemin, J.B.; Crowe, J.E., Jr.; et al. Broad dengue neutralization in mosquitoes expressing an engineered antibody. *PLoS Pathog.* **2020**, *16*, e1008103. [CrossRef] [PubMed]

54. Okorie, P.N.; Marshall, J.M.; Akpa, O.M.; Ademowo, O.G. Perceptions and recommendations by scientists for a potential release of genetically modified mosquitoes in Nigeria. *Malar J.* **2014**, *13*, 154. [CrossRef]

55. Lambrechts, L.; Saleh, M.C. Manipulating Mosquito Tolerance for Arbovirus Control. *Cell Host Microbe* **2019**, *26*, 309–313. [CrossRef]

56. Cirimotich, C.M.; Ramirez, J.L.; Dimopoulos, G. Native microbiota shape insect vector competence for human pathogens. *Cell Host Microbe* **2011**, *10*, 307–310. [CrossRef]

57. Minard, G.; Tran, F.H.; Raharimalala, F.N.; Hellard, E.; Ravelonandro, P.; Mavingui, P.; Valiente Moro, C. Prevalence, genomic and metabolic profiles of Acinetobacter and Asaia associated with field-caught Aedes albopictus from Madagascar. *FEMS Microbiol. Ecol.* **2013**, *83*, 63–73. [CrossRef]

58. Bourtzis, K.; Dobson, S.L.; Xi, Z.; Rasgon, J.L.; Calvitti, M.; Moreira, L.A.; Bossin, H.C.; Moretti, R.; Baton, L.A.; Hughes, G.L.; et al. Harnessing mosquito-Wolbachia symbiosis for vector and disease control. *Acta Trop.* **2014**, *132*, S150–S163. [CrossRef]

59. Dutra, H.L.C.; Rocha, M.N.; Dias, F.B.S.; Mansur, S.B.; Caragata, E.P.; Moreira, L.A. Wolbachia blocks currently circulating Zika virus isolates in Brazilian Aedes aegypti mosquitoes. *Cell Host Microbe* **2016**, *19*, 771–774. [CrossRef]

60. Masui, S.; Kamoda, S.; Sasaki, T.; Ishikawa, H. Distribution and evolution of bacteriophage WO in Wolbachia, the endosymbiont causing sexual alterations in arthropods. *J. Mol. Evol.* **2000**, *51*, 491–497. [CrossRef]

61. Hertig, M.; Wolbach, S.B. Studies on Rickettsia-Like Micro-Organisms in Insects. *J. Med. Res.* **1924**, *44*, 329–374.7.

62. Werren, J.H.; Baldo, L.; Clark, M.E. Wolbachia: Master manipulators of invertebrate biology. *Nat. Rev. Microbiol.* **2008**, *6*, 741–751. [CrossRef]

63. Landmann, F. The Wolbachia Endosymbionts. *Microbiol. Spectr.* **2019**, *7*. [CrossRef]

64. Hilgenboecker, K.; Hammerstein, P.; Schlattmann, P.; Telschow, A.; Werren, J.H. How many species are infected with Wolbachia?—A statistical analysis of current data. *FEMS Microbiol. Lett.* **2008**, *281*, 215–220. [CrossRef]

65. Bonneau, M.; Landmann, F.; Labbe, P.; Justy, F.; Weill, M.; Sicard, M. The cellular phenotype of cytoplasmic incompatibility in Culex pipiens in the light of cidB diversity. *PLoS Pathog.* **2018**, *14*, e1007364. [CrossRef]

66. Beckmann, J.F.; Ronau, J.A.; Hochstrasser, M. A Wolbachia deubiquitylating enzyme induces cytoplasmic incompatibility. *Nat. Microbiol.* **2017**, *2*, 17007. [CrossRef]

67. Pietri, J.E.; DeBruhl, H.; Sullivan, W. The rich somatic life of Wolbachia. *Microbiologyopen* **2016**, *5*, 923–936. [CrossRef]

68. Osborne, S.E.; Leong, Y.S.; O'Neill, S.L.; Johnson, K.N. Variation in antiviral protection mediated by different Wolbachia strains in Drosophila simulans. *PLoS Pathog.* **2009**, *5*, e1000656. [CrossRef]

69. Teixeira, L.; Ferreira, A.; Ashburner, M. The bacterial symbiont Wolbachia induces resistance to RNA viral infections in Drosophila melanogaster. *PLoS Biol.* **2008**, *6*, e2. [CrossRef]

70. Moreira, L.A.; Iturbe-Ormaetxe, I.; Jeffery, J.A.; Lu, G.; Pyke, A.T.; Hedges, L.M.; Rocha, B.C.; Hall-Mendelin, S.; Day, A.; Riegler, M.; et al. A Wolbachia symbiont in Aedes aegypti limits infection with dengue, Chikungunya, and Plasmodium. *Cell* **2009**, *139*, 1268–1278. [CrossRef]

71. Hedges, L.M.; Brownlie, J.C.; O'Neill, S.L.; Johnson, K.N. Wolbachia and virus protection in insects. *Science* **2008**, *322*, 702. [CrossRef]

72. Souza-Neto, J.A.; Powell, J.R.; Bonizzoni, M. Aedes aegypti vector competence studies: A review. *Infect. Genet. Evol.* **2019**, *67*, 191–209. [CrossRef]

73. Hoffmann, A.A.; Montgomery, B.L.; Popovici, J.; Iturbe-Ormaetxe, I.; Johnson, P.H.; Muzzi, F.; Greenfield, M.; Durkan, M.; Leong, Y.S.; Dong, Y.; et al. Successful establishment of Wolbachia in Aedes populations to suppress dengue transmission. *Nature* **2011**, *476*, 454–457. [CrossRef]

74. Zouache, K.; Voronin, D.; Tran-Van, V.; Mousson, L.; Failloux, A.B.; Mavingui, P. Persistent *Wolbachia* and cultivable bacteria infection in the reproductive and somatic tissues of the mosquito vector *Aedes albopictus*. *PLoS ONE* **2009**, *4*, e6388. [CrossRef]

75. Lu, P.; Bian, G.; Pan, X.; Xi, Z. Wolbachia induces density-dependent inhibition to dengue virus in mosquito cells. *PLoS Negl. Trop. Dis.* **2012**, *6*, e1754. [CrossRef]

76. Martinez, J.; Ok, S.; Smith, S.; Snoeck, K.; Day, J.P.; Jiggins, F.M. Should Symbionts Be Nice or Selfish? Antiviral Effects of Wolbachia Are Costly but Reproductive Parasitism Is Not. *PLoS Pathog.* **2015**, *11*, e1005021. [CrossRef]

77. Martinez, J.; Bruner-Montero, G.; Arunkumar, R.; Smith, S.C.L.; Day, J.P.; Longdon, B.; Jiggins, F.M. Virus evolution in Wolbachia-infected Drosophila. *Proc. Biol. Sci.* **2019**, *286*, 20192117. [CrossRef]

78. Kambris, Z.; Cook, P.E.; Phuc, H.K.; Sinkins, S.P. Immune activation by life-shortening Wolbachia and reduced filarial competence in mosquitoes. *Science* **2009**, *326*, 134–136. [CrossRef]

79. Kambris, Z.; Blagborough, A.M.; Pinto, S.B.; Blagrove, M.S.; Godfray, H.C.; Sinden, R.E.; Sinkins, S.P. Wolbachia stimulates immune gene expression and inhibits plasmodium development in Anopheles gambiae. *PLoS Pathog.* **2010**, *6*, e1001143. [CrossRef]

80. Silva, J.B.L.; Magalhaes Alves, D.; Bottino-Rojas, V.; Pereira, T.N.; Sorgine, M.H.F.; Caragata, E.P.; Moreira, L.A. Wolbachia and dengue virus infection in the mosquito Aedes fluviatilis (Diptera: Culicidae). *PLoS ONE* **2017**, *12*, e0181678. [CrossRef]

81. Zhang, G.; Hussain, M.; O'Neill, S.L.; Asgari, S. Wolbachia uses a host microRNA to regulate transcripts of a methyltransferase, contributing to dengue virus inhibition in Aedes aegypti. *Proc. Natl. Acad. Sci. USA* **2013**, *110*, 10276–10281. [CrossRef]

82. Mayoral, J.G.; Etebari, K.; Hussain, M.; Khromykh, A.A.; Asgari, S. Wolbachia infection modifies the profile, shuttling and structure of microRNAs in a mosquito cell line. *PLoS ONE* **2014**, *9*, e96107. [CrossRef]

83. Slonchak, A.; Hussain, M.; Torres, S.; Asgari, S.; Khromykh, A.A. Expression of mosquito microRNA Aae-miR-2940-5p is downregulated in response to West Nile virus infection to restrict viral replication. *J. Virol.* **2014**, *88*, 8457–8467. [CrossRef]

84. Rainey, S.M.; Martinez, J.; McFarlane, M.; Juneja, P.; Sarkies, P.; Lulla, A.; Schnettler, E.; Varjak, M.; Merits, A.; Miska, E.A.; et al. Wolbachia Blocks Viral Genome Replication Early in Infection without a Transcriptional Response by the Endosymbiont or Host Small RNA Pathways. *PLoS Pathog.* **2016**, *12*, e1005536. [CrossRef]

85. Geoghegan, V.; Stainton, K.; Rainey, S.M.; Ant, T.H.; Dowle, A.A.; Larson, T.; Hester, S.; Charles, P.D.; Thomas, B.; Sinkins, S.P. Perturbed cholesterol and vesicular trafficking associated with dengue blocking in Wolbachia-infected Aedes aegypti cells. *Nat. Commun.* **2017**, *8*, 526. [CrossRef]

86. Koh, C.; Audsley, M.D.; Di Giallonardo, F.; Kerton, E.J.; Young, P.R.; Holmes, E.C.; McGraw, E.A. Sustained Wolbachia-mediated blocking of dengue virus isolates following serial passage in Aedes aegypti cell culture. *Virus Evol.* **2019**, *5*, 012. [CrossRef]

87. Teramoto, T.; Huang, X.; Armbruster, P.A.; Padmanabhan, R. Infection of Aedes albopictus Mosquito C6/36 Cells with the wMelpop Strain of Wolbachia Modulates Dengue Virus-Induced Host Cellular Transcripts and Induces Critical Sequence Alterations in the Dengue Viral Genome. *J. Virol.* **2019**, *93*. [CrossRef]

88. Blagrove, M.S.; Arias-Goeta, C.; Di Genua, C.; Failloux, A.B.; Sinkins, S.P. A Wolbachia wMel transinfection in Aedes albopictus is not detrimental to host fitness and inhibits Chikungunya virus. *PLoS Negl. Trop. Dis.* **2013**, *7*, e2152. [CrossRef]

89. Ahmad, N.A.; Vythilingam, I.; Lim, Y.A.L.; Zabari, N.; Lee, H.L. Detection of Wolbachia in Aedes albopictus and Their Effects on Chikungunya Virus. *Am. J. Trop. Med. Hyg.* **2017**, *96*, 148–156. [CrossRef]

90. Mousson, L.; Zouache, K.; Arias-Goeta, C.; Raquin, V.; Mavingui, P.; Failloux, A.B. The native Wolbachia symbionts limit transmission of dengue virus in Aedes albopictus. *PLoS Negl. Trop. Dis.* **2012**, *6*, e1989. [CrossRef]

91. Mousson, L.; Martin, E.; Zouache, K.; Madec, Y.; Mavingui, P.; Failloux, A.B. Wolbachia modulates Chikungunya replication in Aedes albopictus. *Mol. Ecol.* **2010**, *19*, 1953–1964. [CrossRef] [PubMed]

92. Bian, G.; Xu, Y.; Lu, P.; Xie, Y.; Xi, Z. The endosymbiotic bacterium Wolbachia induces resistance to dengue virus in Aedes aegypti. *PLoS Pathog.* **2010**, *6*, e1000833. [CrossRef] [PubMed]

93. Bian, G.; Joshi, D.; Dong, Y.; Lu, P.; Zhou, G.; Pan, X.; Xu, Y.; Dimopoulos, G.; Xi, Z. Wolbachia invades Anopheles stephensi populations and induces refractoriness to Plasmodium infection. *Science* **2013**, *340*, 748–751. [CrossRef]

94. Foo, I.J.; Hoffmann, A.A.; Ross, P.A. Cross-Generational Effects of Heat Stress on Fitness and Wolbachia Density in Aedes aegypti Mosquitoes. *Trop. Med. Infect. Dis.* **2019**, *4*, 13. [CrossRef]

95. Ross, P.A.; Ritchie, S.A.; Axford, J.K.; Hoffmann, A.A. Loss of cytoplasmic incompatibility in Wolbachia-infected Aedes aegypti under field conditions. *PLoS Negl. Trop. Dis.* **2019**, *13*, e0007357. [CrossRef]

96. Ulrich, J.N.; Beier, J.C.; Devine, G.J.; Hugo, L.E. Heat Sensitivity of wMel Wolbachia during Aedes aegypti Development. *PLoS Negl. Trop. Dis.* **2016**, *10*, e0004873. [CrossRef]

97. Ye, Y.H.; Carrasco, A.M.; Dong, Y.; Sgro, C.M.; McGraw, E.A. The Effect of Temperature on Wolbachia-Mediated Dengue Virus Blocking in Aedes aegypti. *Am. J. Trop. Med. Hyg.* **2016**, *94*, 812–819. [CrossRef]

98. Gavotte, L.; Henri, H.; Stouthamer, R.; Charif, D.; Charlat, S.; Bouletreau, M.; Vavre, F. A Survey of the bacteriophage WO in the endosymbiotic bacteria Wolbachia. *Mol. Biol. Evol.* **2007**, *24*, 427–435. [CrossRef]

99. Gavotte, L.; Vavre, F.; Henri, H.; Ravallec, M.; Stouthamer, R.; Bouletreau, M. Diversity, distribution and specificity of WO phage infection in Wolbachia of four insect species. *Insect Mol. Biol.* **2004**, *13*, 147–153. [CrossRef] [PubMed]

100. Chauvatcharin, N.; Ahantarig, A.; Baimai, V.; Kittayapong, P. Bacteriophage WO-B and Wolbachia in natural mosquito hosts: Infection incidence, transmission mode and relative density. *Mol. Ecol.* **2006**, *15*, 2451–2461. [CrossRef]

101. Bordenstein, S.R.; Bordenstein, S.R. Temperature affects the tripartite interactions between bacteriophage WO, Wolbachia, and cytoplasmic incompatibility. *PLoS ONE* **2011**, *6*, e29106. [CrossRef]

102. Tsai, C.H.; Chen, T.H.; Lin, C.; Shu, P.Y.; Su, C.L.; Teng, H.J. The impact of temperature and Wolbachia infection on vector competence of potential dengue vectors Aedes aegypti and Aedes albopictus in the transmission of dengue virus serotype 1 in southern Taiwan. *Parasit Vectors* **2017**, *10*, 551. [CrossRef]

103. Carrington, L.B.; Armijos, M.V.; Lambrechts, L.; Scott, T.W. Fluctuations at a low mean temperature accelerate dengue virus transmission by Aedes aegypti. *PLoS Negl. Trop. Dis.* **2013**, *7*, e2190. [CrossRef]

104. Dohm, D.J.; O'Guinn, M.L.; Turell, M.J. Effect of environmental temperature on the ability of Culex pipiens (Diptera: Culicidae) to transmit West Nile virus. *J. Med. Entomol.* **2002**, *39*, 221–225. [CrossRef]

105. Mourya, D.T.; Yadav, P.; Mishra, A.C. Effect of temperature stress on immature stages and susceptibility of Aedes aegypti mosquitoes to chikungunya virus. *Am. J. Trop. Med. Hyg.* **2004**, *70*, 346–350. [CrossRef]

106. Serbus, L.R.; White, P.M.; Silva, J.P.; Rabe, A.; Teixeira, L.; Albertson, R.; Sullivan, W. The impact of host diet on Wolbachia titer in Drosophila. *PLoS Pathog.* **2015**, *11*, e1004777. [CrossRef]

107. Walker, T.; Johnson, P.H.; Moreira, L.A.; Iturbe-Ormaetxe, I.; Frentiu, F.D.; McMeniman, C.J.; Leong, Y.S.; Dong, Y.; Axford, J.; Kriesner, P.; et al. The wMel Wolbachia strain blocks dengue and invades caged Aedes aegypti populations. *Nature* **2011**, *476*, 450–453. [CrossRef]

108. Balaji, S.; Jayachandran, S.; Prabagaran, S.R. Evidence for the natural occurrence of Wolbachia in Aedes aegypti mosquitoes. *FEMS Microbiol. Lett.* **2019**, *366*. [CrossRef]

109. Teo, C.; Lim, P.; Voon, K.; Mak, J. Detection of dengue viruses and Wolbachia in Aedes aegypti and Aedes albopictus larvae from four urban localities in Kuala Lumpur, Malaysia. *Trop. Biomed.* **2017**, *34*, 14.

110. Bennett, K.L.; Gomez-Martinez, C.; Chin, Y.; Saltonstall, K.; McMillan, W.O.; Rovira, J.R.; Loaiza, J.R. Dynamics and diversity of bacteria associated with the disease vectors Aedes aegypti and Aedes albopictus. *Sci. Rep.* **2019**, *9*, 12160. [CrossRef]

111. Thongsripong, P.; Chandler, J.A.; Green, A.B.; Kittayapong, P.; Wilcox, B.A.; Kapan, D.D.; Bennett, S.N. Mosquito vector-associated microbiota: Metabarcoding bacteria and eukaryotic symbionts across habitat types in Thailand endemic for dengue and other arthropod-borne diseases. *Ecol. Evol.* **2018**, *8*, 1352–1368. [CrossRef] [PubMed]

112. Carvajal, T.M.; Hashimoto, K.; Harnandika, R.K.; Amalin, D.M.; Watanabe, K. Detection of Wolbachia in field-collected Aedes aegypti mosquitoes in metropolitan Manila, Philippines. *Parasit Vectors* **2019**, *12*, 361. [CrossRef]

113. Kulkarni, A.; Yu, W.; Jiang, J.; Sanchez, C.; Karna, A.K.; Martinez, K.J.L.; Hanley, K.A.; Buenemann, M.; Hansen, I.A.; Xue, R.D.; et al. Wolbachia pipientis occurs in Aedes aegypti populations in New Mexico and Florida, USA. *Ecol. Evol.* **2019**, *9*, 6148–6156. [CrossRef] [PubMed]
114. Coon, K.L.; Brown, M.R.; Strand, M.R. Mosquitoes host communities of bacteria that are essential for development but vary greatly between local habitats. *Mol. Ecol.* **2016**, *25*, 5806–5826. [CrossRef]
115. Hegde, S.; Khanipov, K.; Albayrak, L.; Golovko, G.; Pimenova, M.; Saldana, M.A.; Rojas, M.M.; Hornett, E.A.; Motl, G.C.; Fredregill, C.L.; et al. Microbiome Interaction Networks and Community Structure From Laboratory-Reared and Field-Collected Aedes aegypti, Aedes albopictus, and Culex quinquefasciatus Mosquito Vectors. *Front. Microbiol.* **2018**, *9*, 2160. [CrossRef]
116. Crain, P.R.; Crowley, P.H.; Dobson, S.L. Wolbachia re-replacement without incompatibility: Potential for intended and unintended consequences. *J. Med. Entomol.* **2013**, *50*, 1152–1158. [CrossRef]

Article

A Systematic Review: Is *Aedes albopictus* an Efficient Bridge Vector for Zoonotic Arboviruses?

Taissa Pereira-dos-Santos [1,*], **David Roiz** [1], **Ricardo Lourenço-de-Oliveira** [2] **and Christophe Paupy** [1,*]

[1]   MIVEGEC, Univ. Montpellier, IRD, CNRS, 34090 Montpellier, France; david.roiz@ird.fr
[2]   LATHEMA, Instituto Oswaldo Cruz, FIOCRUZ, Rio de Janeiro-RJ 4364, Brazil; lourenco@ioc.fiocruz.br
*   Correspondence: tayssadnz@gmail.com (T.P.-d.-S.); christophe.paupy@ird.fr (C.P.)

Received: 15 March 2020; Accepted: 4 April 2020; Published: 7 April 2020

Abstract: Mosquito-borne arboviruses are increasing due to human disturbances of natural ecosystems and globalization of trade and travel. These anthropic changes may affect mosquito communities by modulating ecological traits that influence the "spill-over" dynamics of zoonotic pathogens, especially at the interface between natural and human environments. Particularly, the global invasion of *Aedes albopictus* is observed not only across urban and peri-urban settings, but also in newly invaded areas in natural settings. This could foster the interaction of *Ae. albopictus* with wildlife, including local reservoirs of enzootic arboviruses, with implications for the potential zoonotic transfer of pathogens. To evaluate the potential of *Ae. albopictus* as a bridge vector of arboviruses between wildlife and humans, we performed a bibliographic search and analysis focusing on three components: (1) The capacity of *Ae. albopictus* to exploit natural larval breeding sites, (2) the blood-feeding behaviour of *Ae. albopictus*, and (3) *Ae. albopictus'* vector competence for arboviruses. Our analysis confirms the potential of *Ae. albopictus* as a bridge vector based on its colonization of natural breeding sites in newly invaded areas, its opportunistic feeding behaviour together with the preference for human blood, and the competence to transmit 14 arboviruses.

Keywords: *Aedes albopictus*; emerging diseases; vector competence; spill-over; blood-feeding; bridge vector; arboviruses; mosquito

---

1. Introduction

The human alteration of Earth's natural systems has become a great concern and a threat to human health. Indeed, these changes are likely to drive most of the global disease burden over the coming century [1]. During the last decades, the burden of emerging infectious diseases has increased to represent a substantial threat to global health, security, and economy growth. About 75% of emerging infectious diseases are zoonotic diseases, mostly of wildlife origin [2,3]. The risk of zoonotic emergences is considered high in tropical forest regions associated with a range of facilitating factors, particularly high vertebrate species diversity and agricultural land use changes [4]. Understanding the mechanisms of disease emergence allows the development of early detection and control programs for reducing disease incidence and economic burden [5].

Zoonotic pathogens can be transmitted from animals to humans directly, or indirectly when arthropod vectors are needed to accomplish their life cycle. Zoonotic vector-borne diseases are maintained in enzootic cycles, but can be transmitted from animal reservoir populations to sympatric human populations or to domestic animals during "spill-over events", and also from humans to animals during "spill-back events" [2,6]. The global emergence of vector-borne diseases is helped by international travel and trade, after their local emergence has been driven by a combination of environmental changes that are not yet completely understood [7]. Therefore, research is needed to determine the potential of these pathogens to emerge in the future, and to identify critical geographic areas where early warning systems must be put in place to mitigate the pathogen's impact on human health [8].

Here, we focused on zoonotic arboviruses (arthropod-borne viruses) transmitted by mosquitoes that are part of enzootic cycles evolving in wildlife or domestic animals, independently of mankind. Animals might act as amplification hosts for spill-over events to humans [9], mainly in tropical forest environments [10]. Some arboviruses, such as those causing epidemic *Aedes*-borne viral diseases (dengue, chikungunya, and zika), have adapted to epidemic cycles in which viremic humans became the source of infection in urban areas where *Aedes aegypti*, and to a lesser extent, *Aedes albopictus* [11,12] ensure person-to-person transmission [13]. The burden of *Aedes*-borne diseases is dramatic. For instance, dengue incidence has increased by 30 times over the last 50 years, with about 390 million infections reported annually worldwide [14,15]. Dengue and chikungunya outbreak waves have resulted in several million cases in the Southwest Indian Ocean region, India, and the Americas [16]. Zika virus (ZIKAV) disease emerged in 87 countries (or territories) [17]. ZIKAV infection during pregnancy can cause microcephaly in newborns and is becoming a major threat due to its long-term sanitary and economic impacts, especially in Latin America [18]. Although these infection outbreaks are caused by independent urban cycles, enzootic cycles still remain essential sources of pathogens and/or vectors that can be introduced, adapt, and disperse, causing new severe threats [2], as exemplified by the recent re-emergence of Yellow Fever Virus (YFV) in Brazil, Angola, and the Democratic Republic of Congo [19,20]. For YFV, spill-over events from non-human primates that involve mosquito bridge vectors have been described in tropical Africa (e.g., involving *Aedes africanus* or *Aedes furcifer*) [21] and in America (involving mosquito species from the *Haemagogus* and *Sabethes* genera) [20]. After its introduction in the Americas, YFV has efficiently spilled back into sylvatic cycles via bridge vectors. In African villages or cities, YFV transmission is supported by epidemic vectors, such as *Ae. aegypti* [21,22]. These data indicate that mosquito bridge vectors play key roles in the early processes leading to the emergence of enzootic viruses, before the urban transmission cycles [6,8].

We define a bridge vector as an "appropriate hematophagous arthropod" that ensures the biological transmission of a pathogen across different landscapes and its circulation between enzootic, domestic animal, and human hosts. In the absence of a bridge vector, pathogen transmission generally remains restricted to a specific area within the enzootic or epidemic cycle and among hosts/reservoirs. Bridge vectors are the key that interconnects animal reservoirs to new vertebrate hosts, including humans, and that allows both spill-over and spill-back events. For this study, we considered that bridge vectors show several bio-ecological traits that influence the shifting risks of pathogen transfer and that are mainly related to their ecological distribution, blood feeding behaviour, and vector competence. Regarding ecological habitats, high ecological and physiological plasticity favours the vector dispersal and its establishment (breeding in specific microhabitats) across different ecosystems, landscapes, or habitats (e.g., forest/rural/urban, forest/savannah, ground/canopy, natural/anthropic larval breeding and adult resting sites). Regarding blood feeding behaviour, low specificity in blood-meal sources and opportunistic feeding behaviour involving multiple hosts increases the probability of contact between the vector and different animal reservoirs, and thus interspecies pathogen transfer. This probability also depends on the vector and host density and on the host's defensive behaviour. Regarding vector competence, for the biological transmission of a pathogen after its acquisition on an infected vertebrate, a bridge vector must be able to ensure its replication/multiplication, dissemination, and transmission to subsequently bitten vertebrates. Arthropod species competent for a large panel of pathogens or with high vector competence for one pathogen represent particularly suitable candidates to act as (bridge) vectors.

Here, we evaluated the potential role of the Asian tiger mosquito *Ae. albopictus* as a bridge vector. This invasive vector species originates from Asian tropical forests, but nowadays is present in all continents [23], and has become a major public health issue. In its native area, sylvatic *Ae. albopictus* populations complete their biological cycle by exploiting wild animals as blood sources, and natural water collection points (e.g., tree holes, bamboo stumps, or rock holes) as oviposition sites in the woods [24], particularly at the forest edge [24]. The capacity to colonize artificial man-made containers (together with desiccation-resistant and diapausing eggs) led to its "domestication". Its ecological plasticity to several habitats, its passive dispersion through the global transport of tires and inside cars [25], and the inefficiency of control programmes have allowed *Ae. albopictus* to become one of most invasive species worldwide [23,24]. In native and newly colonized areas, it has been found in urban, rural, and forest habitats; however, unclear information is available on its natural breeding sites and its presence in forested environments. Moreover, its presence in natural breeding sites in the invaded territories has not been analysed. In general, *Ae. albopictus* is considered an opportunistic feeder that is attracted to mammals, particularly humans, rather than other hosts [26,27]. However, to our knowledge, a detailed and quantified analysis of its host preferences has never been done.

In relation to epidemic virus transmission, *Ae. albopictus* has been considered the vector for the chikungunya virus (CHIKV), dengue virus (DENV), and ZIKAV in Gabon and Central Africa [28,29], for DENV and CHIKV in la Réunion island [30], and for CHIKV in Madagascar and Mayotte [30,31]. In Europe, it has been incriminated in Italy and France during CHIKV and DENV outbreaks [32,33] and in Japan in DENV transmission [34]. Moreover, this mosquito represents a potential risk of outbreaks in many other areas, for example, in Brazil and USA where *Ae. albopictus* is widespread [35–38]. Different studies have shown that *Ae. albopictus* can develop infection from up to 32 arboviruses [16,23,36]; however, to our knowledge, its ability to transmit any of them has not been clearly demonstrated yet.

In this work, we hypothesized that *Ae. albopictus* may have an active role as a bridge vector for the transfer from vertebrate hosts to humans (spill-over events) and therefore, in the emergence of enzootic arboviruses. To test this hypothesis, we reviewed and quantified: (1) *Ae. albopictus'* capacity to exploit natural water collections as larval breeding sites (as a proxy for its establishment in rural/sylvatic/forested areas) in native or invaded regions; (2) its feeding behaviour with regard to humans, domestic, or wild animals (as a proxy for the contact between vertebrate hosts and humans); and (3) its vector competence, tested experimentally for different arboviruses and natural infections reported from the field in mosquitoes (as a proxy for its potential for virus transmission in the field). Finally, we discuss the potential spill-over transmission risk from vertebrate hosts to humans and the methodological issues and knowledge gaps that need to be tackled.

2. Results

2.1. Natural Breeding Sites

Based on the literature (see Methods and Supplementary Tables S1 and S2), we found 27 articles that quantified the number and type of natural breeding sites exploited by *Ae. albopictus* in areas where the species is considered native (n = 10 articles) or invasive (i.e., colonized areas) (n = 17 articles). Preimaginal stages of *Ae. albopictus* were mainly detected in coconut shells (54.7%) [37–45], bromeliads (19%) [46–50], bamboo stumps (8.3%) [39,40,51–54], tree holes (8.2%) [37,42,43,51,53–59], palm leaves (3.6%) [51], rock holes (3.2%) [37,42,43,51,53,57,60], leaf axils (1%) [39,40,42,61], and sporadically (<1%) in other natural breeding sites, such as snail shells [43,53], palm bracts [53], dead leaves [37,43], cacao pods on the ground [43], dead cow horns [43], puddles [62], ground cavities [46], and hollow logs [53] (Figure 1).

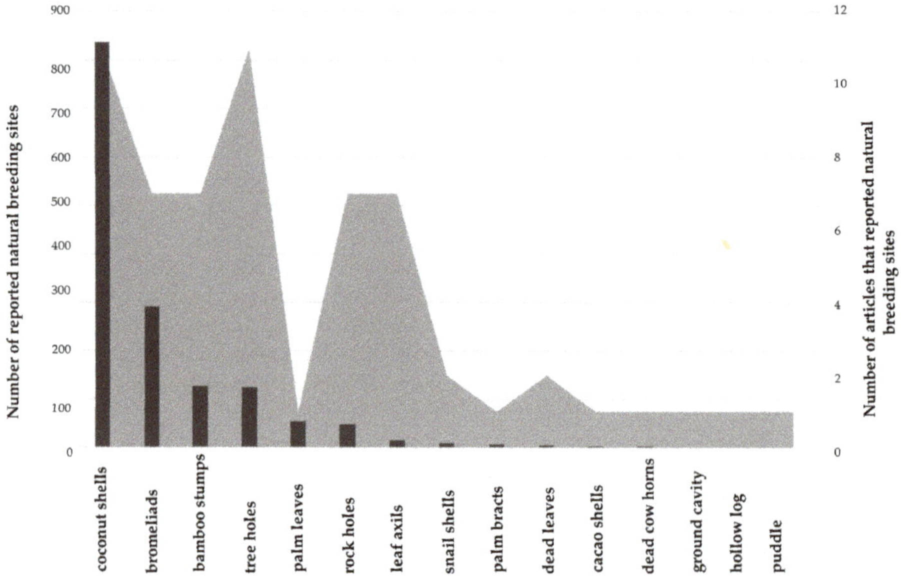

Figure 1. Natural larval breeding sites exploited by *Ae. albopictus*. Number of reported natural breeding sites (black bars) and number of articles that reported natural breeding sites (grey areas).

Coconut shells and tree holes were more often reported (11 articles each), followed by bamboo stumps, bromeliads, rock holes, and leaf axils (7 articles each), and finally, the other natural breeding sites (1–2 articles each). In native areas, most of the reported natural breeding sites were coconut shells (83%), followed by bamboo stumps (11%), tree holes (5%), leaf axils (1%), and rock holes (1%). In colonized areas, a great diversity of breeding sites was reported: bromeliads (50.8%), tree holes (13%), palm leaves (9.6%), rock holes (8.4%), coconut shells (8%), bamboo stumps (3.8%), leaf axils (1%), palm bracts (1.2 %), snail shell (1.7%), and others (<1% each).

2.2. Feeding Behaviour

Our quantification of the feeding behaviour indicates that *Ae. albopictus* has been mainly reported as a species that prefers mammals (including humans) as blood sources (mean and standard deviation: 92.0% ± 8), compared with birds (8% ± 8) and other animals (3.7% ± 1.7) [26,27,63–82] (Figure 2A). Among mammals, blood meals were mainly from humans (60%) than non-human species (30%). From an "interspecies risk of transfer" perspective, it is relevant to note that *Ae. albopictus* seems to be biting domestic animals (25%) more frequently than wildlife animals (10%) (Figure 2B). Importantly, there is huge variability in the percentage of human blood meals in the different studies.

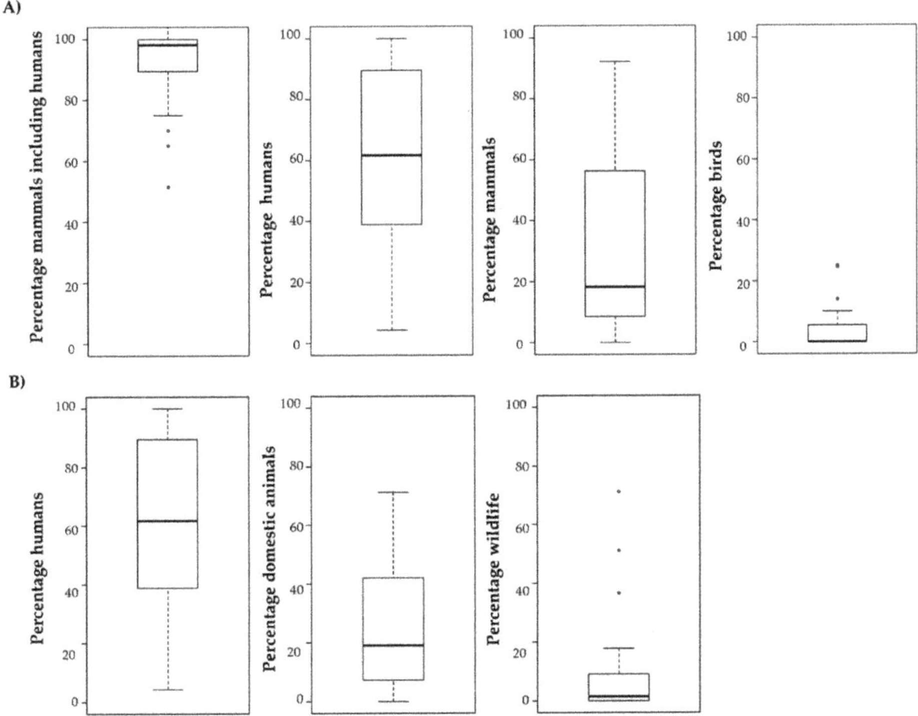

Figure 2. Boxplots showing the host feeding preferences (i.e., percentage of bites) of *Ae. albopictus* without taking into account host availability. (**A**) Mammals, humans, non-human mammals, and birds; (**B**) Humans, domestic animals, and wildlife. Black line: median.

Among domestic and peri-domestic animals, dogs, rodents, and rabbits were reported as the main blood sources for *Ae. albopictus*, followed by cats, bovines, chickens, horses, and pigs (Supplementary Figure S1). When classified according to the biological family of blood sources, *Ae. albopictus* can feed on 28 different host biological families, and preferentially on animals belonging to Hominidae (60%), Muridae (15%), Canidae (12%), and Phasianidae (10%) (see Table 1 for detailed information and Supplementary Table S3 for bibliographical information).

Table 1. Mean biting frequency by *Aedes albopictus* in animals classified according to biological class and family.

Biological Class	Biological Family	Mean Frequency (%)
	Phasianidae	10.08
	Passeridae	7.78
	Anatidae	7.5
	Columbidae	5.83
Aves	Sulidae	2.33
	Thamnophilidae	1.49
	Pycnonotidae	1.39
	Corvidae	1.11
	Ciconiidae	1.0
	Hominidae (Humans)	59.83
	Muridae	15.34
	Canidae	11.6
	Herpestidae	9.53
	Bovidae	8.9
	Felidae	8.49
	Leporidae	8.27
	Sciuridae	5.07
Mammalia	Suidae	4.99
	Didelphidae	4.6
	Equidae	4.39
	Cervidae	4.15
	Muridae/Soricidae	3.43
	Phyllostomidae	2.99
	Procyonidae	2.71
	Furipteridae	1.49
	Cricetidae	0.61
Actinopterygii	Cobitidae	1.11
Amphibia	Salamandridae	2.22

The mean frequencies were calculated using the data found in articles that described different *Ae. albopictus* populations biting different animals in different locations. As these articles do not describe the same biological families, the total mean bite frequency does not correspond to 100%.

2.3. Arbovirus Transmission

In the literature search, in addition to the epidemic DENV (serotypes 1, 2, 3, and 4), CHIKV and ZIKV, we found reports on experimental infections of *Ae. albopictus* with the following 36 arboviruses: Arumowot (AMTV) [83], Bujaru (BUJV) [83], Bussuquara (BSQV) [84], Cache Valley (CVV) [85], Chandipura (CHPV) [86], Chilibre (CHIV) [83], Eastern Equine Encephalomyelitis (EEEV) [87–90], Getah (GETV) [91], Icoaraci (ICOV) [83], Ilheus (ILHV) [92], Itaporanga (ITPV) [83], Jamestown Canyon (JCV) [93], Japanese Encephalitis (JEV) [92,94–96], Karimabad (KARV) [83], Keystone (KEYV) [92,93], Kokobera (KOKV) [92], Kunjin (KUNV) [92], La Crosse (LACV) [92,93,97–99], Mayaro (MAYV) [100], Oropuche (OROV) [100], Orungo (ORUV) [101], Pacui (PACV) [83], Potosi (POTV) [102–104], Rift Valley fever (RVFV) [105,106], Ross River (RRV) [107,108], Salehabad (SALV) [83], San Angelo (SA) [84,92,109], St. Louis encephalitis (SLEV) [110], Tensaw (TENV) [111], Trivittatus (TVTV) [93], Uganda S. (UGSV) [92], Urucuri (URUV) [83], Usutu virus (USUV) [112], Venezuelan equine encephalitis (VEEV) [113–115], West Nile virus (WNV) [116–125], and YFV [108,126–130] (see Supplementary Table S4 for bibliographical information). However, besides the addition to the epidemic DENV (serotypes 1, 2, 3 and 4) [131], CHIKV, and ZIKV [28], natural infections of *Ae. albopictus* were only reported for eight viruses: CCV [85,132], EEEV [133], KEYV [133], LACV [99,132,134,135], POTV [102,132,136], TENV [133], USUV [112], and WNV [118–120] (see Supplementary Table S5 for bibliographical information). These infections were detected by virus isolation on cell lines, immunological or molecular methods (Vero cells, direct or indirect immunofluorescence, polymerase chain reaction). These infections provide evidence of contact between *Ae. albopictus* and the hosts of these viruses, but do not necessarily indicate their biological transmission by this mosquito. On the other hand, for the BSQV, ILHV, KOKV, KUNV, and UGSV arboviruses, only intrathoracic injection experiments were carried out to investigate transovarian transmission between different generations. Supplementary Table S6 gives information on the taxonomic classification of these viruses, their geographic distribution, their natural host family (i.e., vertebrate host family in which the virus was isolated or in which serological evidence was found), the mosquito species from which the virus was isolated, and the detection method in *Ae. albopictus*.

Among studies on *Ae. albopictus* vector competence, we found important variations concerning the methodology used to perform the infection (intra-thoracic inoculation of viruses, oral challenge using infected blood meals or infected animals), the mosquito strains, the viral strains and the virus loads used, the conditions of mosquito incubation (e.g., time, temperature), and the methods used to determine mosquito infection and transmission efficiency. Concerning the virus inoculation methodology, intra-thoracic injection was used for 11 viruses to assess vector infection, and oral infection was performed using infected hosts (n = 11 arboviruses), or membrane feeding methods (n = 11 arboviruses).

The mean infection values in *Ae. albopictus* after infection by intrathoracic injection greatly varied in function of the tested virus, and ranged from 100% ± 0 (AMTV, BUJV, and PACV) to 37.5% ± 17.67 (ORUV). Among these viruses, the transmission rate after intrathoracic injection was estimated only for ORUV (37.5% ± 17.67) and RVFV (15.9% ± 7.3). The mean infection rate (IR) in *Ae. albopictus* that fed directly on infected vertebrate hosts or on an infectious artificial blood-meal through a membrane also hugely varied, from 100% ± 0 for GETV to 6.6% ± 5.2 for OROV. The mean Dissemination Efficiency (DE) in *Ae. albopictus* varied from 89.85% ± 5.9 for POTV to 4.06% ± 1.32 for MAYV. The mean Transmissions Rates (TR) in *Ae. albopictus* varied from 82.7% ± 11.5 (WNV) to 7.7% ± 0 (JCV). Finally, the mean Transmission Efficiency (TE) by *Ae. albopictus* varied from 68.6% ± 18.6 (WNV) to 3.5% ± 0.69 (MAYV).

Whatever the methodology used for the experimental infection, transmission was confirmed for 14 viruses. Six displayed a mean TE higher than 30% (WNV, EEEV, RRV, JEV, VEEV, and ORUV), and five had a mean TE between 10% and 30% (LACV, CVV, POTV, CHPV, and RVFV). The mean TE for YFV, JCV, and MAYV was below 10%. All TE rates in *Ae. albopictus* (using both experimental infections with infectious animals and infectious artificial blood meals) are summarized in Figure 3, without taking into account the different mosquito populations used, the viral loads, or genotypes. For more details on the infection parameters (IR, DE, TR, and TE) obtained using the different inoculation methods, see Table 2.

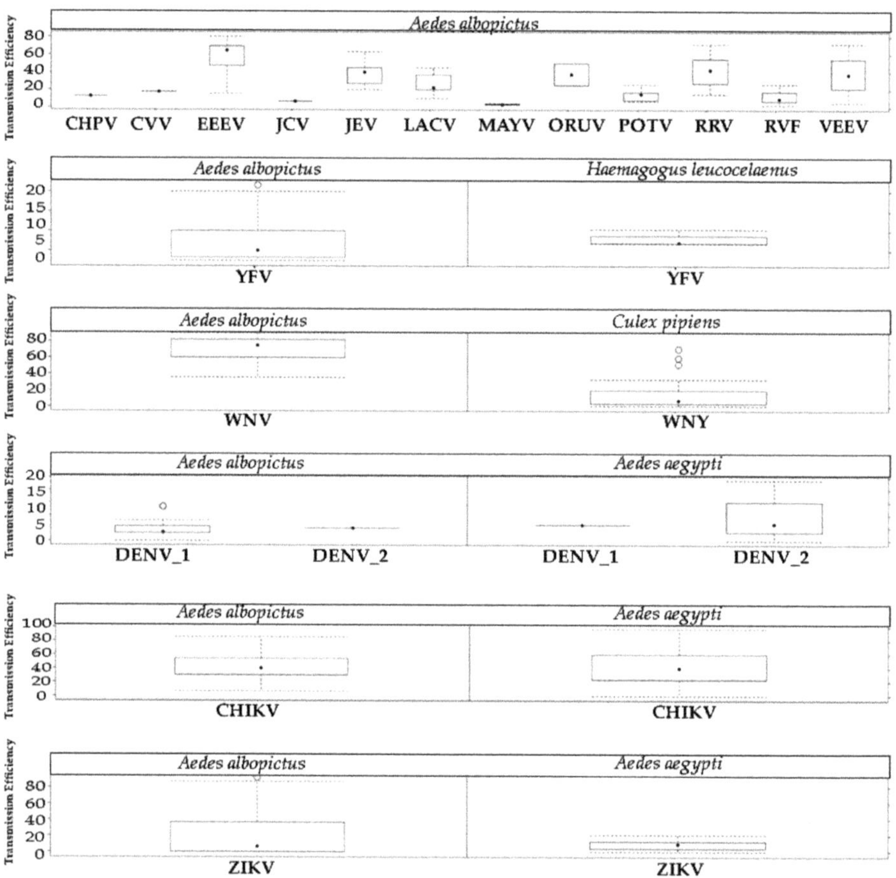

Figure 3. Transmission efficiency across studies that evaluated *Ae. albopictus* vector competence for different arboviruses. Bridge Vector*Virus and Epidemic Vector*Virus pairs were added to compare the transmission efficiency.

Table 2. Infection rate, dissemination rate, dissemination efficiency, transmission rate, and transmission efficiency (mean and standard deviation) of *Aedes albopictus* for the indicated arboviruses, according to the inoculation method.

Infection Method	Virus	Infection or Infection Rate		Dissemination Rate		Dissemination Efficiency		Transmission Rate		Transmission Efficiency	
		Mean	SD	Mean	SD	Mean	SD	Mean	SD	Mean	SD
Host feeding	CHPV	25.00	0.00	ND	ND	ND	ND	ND	ND	12.50	0.00
	EEEV	75.36	35.35	85.19	25.66	76.99	33.58	40.00	0	57.17	20.15
	JEV	ND	ND	ND	ND	ND	ND	ND	ND	37.00	9.17
	LACV	ND	ND	ND	ND	ND	ND	ND	ND	23.86	6.69
	MAYV	11.88	3.31	20.00	0.00	4.07	1.32	ND	ND	3.46	0.69
	OROV	6.67	5.20	ND	ND	ND	ND	ND	ND	ND	ND
	POTV	26.26	17.06	ND	ND	ND	ND	ND	ND	ND	ND
	RRV	80.66	23.02	ND	ND	ND	ND	ND	ND	41.40	16.57
	RVFV	69.26	27.24	60.04	6.34	40.72	11.96	15.00	7.07	6.54	4.67
	VEEV	71.20	20.49	89.48	10.48	64.78	22.53	59.94	26.57	38.09	23.53
	WNV	73.41	23.81	94.39	3.91	69.80	23.98	82.72	11.49	68.63	18.62
Intrathoracic injection	AMTV	100.00	ND	ND	ND	ND	ND	ND	ND	ND	ND
	BUJV	100.00	ND	ND	ND	ND	ND	ND	ND	ND	ND
	CHIV	96.88	ND	ND	ND	ND	ND	ND	ND	ND	ND
	ICOV	40.91	ND	ND	ND	ND	ND	ND	ND	ND	ND
	ITPV	81.25	ND	ND	ND	ND	ND	ND	ND	ND	ND
	KARV	94.12	ND	ND	ND	ND	ND	ND	ND	ND	ND
	ORUV	37.50	17.68	ND	ND	ND	ND	ND	ND	37.50	17.68
	PACV	100.00	0.00	ND	ND	ND	ND	ND	ND	ND	ND
	RVFV	ND	ND	ND	ND	ND	ND	ND	ND	15.93	7.35
	SALV	92.86	0.00	ND	ND	ND	ND	ND	ND	ND	ND
	URUV	94.12	0.00	ND	ND	ND	ND	ND	ND	ND	ND
Membrane feeding	CHIKV	58.92	28.23	77.58	22.60	79.06	23.45	53.49	33.98	42.68	23.78
	CVV	56.50	0.00	100.00	0.00	ND	ND	29.60	0.00	17.39	0.00
	DENV-1	60.18	16.01	63.79	23.97	39.56	23.90	8.33	0.00	6.25	0.00
	DENV-2	58.10	30.93	53.12	22.93	34.83	18.81	12.47	13.20	10.13	12.29
	GETV	100.00	0.00	ND	ND	ND	ND	ND	ND	ND	ND
	JCV	96.67	0.00	89.66	0.00	86.67	0.00	7.69	0.00	6.67	0.00
	JEV	91.98	10.72	90.79	14.56	84.63	19.92	ND	ND	40.50	15.98
	KEYV	91.89	0.00	91.18	0.00	83.78	0.00	ND	ND	ND	ND
	LACV	89.72	7.38	86.83	13.70	71.03	22.93	35.84	14.25	29.93	16.75
	POTV	93.55	6.59	96.13	3.21	89.86	5.96	ND	ND	14.67	7.00
	RVFV	10.53	0.00	25.00	0.00	2.63	0.00	100.00	0.00	2.63	0.00
	TVTV	28.00	0.00	85.71	0.00	24.00	0.00	ND	ND	ND	ND
	USUV	64.40	31.2	0.00	0.00	0.00	0.00	0.00	0.00	0.00	0.00
	WNV	32.61	24.53	64.59	25.58	20.33	16.96	ND	ND	ND	ND
	YFV	33.18	21.18	55.52	20.97	20.86	10.90	36.52	26.17	7.68	5.94
	ZIKV	67.19	23.70	38.71	21.76	29.25	22.80	24.62	22.46	9.21	6.91

Infection rate: number of mosquitoes showing virus infection in the gut divided by the number of mosquitoes fed with infected blood x 100. Infection: percentage of mosquitoes in which the virus was detected after 7–10 day of incubation following intrathoracic injection of the indicated virus. For this test, the ground mosquito suspension was inoculated in rats, or the virus presence was quantified by assays in Vero cells. ND: Not described SD: standard deviation AMTV, Arumowot virus; BUJV, Bujaru virus; CHIKV, Chikungunya virus; CVV, Cache Valley virus; CHPV, Chandipura virus; CHIV, Chilibre virus; DENV-1, Dengue virus serotype 1; DENV-2, Dengue virus serotype 2; EEEV, Eastern Equine Encephalomyelitis virus; GETV, Getah virus; ICOV, Icoaraci virus; ITPV, Itaporanga virus; JCV, Jamestown Canyon virus; JEV, Japanese Encephalitis virus; KARV, Karimabad virus; KEYV, Keystone virus; LACV, La Crosse virus; MAYV, Mayaro virus; OROV, Oropuche virus; ORUV, Orungo virus; PACV, Pacui virus; POTV, Potosi virus; RVFV, Rift Valley fever virus; RRV, Ross River virus; SALV, Salehabad virus; SAV, San Angelo virus; SLEV, St. Louis encephalitis virus; TENV, Tensaw virus; TVTV, Trivittatus virus; URUV, Urucuri virus; USUV, Usutu virus; VEEV, Venezuelan equine encephalitis virus; WNV, West Nile virus; YFV, Yellow fever virus; and ZIKV, Zika virus. For BSQV, Bussuquara virus, ILHV, Ilheus virus, KOKV, Kokobera virus, KUNV, Kunjin virus, UGSV, and Uganda S. virus, only transovarial transmission tests were described.

Comparison of IR, DE, and TE (see Methods and Supplementary Table S7) values calculated for known efficient bridge vectors infected with different arboviruses, and those for *Ae. albopictus* (Table 3) showed that the YFV TE rate for *Ae. albopictus* (7.68% ± 5.9) was similar to the rate calculated for *Haemagogus leucocelenus* [127] (8.08% ± 2.0). Conversely, the TE rates varied more for WNV: 68.6% ± 18.6 for *Ae. albopictus* and 13.49 ± 14.8 for *Culex pipiens* (a primary vector of WNV in the field) [117,124,137–146]. Moreover, *Ae. albopictus* and *Ae. aegypti* (a recognized epidemic vector) showed similar TE rates for CHIKV [126,147–152], DENV-1 [153–157], and DENV-2 [126,147,154,155,157–159], but *Ae. aegypti* was more efficient at transmitting ZIKV [126,160–171] and YFV [127,128,172–175].

Table 3. Comparison of the infection rate, dissemination efficiency, and transmission efficiency (mean and standard deviation) for *Aedes albopictus* and other mosquito vectors.

Mosquito Species	Virus	IR (%)	DE (%)	TE (%)
Aedes aegypti	CHIKV	NA	98.3 ± 3.8	42.92 ± 20.19
	DENV-1	37.7 ± 27	34.4 ± 24.9	4.9 ± 4.6
	DENV-2	44.4 ± 33.4	33.3 ± 24.2	5 ± 0
	ZIKV	69.0 ± 27.4	44.0 ± 28.3	20.48 ± 26.87
	YFV	46.4.0±23.6	21.3 ± 19.0	16.5 ± 17.7
Aedes albopictus	CHIKV	58.9 ± 28.2	79.0 ± 23.4	42.68 ± 23.7
	DENV-1	60.2 ± 16	39.5 ± 24.2	6.25 ± 0
	DENV-2	58.0 ± 30.9	34.8 ± 18.8	10.13 ± 12.28
	WNV	63.8 ± 29.2	58.1 ± 30.8	68.6 ± 18.6
	YFV	33.1 ± 21.1	20.8 ± 10.8	7.68 ± 5.9
	ZIKV	67.1 ± 23.7	29.2 ± 22.8	9.21 ± 6.9
Culex pipiens	WNV	47.7 ± 33.7	30.4 ± 29.7	13.49 ± 14.8
Haemagogus leucocelenus	YFV	50.9 ± 4.0	30.06 ± 1.6	8.08 ± 2.0

IR, infection rate; DE, dissemination efficiency; TE, transmission efficiency; CHIKV, Chikungunya virus; DENV-1, Dengue serotype 1; DENV-2, Dengue serotype 2; WNV, West Nile virus; YFV, Yellow fever virus; ZIKV, Zika vírus.

3. Discussion

In the present work, we tried to understand the potential role of the Asian tiger mosquito *Ae. albopictus* as a bridge vector that might favour the transfer of zoonotic arboviruses from enzootic or domestic hosts to humans and vice-versa. To this aim, we evaluated its ability to colonize natural breeding sites in newly invaded and native areas, its appetence for animal blood sources, and its global efficiency for transmitting arboviruses. This mosquito species was described as capable of developing infection from a large number of arboviruses in laboratory conditions [36]. However, based on the published evidences of vector competence, we found that transmission by *Ae. albopictus* is proven only for 14 of them, without considering the epidemic *Aedes*-borne CHIKV, DENV (4 serotypes), and ZIKAV.

In relation to the capacity of *Ae. albopictus* to establish in natural areas (rural/sylvan environments), tree holes were described as the most common natural breeding sites, although it has been detected also in bamboo stumps, and more sporadically in rock holes and plant axils [24]. Our analysis indicates that coconut shells, bromeliads, and bamboo stumps might be as common as tree holes, whereas rock holes and leaf axils of other plants are less frequently used. These results might be biased due to differences across studies related to sampling efforts and the environmental characteristics of sampled areas. Therefore, they should be confirmed by comparisons with larval sampling in natural and artificial breeding sites in natural areas and forest edges. Moreover, when possible, the productivity in these habitats should be described and compared by pupal sampling, with the same methodology used for quantifying the productivity of anthropic containers in urban areas [176]. For example, a study in Rio de Janeiro showed that the percentage of *Ae. albopictus* larvae in bromeliads corresponded to 0.18% of all sampled larva, demonstrating the low productivity of this breeding place [48]. However, studies describing the productivity of natural breeding sites in the natural environment or at an interface between natural and man-modified environments are lacking. In native forested areas, natural containers of larvae (tree holes, bamboo stumps, rock holes) were observed at the forest edge, like in a colonized forested area. Breeding sites in the deep forest have never been detected for this species [24,27].

Our results also confirmed the opportunistic feeding behaviour of *Ae. albopictus* and its strong preference for mammals, especially humans (humans = 60%, non-humans = 30%) compared with other groups, such as birds (4%). *Ae. albopictus* can feed on 28 different biological families. Reports on *Ae. albopictus* biting on any primates other than man were lacking until very recently. Specifically, a study described *Ae. albopictus* probing on a howler monkey that had just died due to YFV and was lying on the forest edge in Brazil [177]. This mosquito also bites domesticated animals—Muridae, Canidae, Phasianidae, Herpestidae, and Bovinae. Several studies suggested this opportunism. For instance, laboratory experiments on the host choice showed that this mosquito preferentially bites humans compared with other animals [30]. This opportunism was confirmed in studies on blood-fed mosquitoes collected in the field [27,30,69,78]. From our literature analysis, birds appeared as a non-preferential host group. Based on the reported proportion of blood meals, domestic and peri-domestic animals (25%) should be considered more relevant than wildlife (10%) as sources of zoonotic pathogens for *Ae. albopictus*. However, a limited number of studies were carried out in natural habitats where wildlife is abundant. Therefore, additional research is needed in natural areas to precisely describe the blood feeding patterns of *Ae. albopictus* and its interaction with wildlife. If possible, the availability of vertebrate hosts should be taken into account by using field census procedure and by calculating indexes of feeding preferences [178]. Such approaches should prevent the underestimation of the *Ae. albopictus'* potential to transmit pathogens from domestic/sylvatic vertebrate hosts to humans, but also from domestic to sylvatic vertebrate hosts, and vice versa. Our analysis also highlighted a huge variability in the proportion of human blood meals. This is a relevant factor for calculating the vector capacity, the disease reproduction rate (Ro), and the spill-over risk that may be determined by several parameters [178].

Concerning vector competence, this species was suggested as a potential vector for many viruses. It is important to emphasize that the mean TE values of enzootic viruses, such as WNV (68.6% ± 18.6), EEEV (57.16% ± 20.14), RRV (41.39% ± 16.5), JEV (39.3% ± 13.5), VEEV (38.1% ±), LACV (27.3% ± 12.87), CVV (17.4% ± 0), and POTV (14.6% ± 7), were higher or comparable with those reported for epidemic viruses, such as DENV-1 (6.25 ± 0), DENV-2 (10.13 ± 12.28), YFV (7.68 ± 5.9), ZIKV (9.21 ± 6.9), CHPV (12.5% ± 0), YFV (8.2% ± 6), JCV (6.6% ± 0), RVFV (5.2% ± 3.9), and MAYV (3.5% ± 0.69). The large difference in TE rates between enzootic and epidemic viruses is a reflection of the techniques employed to assess parameters. Most of the analysis on enzootic viruses were performed mainly in the 1990s and up to the beginning of the 2000s. Conversely, epidemic viruses were analysed using more precise techniques during the last 5 years. Despite the biases of the older methodologies, *Ae. albopictus* presented a high TE rate for enzootic arboviruses; therefore, it might transmit these viruses if taken from viremic natural vertebrates.

Comparing the vector competence of *Ae. aegypti* and *Ae. albopictus* for different epidemic viruses did not allow for a conclusion that there is a difference in their TE rates for ZIKV, CHIKV, DENV-1, and DENV-2. However, for bridge vectors*virus pairs, WNV TE was higher for *Ae. albopictus* than for *Cx. pipiens*, contrary to what was expected. Although the WNV transmission efficiency rate by *Ae. albopictus* is high in experimental conditions, this species has never been incriminated as a WNV vector in the field, possibly due to its low propensity to bite birds. *Ae. albopictus* presented similar TE rates as *Hg. leucocelenus*, a primary YFV vector within and at the edges of Brazilian forests [27,179]. However, few studies have been carried out to assess *Hg. leucocelenus* vector competence. In general, the contribution of laboratory studies for assessing the role of vector(s) in natural environments is limited.

Based on vector competence and blood meal studies, we conclude that *Ae. albopictus* could act as a bridge vector for many viruses (e.g., WNV, EEEV, ORUV, RRV, YFV, JEV, VEEV, LACV, RVFV, CVV, CHPV, JCV, and MAYV) with a potential risk for disease emergence. One of our goals was to identify in a quantitative way the viruses with a higher risk of emergence, and to develop an analysis to quantify the relative risk of transfer to humans of each enzootic arbovirus that can be efficiently transmitted by *Ae. albopictus* in laboratory conditions. The methodology used was based on two previous published works [180,181] that quantified the risk of WNV transfer by *Culex* mosquitoes. We then calculated the relative risk of *Ae. albopictus*-mediated virus transfer from its natural hosts to humans using a simplified version of Kilpatrick's equation (see Supplementary Information for more details concerning the methodology used and Figure S2) that takes into account *Ae. albopictus* vector competence for a given virus (i.e., TE), and the mean relative feeding frequencies on humans (FHi) and on animal hosts (FAi). Unfortunately, this analysis was hindered by the limited information available on the enzootic/sylvatic reservoirs of several of these arboviruses (some hosts remain unknown or are not sufficiently identified). Moreover, some viruses have many potential reservoirs, and their objective weighting is difficult. Additionally, data on *Ae. albopictus* propensity to bite a given animal reservoir species are often lacking (e.g., primates). Consequently, only biting frequencies at animal family levels could be used, leading to overly unreliable and speculative risk transfer estimates. Therefore, we chose not to include them here, although these estimates are crucial to better assess the risk of spill-over and emergence of enzootic arboviruses in relation with the secondary invasion of *Ae. albopictus* in forested areas.

Another important limitation of the present work is the great methodological variation and the lack of standardization of the protocols used to assess the vector competence of *Ae. albopictus*. Vector competence for arboviruses is influenced by genetic factors in the mosquito population and in the virus strain, such as the geographical genetic origin of the vector population or the interaction between the vector and arbovirus genotype [182,183]. Therefore, the intraspecific genetic variability in mosquito species/populations, as well as the intra- and inter-specific variability of arboviruses can affect vector competence and risk estimations. External factors, such as the incubation temperature, can also affect vector competence, and consequently the transmission and analysis of the risk [184].

Other factors interfering with the vector competence results are the way of ingesting the virus-infected blood (in vivo or in vitro), the viral load concentration, and the sensibility of the method used to detect the virus in the mosquito body or saliva. We are aware that our study is limited due to the methodological differences of the analysed articles, and also because the risk of arbovirus emergence is a multifactorial process and it is actually impossible to estimate the interactions of all factors with the limited evidences available. Thus, more standardized studies of vector competence and blood feeding preferences are necessary. In this sense, the project Infravec2 (https://infravec2.eu) is an important international initiative, and one of its themes is the standardization of methods.

In conclusion, data from the literature show that *Ae. albopictus* can colonize forest environments, and has possible interactions with domestic animals and wildlife, suggesting a risk for interaction with animal viruses. Such a risk is particularly high in areas that are considered to be biodiversity hotspots, such as the Congo and Amazon Basin forests. The presence of *Ae. albopictus* in small towns and hamlets in the Amazon Forest highlights the risk of spill-over of some arboviruses that cause human diseases, such as OROV, YFV, and MAYV [27]. In Brazil, *Ae. albopictus* populations are experimentally competent for YFV transmission, but this has not been confirmed by infecting *Ae. albopictus* [127,185]. In Africa, many arboviruses could be investigated to elucidate their potential transmission and emergence facilitated by *Ae. albopictus*, as done for CHIKV [152]. In the United States, where this mosquito species is widespread, its potential role in LACV, EEEV, WNV, and POTV transmission must be investigated [36,133,135]. In Asia and Oceania, the potential for inter-species transmission of JEV and RRV must be evaluated. It is important to take into account that the risk of arbovirus emergence is dynamic and in continuous evolution because mosquito populations, virus genetics, and the possibility of their contact varies according to time and place, and adaptations could be expected, particularly for invasive pathogens and vectors [186]. For instance, in the Indian Ocean region, the interaction between *Ae. albopictus* and CHIKV led to the selection of a virus strain that infects vectors and can spread around the world more easily. Studies on mutation selection for more susceptible arbovirus strains are still limited, but can be useful for predicting spill-over events [187]. Also, vector competence must be evaluated with as many strains as possible to maximize viral diversity, if possible using strains recently isolated from animals.

Our literature review showed that *Ae. albopictus* is competent for many different arboviruses, is present in natural habitats and forest edges, and can feed on several animal groups [30]. All these features make of *Ae. albopictus* a potential bridge vector of several emerging arboviruses (at least 14 viruses [23,36]), thus increasing the risk of spill-over and spill-back events. We hope that our approach will encourage more research to disentangle this risk in the field and the laboratory, with the aim of preventing the emergence of zoonotic diseases and reducing potential health and economic burdens, particularly for vulnerable populations.

4. Material and Methods

4.1. Natural Breeding Sites

First, a literature search was done in Google Scholar to identify articles reporting the presence of *Ae. albopictus* in natural larval breeding sites and their types, using the keywords "Natural Breeding sites *Aedes albopictus*" or "Oviposition sites *Aedes albopictus*" or "Larval habitats *Aedes albopictus*". This allowed for the identification of 16 articles [43,44,46,52,54,55,61,62,188–194] (Supplementary Table S1). From these articles, the main natural breeding sites were listed: bamboo stumps, bromeliads, coconut shells, leaf axils, rock holes, tree holes, snail shells, cacao shells, puddles, dead cow horns, dead leaves, ground cavity, hollow log, palm bracts, and palm leaves. Then, a search on each type of natural breeding site was carried out using PubMed, using the following words: (*Aedes albopictus* [Title/Abstract] AND "Breeding type" [Title/Abstract]). The aim of this search was to quantify the number of articles and the number of detections that described the presence of this mosquito in each of the identified natural breeding sites (Supplementary Table S2). Articles that did not quantify the number of times the breeding sites were found positive were excluded. The bibliographic search was done between August and December 2018.

4.2. Feeding Behaviour

A literature search was done in Google Scholar with the key words "blood meal" and "host feeding", followed by "*Aedes albopictus*" until December 2018. Three studies were excluded because they were considered unreliable: (i) the study by Gingrich and Williams, 2005 [67], which did not test for human blood meals, thus bringing a potential bias into the results; (ii) the study performed in a zoo by Tuten et al., 2012 [195]; and (iii) the study by Hess et al., 1968 [196] that was exclusively carried out in a bird area on Hawaii Island. Finally, 22 studies were selected (see references and details for each of them in Supplementary Table S3) to build a database of blood feeding preferences, based on the *Ae. albopictus* biting frequency for each host species, biological family, or group of vertebrate hosts (human, mammals, birds, domestic animals, wild animals). The database was used to quantify the relative importance as a blood meal of each host group and of specific hosts, based on the reported blood meal sources identified using different techniques (DNA sequencing, ELISA blond meal analyses, agarose gel precipitin). Then, these preferences were analysed independently of the host availability, which was quantified in very few studies.

4.3. Arbovirus Transmission

First, all referenced arboviruses that might be transmitted by *Ae. albopictus* were selected using the arbocat database from Centers for Disease Control and Prevention (CDC) (https://wwwn.cdc.gov/arbocat/VirusBrowser.aspx). Then, Google Scholar and PubMed were searched with the key words "Virus name" and "Vector Competence", followed by "*Aedes albopictus*". Among the 49 articles obtained with this search, articles containing data on virus detection/isolation from field-collected mosquitoes, and data on vector competence parameters, including "susceptibility", "infection, dissemination", or "transmission rates" were selected (see Supplementary Table S4 showing the viruses and the bibliographic references). Data from each article were used to calculate the infection rates as the number of mosquitoes showing virus infection in the gut divided by the number of mosquitoes fed with infected blood x 100. Dissemination efficiency was calculated as the number of mosquitoes with viruses disseminated in the legs, wings, or head divided by the number of mosquitoes fed with infected blood x 100. Transmission rates were calculated as the number of mosquitoes that could deliver the virus with saliva (detection of virus in mosquito saliva, or demonstration of transmission using animal hosts exposed to infected mosquito bites) divided by the number of mosquitoes with viruses disseminated in the legs, wings, or head (body) × 100. Transmission efficiency was calculated as the number of mosquitoes that could deliver the virus with saliva (detection of living viruses or viral genome in mosquito saliva, or demonstration of transmission using animal hosts exposed to infected mosquito bites) divided by the number of mosquitoes fed with infected blood [168]. In the present work, infection performed from intrathoracic assays corresponds to mosquitoes that after intrathoracic injection, were detected with the virus after a 7–10 day incubation period. For this detection, the ground mosquito suspension was inoculated in rats, or the presence of the virus quantified by assays in Vero cell cultures. After intrathoracic injection, infected mosquitoes may transmit the virus to another animal. Some articles only described transovarial transmission tested after intrathoracic infection. These works demonstrated *Ae. albopictus* susceptibility to develop infection by a given arbovirus. However, these articles did not quantify the infection and transmission rates.

To compare the results, the same bibliographic search was performed to find the vector competence values reported for efficient bridge vector–virus pairs, such as *Culex pipiens* * WNV and *Haemagogus leucocelenus* * YFV, and for epidemic vector–virus pairs, such as *Aedes aegypti* * YFV, *Aedes albopictus* *DENV_1, *Aedes albopictus* *DENV_2, *Aedes aegypti* *DENV_1, *Aedes aegypti* *DENV_2, *Aedes albopictus* *CHIKV, *Aedes aegypti* * CHIKV, *Aedes albopictus* *ZIKV virus, and *Aedes aegypti* * ZIKV (Supplementary Table S7). The bibliographic search was done between August 2018 and November 2019.

Supplementary Materials: The following are available online at http://www.mdpi.com/2076-0817/9/4/266/s1, Table S1. List of the 16 articles found by searching Google Scholar to characterize the types of natural breeding sites exploited by *Ae. albopictus*, Table S2. Typology and number of reported natural containers exploited by *Ae. albopictus* from articles found in PubMed, Table S3: List of references used to analyse the host feeding preferences of *Aedes albopictus*, Table S4: List of references that reported infection, infections rate, dissemination rate, dissemination efficiency, transmissions rate or transmission efficiency in *Ae. albopictus* for the indicated arboviruses, Table S5: List of references used to analyse the vector competence of several mosquito-virus pairs: *Aedes aegypti**CHIKV, *Aedes aegypti**DENV-1, *Aedes aegypti**DENV-2, *Aedes aegypti**ZIKV, *Aedes albopictus**CHIKV, *Aedes albopictus**DENV-1, *Aedes albopictus**DENV_2, *Aedes albopictus**ZIKV, *Culex pipiens**WNV, and *Haemagogus leucocelenus**YFV, Table S6: Natural detection or isolation of arboviruses in *Ae. albopictus* from field-collected mosquitoes. CCV, Cache Valley virus; EEEV, Eastern Equine Encephalomyelitis virus; KEYV, Keystone virus; LACV, La Crosse virus; POTV, Potosi virus; TENV, Tensaw virus; WNV, West Nile virus, Table S7: Geographic distribution, vertebrate hosts and potential vectors of arboviruses isolated or tested for vector competence in *Ae. albopictus*. Figure S1. Analysis of the host feeding patterns of *Ae. albopictus* for the different species of domestic animals without taking into account the host availability.

Author Contributions: T.P.-d.-S., D.R., and C.P. conceived the study and designed the methodology. T.P.-d.-S., D.R., R.L.-d.-O. and C.P. wrote the manuscript. All authors have read and agreed to the published version of the manuscript.

Funding: This study was funded by the French Government Investissement d'Avenir program, Laboratoire d'Excellence "Centre d'Etude de la Biodiversité Amazonienne" (grant ANR-10-LABX-25-01), by the European Union Horizon 2020 Research and Innovation Programme under ZIKAlliance (Grant Agreement no. 734548), the Pasteur Institute via the PTR (grant n°528) and the ANR PRC TIGERBRIDGE (grant ANR-16-CE35-0010-01). Taissa Pereira dos Santos received a PhD mobility grant (201927/2014-4) from the CNPq "Science without Borders" programme.

Conflicts of Interest: The authors declare that there is no conflict of interest.

References

1. Whitmee, S.; Haines, A.; Beyrer, C.; Boltz, F.; Capon, A.G.; De Souza Dias, B.F.; Ezeh, A.; Frumkin, H.; Gong, P.; Head, P.; et al. Safeguarding human health in the Anthropocene epoch: Report of the Rockefeller Foundation-Lancet Commission on planetary health. *Lancet* **2015**, *386*, 1973–2028. [CrossRef]
2. Taylor, L.H.; Latham, S.M.; Woolhouse, M.E.J.; Lathamt, S.M.; Bush, E. Risk factors for human disease emergence. *R. Soc.* **2001**, *356*, 983–989. [CrossRef] [PubMed]
3. Jones, K.K.E.; Patel, N.G.N.; Levy, M.A.; Storeygard, A.; Balk, D.; Gittleman, J.L.; Daszak, P. Global trends in emerging infectious diseases. *Nature* **2008**, *451*, 990–993. [CrossRef] [PubMed]

4. Allen, T.; Murray, K.A.; Zambrana-Torrelio, C.; Morse, S.S.; Rondinini, C.; Di Marco, M.; Breit, N.; Olival, K.J.; Daszak, P. Global hotspots and correlates of emerging zoonotic diseases. *Nat. Commun.* **2017**, *8*, 1124. [CrossRef] [PubMed]

5. Karesh, W.B.; Dobson, A.; Lloyd-Smith, J.O.; Lubroth, J.; Dixon, M.A.; Bennett, M.; Aldrich, S.; Harrington, T.; Formenty, P.; Loh, E.H.; et al. Ecology of zoonoses: Natural and unnatural histories. *Lancet* **2012**, *380*, 1936–1945. [CrossRef]

6. Vasilakis, N.; Cardosa, J.; Hanley, K.A.; Holmes, E.C.; Weaver, S.C. Fever from the forest: Prospects for the continued emergence of sylvatic dengue virus and its impact on public health. *Nat. Rev. Microbiol.* **2011**, *9*, 532–541. [CrossRef] [PubMed]

7. Kilpatrick, A.; Randolph, S. Drivers, dynamics, and control of emerging vector-borne zoonotic diseases. *Lancet* **2012**, *380*, 1946–1955. [CrossRef]

8. Weaver, S.C.; Reisen, W.K. Present and future arboviral threats. *Antivir. Res.* **2010**, *85*, 328–345. [CrossRef]

9. Go, Y.Y.; Balasuriya, U.B.R.; Lee, C. Zoonotic encephalitides caused by arboviruses: Transmission and epidemiology of alphaviruses and flaviviruses. *Clin. Exp. Vaccine Res.* **2014**, *3*, 58–77. [CrossRef]

10. Weaver, S.C.; Barrett, A.D.T. Transmission cycles, host range, evolution and emergence of arboviral disease. *Nat. Rev. Microbiol.* **2004**, *2*, 789–801. [CrossRef]

11. Gubler, D. Vector-borne diseases. *Rev. Sci. Tech.* **2009**, *28*, 583–588. [CrossRef] [PubMed]

12. Kamgang, B.; Happi, J.Y.; Boisier, P.; Njiokou, F.; Hervé, J.P.; Simard, F.; Paupy, C. Geographic and ecological distribution of the dengue and chikungunya virus vectors Aedes aegypti and Aedes albopictus in three major Cameroonian towns. *Med. Vet. Entomol.* **2010**, *24*, 132–141. [CrossRef] [PubMed]

13. Weaver, S.C. Host range, amplification and arboviral disease emergence. *Arch. Virol. Suppl.* **2005**, *19*, 33–44.

14. Bhatt, S.; Gething, P.; Brady, O.; Messina, J.; Farlow, A.; Moyes, C. The global distribution and burden of dengue. *Nature* **2012**, *496*, 504–507. [CrossRef] [PubMed]

15. Guzman, M.G.; Harris, E. Dengue. *Lancet* **2015**, *385*, 453–465. [CrossRef]

16. Staples, E.; Fischer, M. Chikungunya virus in the Americas-what a vectorborne pathogen can do. *N. Engl. J. Med.* **2014**, *371*, 887–889. [CrossRef] [PubMed]

17. Hills, S.; Fischer, M.; Petersen, L. Epidemiology of Zika Virus Infection. *J. Infect. Dis.* **2017**, *216*, 868–874. [CrossRef]

18. Zanluca, C.; De Melo, V.C.A.; Mosimann, A.L.P.; Dos Santos, G.I.V.; dos Santos, C.N.D.; Luz, K. First report of autochthonous transmission of Zika virus in Brazil. *Mem. Inst. Oswaldo Cruz* **2015**, *110*, 569–572. [CrossRef]

19. Ahmed, Q.; Memish, Z. Yellow fever from Angola and Congo: A storm gathers. *Trop. Doct.* **2017**, *47*, 92–96. [CrossRef]

20. Possas, C.; Martins, R.M.; de Oliveira, R.L.; Homma, A. Urgent call for action: Avoiding spread and re-urbanisation of yellow fever in Brazil. *Mem. Inst. Oswaldo Cruz* **2018**, *113*, 1. [CrossRef]

21. Cordellier, R. L'epidemiologie de la fièvre jaune en Afrique de l'Ouest. *Bull. World Health Organ.* **1991**, *69*, 73–84. [PubMed]

22. Klitting, R.; Fischer, C.; Drexler, J.; Gould, E.; Roiz, D.; Paupy, C.; de Lamballerie, X. What Does the Future Hold for Yellow Fever Virus? (II). *Genes (Basel)* **2018**, *9*, 425. [CrossRef] [PubMed]

23. Paupy, C.; Delatte, H.; Bagny, L.; Corbel, V.; Fontenille, D. Aedes albopictus, an arbovirus vector: From the darkness to the light. *Microbes Infect.* **2009**, *11*, 1177–1185. [CrossRef] [PubMed]

24. Hawley, W. The biology of Aedes albopictus. *J. Am. Mosq. Control Assoc.* **1988**, *4*, 1–39.

25. Eritja, R.; Palmer, J.; Roiz, D.; Sanpera-Calbet, I.; Bartumeus, F. Direct Evidence of Adult Aedes albopictus Dispersal by Car. *Sci. Rep.* **2017**, *7*, 1e4399. [CrossRef] [PubMed]

26. Savage, H.; Niebylski, M.; Smith, G.; Mitchell, C.; Craig, G. Host-feeding patterns of Aedes albopictus (Diptera: Culicidae) at a temperate North American site. *J. Med. Entomol.* **1993**, *30*, 27–34. [CrossRef]

27. Pereira dos Santos, T.; Roiz, D.; de Abreu, F.; Luz Bessa, S.; Santa Lucia, M.; Jiolle, D.; Neves, N.; Simard, F.; Lourenço De Oliveira, R.; Paupy, C. Potential of Aedes albopictus as a bridge vector for zoonotic pathogens at the urban-forest interface of Brazil. *Emerg. Microbes Infect.* **2018**, *7*, 191. [CrossRef]

28. Grard, G.; Caron, M.; Mombo, I.M.; Nkoghe, D.; Mboui Ondo, S.; Jiolle, D.; Fontenille, D.; Paupy, C.; Leroy, E.M. Zika Virus in Gabon (Central Africa) - 2007: A New Threat from Aedes albopictus? *PLoS Negl. Trop. Dis.* **2014**, *8*, e2681. [CrossRef]

29. Leroy, E.M.; Nkoghe, D.; Ollomo, B.; Nze-Nkogue, C.; Becquart, P.; Grard, G.; Pourrut, X.; Charrel, R.; Moureau, G.; Ndjoyi-Mbiguino, A.; et al. Concurrent chikungunya and dengue virus infections during simultaneous outbreaks, Gabon, 2007. *Emerg. Infect. Dis.* **2009**, *15*, 591–593. [CrossRef]

30. Delatte, H.; Desvars, A.; Bouétard, A.; Bord, S.; Gimonneau, G.; Vourc'h, G.; Fontenille, D. Blood-Feeding Behavior of Aedes albopictus, a Vector of Chikungunya on La Réunion. *Vector-Borne Zoonotic Dis.* **2010**, *10*, 249–258. [CrossRef]

31. Flahault, A.; Aumont, G.; Boisson, V.; Lamballerie, X.; Favier, F.; Fontenille, D.; Gauzere, B.; Journeaux, S.; Lotteau, V.; Paupy, C.; et al. Chikungunya, La Réunion et Mayotte, 2005–2006: Une épidémie sans histoire? *Sante Publique (Paris)* **2007**, *19*, 165–195.

32. Rezza, G.; Nicoletti, L.; Angelini, R.; Romi, R.; Finarelli, A.; Panning, M.; Cordioli, P.; Fortuna, C.; Boros, S.; Magurano, F.; et al. Infection with chikungunya virus in Italy: An outbreak in a temperate region. *Lancet* **2007**, *370*, 1840–1846. [CrossRef]

33. Venturi, G.; Di Luca, M.; Fortuna, C.; Elena Remoli, M.; Riccardo, F.; Severini, F.; Toma, L.; Del Manso, M.; Benedetti, E.; Grazia Caporali, M.; et al. Detection of a chikungunya outbreak in Central Italy Detection of a chikungunya outbreak in Central. *Euro Surveill.* **2017**, *22*, 17–00646.

34. Sawabe, K. Autochthonous dengue outbreak in Japan after a blank of 70 years and the future prediction of such cases. *Med. Entomol. Zool.* **2015**, *66*, 203–205. [CrossRef]

35. Carvalho, R.G.; Lourenço-De-Oliveira, R.; Braga, I.A. Updating the geographical distribution and frequency of Aedes albopictus in Brazil with remarks regarding its range in the Americas. *Mem. Inst. Oswaldo Cruz* **2014**, *109*, 787–796. [CrossRef]

36. Vanlandingham, D.; Higgs, S.; Huang, Y.-J. Aedes albopictus (Diptera: Culicidae) and Mosquito-Borne Viruses in the United States. *J. Med. Entomol.* **2016**, *53*, 1024–1028. [CrossRef]

37. Bagny, L.; Delatte, H.; Elissa, N.; Quilici, S.; Fontenille, D.; Adhami, J.; Murati, N.; Adhami, J.; Reiter, P.; Beltrame, A.; et al. Aedes (Diptera: Culicidae) vectors of arboviruses in Mayotte (Indian Ocean): Distribution area and larval habitats. *J. Med. Entomol.* **2009**, *46*, 198–207. [CrossRef]

38. Banerjee, S.; Aditya, G.; Saha, G.K. Household disposables as breeding habitats of dengue vectors: Linking wastes and public health. *Waste Manag.* **2013**, *33*, 233–239. [CrossRef]

39. Dev, V. Dengue vectors in urban and suburban Assam, India: Entomological observations. *Who South-East Asia J. Public Heal.* **2014**, *3*, 5838. [CrossRef]

40. Edillo, F.E.; Roble, N.D.; Otero, N.D. The key breeding sites by pupal survey for dengue mosquito vectors, Aedes aegypti (Linnaeus) and Aedes albopictus (Skuse), in Guba, Cebu City, Philippines. *Southeast Asian J. Trop. Med. Public Health* **2012**, *43*, 1365–1374.

41. Rao, B.B.; George, B. Breeding patterns of aedes stegomyia albopictus in periurban areas of Calicut, Kerala, India. *Southeast Asian J. Trop. Med. Public Health* **2010**, *41*, 536–540. [PubMed]

42. Shriram, A.N.; Sivan, A.; Sugunan, A.P. Spatial distribution of Aedes aegypti and Aedes albopictus in relation to geo-ecological features in South Andaman, Andaman and Nicobar Islands, India. *Bull. Entomol. Res.* **2018**, *108*, 166–174. [CrossRef] [PubMed]

43. Simard, F.; Nchoutpouen, E.; Toto, J.C.; Fontenille, D. Geographic distribution and breeding site preference of Aedes albopictus and Aedes aegypti (Diptera: Culicidae) in Cameroon, Central Africa. *J. Med. Entomol.* **2005**, *42*, 726–731. [CrossRef] [PubMed]

44. Thavara, U.; Tawatsin, A.; Chansang, C.; Kong-ngamsuk, W.; Paosriwong, S.; Boon-Long, J.; Rongsriyam, Y.; Komalamisra, N. Larval occurrence, oviposition behavior and biting activity of potential mosquito vectors of dengue on Samui Island, Thailand. *J. Vector Ecol.* **2001**, *26*, 172–180. [PubMed]

45. Vijayakumar, K.; Sudheesh Kumar, T.K.; Nujum, Z.T.; Umarul, F.; Kuriakose, A. A study on container breeding mosquitoes with special reference to Aedes (Stegomyia) aegypti and Aedes albopictus in Thiruvananthapuram district, India. *J. Vector Borne Dis.* **2014**, *51*, 27–32.

46. Forattini, O.P.; Brito, M. Brief Communication an Unusual Grround Larval Habitat of Aedes albopictus. *Rev. Inst. Med. Trop. Sao Paulo* **1998**, *40*, 121–122. [CrossRef]

47. Marques, G.R.; Forattini, O.P. Aedes albopictus in soil bromeliads in Ilhabela, coastal area of Southeastern Brazil. *Rev. Saude Publica* **2005**, *39*, 548–552. [CrossRef]

48. Mocellin, M.G.; Simões, T.C.; do Nascimento, T.F.S.; Teixeira, M.L.F.; Lounibos, L.P.; de Oliveira, R.L. Bromeliad-inhabiting mosquitoes in an urban botanical garden of dengue endemic Rio de Janeiro - Are bromeliads productive habitats for the invasive vectors Aedes aegypti and Aedes albopictus? *Mem. Inst. Oswaldo Cruz* **2009**, *104*, 1171–1176. [CrossRef]

49. de Oliveira, V.C.; de Almeida Neto, L.C. Ocorrência de Aedes aegypti e Aedes albopictus em bromélias cultivadas no Jardim Botânico Municipal de Bauru, São Paulo, Brasil. *Cad. Saude Publica* **2017**, *33*, e00071016. [CrossRef]

50. O'Meara, G.F.; Cutwa, M.M.; Evans, L.F. Bromeliad-inhabiting mosquitoes in south Florida: Native and exotic plants differ in species composition. *J. Vector Ecol.* **2003**, *28*, 37–46.

51. Delatte, H.; Dehecq, J.S.; Thiria, J.; Domerg, C.; Paupy, C.; Fontenille, D. Geographic Distribution and Developmental Sites of Aedes albopictus (Diptera: Culicidae) During a Chikungunya Epidemic Event. *Vector Control Southeast Asia* **2008**, *8*, 25–34. [CrossRef] [PubMed]

52. Gilotra, S.K.; Rozeboom, L.E.; Bhattacharya, N.C. Observations on possible competitive displacement between populations of Aedes aegypti Linnaeus and Aedes albopictus Skuse in Calcutta. *Bull. World Health Organ.* **1967**, *37*, 437–446. [PubMed]

53. Rozeboom, L.E.; Bridges, J.R. Relative population densities of Aedes albopictus and A. guamensis on Guam. *Bull. World Health Organ.* **1972**, *46*, 477–483. [PubMed]

54. Sota, T.; Mogi, M.; Hayamizu, E. Seasonal distribution and habitat selection by Aedes albopictus and Aedes riversi (Diptera: Culicidae) in Northern Kyushu, Japan. *J. Med. Entomol.* **1992**, *29*, 296–304. [CrossRef]

55. Gomes, A.; Forattini, O.; Kakitani, I.; Marques, G.; Marques, C.; Marucci, D.; Brito, M. Microhabitats de Aedes albopictus (Skuse) na região do Vale do Paraíba, Estado de São Paulo, Brasil. *Rev. Saúde Pública S. Paulo* **1992**, *26*, 108–118. [CrossRef]

56. Müller, G.C.; Kravchenko, V.D.; Junnila, A.; Schlein, Y. Tree-hole breeding mosquitoes in Israel. *J. Vector Ecol.* **2012**, *37*, 102–109. [CrossRef]

57. O'Meara, G.F.; Evans, L.F., Jr.; Womack, M.L. Colonization of rock holes by Aedes albopictus in the southeastern United States. *J. Am. Mosq. Control Assoc.* **1997**, *13*, 270–274.

58. Sivan, A.; Shriram, A.N.; Sugunan, A.P.; Anwesh, M.; Muruganandam, N.; Kartik, C.; Vijayachari, P. Natural transmission of dengue virus serotype 3 by Aedes albopictus (Skuse) during an outbreak in Havelock Island: Entomological characteristics. *Acta Trop.* **2016**, *156*, 122–129. [CrossRef]

59. Urbinatti, P.R.; Menezes, R.M.T.; Natal, D. Sazonalidade de Aedes albopictus em área protegida na cidade de São Paulo, Brasil. *Rev. Saude Publica* **2007**, *41*, 478–481. [CrossRef]

60. Pena, C.J.; Gonzalvez, G.; Chadee, D.D. Seasonal prevalence and container preferences of Aedes albopictus in Santo Domingo City, Dominican Republic. *J. Vector Ecol.* **2003**, *28*, 208–212.

61. Kamgang, B.; Ngoagouni, C.; Manirakiza, A.; Nakouné, E.; Paupy, C.; Kazanji, M. Temporal Patterns of Abundance of Aedes aegypti and Aedes albopictus (Diptera: Culicidae) and Mitochondrial DNA Analysis of Ae. albopictus in the Central African Republic. *PLoS Negl. Trop. Dis.* **2013**, *7*, e2590. [CrossRef] [PubMed]

62. Marquetti, M.; Bisset, J.; Leyva, M.; Garcia, A.; Rodriguez, M. Comportamiento estacional y temporal de Aedes aegypti y Aedes albopictus en La Habana, Cuba. *Rev. Cuba. De Med. Trop.* **2008**, *60*, 62–67.

63. Almeida, P.; Baptista, S.; Sousa, C.; Novo, T.; Ramos, H.; Panella, N.; Godsey, M.; Simões, M.J.; Anselmo, M.L.; Komar, N.; et al. Bioecology and Vectorial Capacity of Aedes albopictus (Diptera: Culicidae) in Macao, China, in Relation to Dengue Virus Transmission. *J. Med. Entomol.* **2005**, *42*, 419–428. [CrossRef] [PubMed]

64. Colless, D.H. Notes on the culicine mosquitoes of singapore. *Ann. Trop. Med. Parasitol.* **1959**, *51*, 87–101. [CrossRef]

65. Dennett, J.A.; Bala, A.; Wuithiranyagool, T.; Randle, Y.; Sargent, C.B.; Guzman, H.; Siirin, M.; Hassan, H.K.; Reyna-Nava, M.; Unnasch, T.R.; et al. Associations between two mosquito populations and West Nile virus in Harris County, Texas, 2003-06. *J. Am. Mosq. Control Assoc.* **2007**, *23*, 264–275. [CrossRef]

66. Faraji, A.; Egizi, A.; Fonseca, D.M.; Unlu, I.; Crepeau, T.; Healy, S.P.; Gaugler, R. Comparative Host Feeding Patterns of the Asian Tiger Mosquito, Aedes albopictus, in Urban and Suburban Northeastern USA and Implications for Disease Transmission. *PLoS Negl. Trop. Dis.* **2014**, *8*, e3037. [CrossRef] [PubMed]

67. Gingrich, J.B.; Williams, G.M. Host-feeding patterns of suspected West Nile virus mosquito vectors in Delaware, 2001-2002. *J. Am. Mosq. Control Assoc.* **2005**, *21*, 194–200. [CrossRef]

68. Gomes, A.; Silva, N.; Marques, G.; Brito, M. Host-feeding patterns of potential human disease vectors in the Paraiba Valley Region, State of Sao Paulo, Brazil. *J. Vector Ecol.* **2002**, *28*, 74–78.

69. Kamgang, B.; Nchoutpouen, E.; Simard, F.; Paupy, C. Notes on the blood-feeding behavior of Aedes albopictus (Diptera: Culicidae) in Cameroon. *Parasit. Vectors* **2012**, *5*, 57. [CrossRef]

70. Kek, R.; Hapuarachchi, H.C.; Chung, C.-Y.; Humaidi, M.B.; Razak, M.A.B.A.; Chiang, S.; Lee, C.; Tan, C.-H.; Yap, G.; Chong, C.-S.; et al. Feeding Host Range of *Aedes albopictus* (Diptera: Culicidae) Demonstrates Its Opportunistic Host-Seeking Behavior in Rural Singapore. *J. Med. Entomol.* **2014**, *51*, 880–884. [CrossRef]

71. Kim, K.S.; Tsuda, Y.; Yamada, A. Bloodmeal Identification and Detection of Avian Malaria Parasite From Mosquitoes (Diptera: Culicidae) Inhabiting Coastal Areas of Tokyo Bay, Japan. *J. Med. Entomol.* **2009**, *46*, 1230–1234. [CrossRef] [PubMed]

72. Kim, H.; mi Yu, H.; Lim, H.W.; Yang, S.C.; Roh, J.Y.; Chang, K.S.; Shin, E.H.; Ju, Y.R.; Lee, W.G. Host-feeding pattern and dengue virus detection of Aedes albopictus (Diptera: Culicidae) captured in an urban park in Korea. *J. Asia. Pac. Entomol.* **2017**, *20*, 809–813. [CrossRef]

73. Muñoz, J.; Eritja, R.; Alcaide, M.; Montalvo, T.; Soriguer, R.C.; Figuerola, J. Host-Feeding Patterns of Native Culex pipiens and Invasive Aedes albopictus Mosquitoes (Diptera: Culicidae) in Urban Zones From Barcelona, Spain. *J. Med. Entomol.* **2011**, *48*, 956–960. [CrossRef] [PubMed]

74. Niebylski, M.L.; Savage, H.M.; Nasci, R.S.; Craig, G.B. Blood hosts of Aedes albopictus in the United States. *J. Am. Mosq. Control Assoc.* **1994**, *10*, 447–450. [PubMed]

75. Ponlawat, A.; Harrington, L.C. Blood Feeding Patterns of Aedes aegypti and Aedes albopictus in Thailand. *J. Med. Entomol.* **2005**, *42*, 844–849. [CrossRef] [PubMed]

76. Richards, S.L.; Ponnusamy, L.; Unnasch, T.R.; Hassan, H.K.; Apperson, C.S. Host-Feeding Patterns of Aedes albopictus (Diptera: Culicidae) in Relation to Availability of Human and Domestic Animals in Suburban Landscapes of Central North Carolina. *J. Med. Entomol.* **2006**, *43*, 543–551. [CrossRef]

77. Sawabe, K.; Isawa, H.; Hoshino, K.; Sasaki, T.; Roychoudhury, S.; Higa, Y.; Kasai, S.; Tsuda, Y.; Nishiumi, I.; Hisai, N.; et al. Host-Feeding Habits of Culex pipiens and Aedes albopictus (Diptera: Culicidae) Collected at the Urban and Suburban Residential Areas of Japan. *J. Med. Entomol.* **2010**, *47*, 442–450. [CrossRef]

78. Sivan, A.; Shriram, A.; Sunish, I.; Vidhya, P. Host-feeding pattern of Aedes aegypti and Aedes albopictus (Diptera: Culicidae) in heterogeneous landscapes of South Andaman, Andaman and Nicobar Islands, India. *Parasitol. Res.* **2015**, *114*, 3539–3546. [CrossRef]

79. Tandon, N.; Ray, S. Host feeding pattern of Aedes aegypti and Aedes albopictus in Kolkata, India. *Denuge Bull.* **2000**, *24*, 117–120.

80. Tempelis, C.H.; Hayes, R.O.; Hess, A.D.; Reeves, W.C. Blood-feeding habits of four species of mosquito found in Hawaii. *Am. J. Trop. Med. Hyg.* **1970**, *19*, 335–341. [CrossRef]

81. Valerio, L.; Francesca, M.; Bongiorno, G.; Facchinelli, L.; Pombi, M.; Caputo, B.; Maroli, M.; Torre, A. Della Host-Feeding Patterns of Aedes albopictus. *Vector-Borne Zoonotic Dis.* **2010**, *10*, 291–294. [CrossRef] [PubMed]

82. Egizi, A.; Healy, S.P.; Fonseca, D.M. Rapid blood meal scoring in anthropophilic Aedes albopictus and application of PCR blocking to avoid pseudogenes. *Infect. Genet. Evol.* **2013**, *16*, 122–128. [CrossRef]

83. Tesh, R.B. Multiplication Phlebotomus phlebotomus fever group arboviruses in mosquitoes after intrathoracic inoculation. *J. Med. Entomol.* **1975**, *12*, 1–4. [CrossRef] [PubMed]

84. Tesh, R.B.; Shroyer, D.A. The mechanism of arbovirus transovarial transmission in mosquitoes: San Angelo virus in Aedes albopictus. *Am. J. Trop. Med. Hyg.* **1980**, *29*, 1394–1404. [CrossRef] [PubMed]

85. Armstrong, P.M.; Anderson, J.F.; Farajollahi, A.; Healy, S.P.; Unlu, I.; Crepeau, T.N.; Gaugler, R.; Fonseca, D.M.; Andreadis, T.G. Isolations of Cache Valley virus from Aedes albopictus (Diptera: Culicidae) in New Jersey and evaluation of its role as a regional arbovirus vector. *J. Med. Entomol.* **2013**, *50*, 1310–1314. [CrossRef]

86. Ramachandra, R.; Sing, K.; Dhanda, V.; Bhattacharya, N. Experimental transmission of Chandipyura Virus by mosquitoes. *Ind. J. Med. Res.* **1967**, *55*, 1306–1310.

87. Mitchell, C.J.; McLean, R.G.; Nasci, R.S.; Crans, W.J.; Smith, G.C.; Caccamise, D.F. Susceptibility parameters of Aedes albopictus to per oral infection with eastern equine encephalitis virus. *J. Med. Entomol.* **1993**, *30*, 233–235. [CrossRef]

88. Moncayo, A.C.; Edman, J.D.; Turell, M.J. Effect of eastern equine encephalomyelitis virus on the survival of Aedes albopictus, Anopheles quadrimaculatus, and Coquillettidia perturbans (Diptera: Culicidae). *J. Med. Entomol.* **2000**, *37*, 701–706. [CrossRef]

89. Scott, T.W.; Lorenz, L.H.; Weaver, S.C. Susceptibility of Aedes albopictus to infection with eastern equine encephalomyelitis virus. *J. Am. Mosq. Control Assoc.* **1990**, *6*, 274–278.

90. Turell, M.; Beaman, J.; Neely, G. Experimental transmission of Eastern equine encephalitis virus by strains of Aedes albopictus and A. taeniorhynchus (Diptera: Culicidae). *J. Med. Entomol.* **1994**, *31*, 287–290. [CrossRef]

91. Takashima, I.; Hashimoto, N. Getah virus in several species of mosquitoes. *Trans. R. Soc. Trop. Med. Hyg.* **1985**, *79*, 546–550. [CrossRef]

92. Tesh, R.B. Experimental studies on the transovarial transmission of Kunjin and San Angelo viruses in mosquitoes. *Am. J. Trop. Med. Hyg.* **1980**, *29*, 657–666. [CrossRef] [PubMed]

93. Grimstad, P.R. Recently Introduced Aedes albopictus in the United States: Potential vector of La Crosse Virus (Bunyaviridae: California Serogroup). *J. Am. Mosq. Control Assoc.* **1989**, *5*, 422–427.

94. Rosen, L.; Tesh, R.B.; Lien, J.; John, H. Transovarial Transmission of Japanese Encephalitis Virus by Mosquitoes. *Am. Assoc. Adv. Sci.* **1978**, *199*, 909–911. [CrossRef] [PubMed]

95. Weng, M.H.; Lien, J.C.; Wang, Y.M.; Wu, H.L.; Chin, C. Susceptibility of three laboratory strains of Aedes albopictus (Diptera: Culicidae) to Japanese encephalitis virus from Taiwan. *J. Med. Entomol.* **1997**, *34*, 745–747. [CrossRef] [PubMed]

96. Wispelaere, M.; Després, P.; Choumet, V. European Aedes albopictus and Culex pipiens Are Competent Vectors for Japanese Encephalitis Virus. *PLoS Negl. Trop. Dis.* **2017**, *11*, e0005294. [CrossRef]

97. Cully, J.F., Jr.; Streit, T.G.; Heard, P.B. Transmission of La Crosse virus by four strains of Aedes albopictus to and from the eastern chipmunk (Tamias striatus). *J. Am. Mosq. Control Assoc.* **1992**, *8*, 237–240.

98. Hughes, M.T.; Gonzalez, J.A.; Reagan, K.L.; Carol, D.; Beaty, B.J.; Beaty, B.J. Comparative Potential of Aedes triseriatus, Aedes albopictus, and Aedes aegypti (Diptera: Culicidae) to Transovarially Transmit La Crosse Virus Comparative Potential of Aedes triseriatus, Aedes albopictus, and Aedes aegypti (Diptera: Culicidae). *Entomol. Soc. Am.* **2006**, *43*, 757–761.

99. Lambert, A.J.; Blair, C.D.; D'Anton, M.; Ewing, W.; Harborth, M.; Seiferth, R.; Xiang, J.; Lanciotti, R.S. La Crosse Virus in Aedes albopictus Mosquitoes, Texas, USA, 2009. *Emerg. Infect. Dis.* **2010**, *16*, 856–858. [CrossRef]

100. Smith, G.C.; Francy, D.B. Laboratory studies of a Brazilian strain of Aedes albopictus as a potential vector of Mayaro and Oropouche viruses. *J. Am. Mosq. Control Assoc.* **1991**, *7*, 89–93.

101. Tomori, O.; Aitken, T.H. Orungo virus: Transmission studies with Aedes albopictus and Aedes aegypti (Diptera: Culicidae). *J. Med. Entomol.* **1978**, *14*, 523–526. [PubMed]

102. Francy, D.; Krabatsos, N.; Wesson, C.; Moore, J.; Lazuick, J.; Niebylski, T.; Tsai, T.; Craig, J. A new Arbovirus from Aedes albopictus, an Asian mosquito Established in the United States. *Science* **1990**, *250*, 1738–1740. [CrossRef] [PubMed]

103. Heard, P.B.; Niebylski, M.L.; Francy, D.B. Transmission of a Newly Recognized Virus (Bunyaviridae, Bunyavirus) Isolated from Aedes albopictus (Diptera: Culicidae) in Potosi, Missouri. *J. Med. Entomol.* **1991**, *28*, 601–605. [CrossRef] [PubMed]

104. Mitchell, C.J.; Smith, G.C.; Miller, B.R. Vector Competence of Aedes Albopictus for a Newly Recognized Bunyavirus From Mosquitoes Collected in Potosi, Missouri1. *J. Am. Mosq. Control Assoc.* **1990**, *6*, 523–527. [PubMed]

105. Brustolin, M.; Talavera, S.; Nunez, A.; Santamaría, C.; Rivas, R.; Pujol, N.; Valle, M.; Verdun, M.; Brun, A.; Pages, N.; et al. Rift Valley fever virus and European mosquitoes: Vector competence of Culex pipiens and Stegomyia albopicta (= Aedes albopictus). *Med. Vet. Entomol.* **2017**, *31*, 365–372. [CrossRef] [PubMed]

106. Turell, M.J.; Bailey, C.L.; Beaman, J.R. Vector competence of a Houston, Texas strain of Aedes albopictus for Rift Valley fever virus. *J. Am. Mosq. Control Assoc.* **1988**, *4*, 94–96.

107. Mitchell, C.J.; Gubler, D.J. Vector competence of geographic strains of Aedes albopictus and Aedes polynesiensis and certain other Aedes (Stegomyia) mosquitoes for Ross River virus. *J. Am. Mosq. Control Assoc.* **1987**, *3*, 142–147.

108. Mitchell, C.J.; Miller, B.R.; Gubler, D.J. Vector competence of Aedes albopictus from Houston, Texas, for dengue serotypes 1 to 4, yellow fever and Ross River viruses. *J. Am. Mosq. Control Assoc.* **1987**, *3*, 460–465.

109. Shroyer, D.A. Transovarial Maintenance of San Angelo Virus in Sequenctial Generations of Aedes albopictus. *Am J Trop Med Hyg.* **1986**, *35*, 408–417. [CrossRef]

110. Hardy, J.L.; Rosen, L.; Kramer, L.D.; Presser, S.B.; Shroyer, D.A.; Turell, M.J. Effect of rearing temperature on transovarial transmission of St. Louis encephalitis virus in mosquitoes. *Am. J. Trop. Med. Hyg.* **1980**, *29*, 963–968. [CrossRef]

111. Mitchell, C.J. Vector competence of North and South American strains of Aedes albopictus for certain arboviruses: A review. *J. Am. Mosq. Control Assoc.* **1991**, *7*, 446–451. [PubMed]

112. Puggioli, A.; Bonilauri, P.; Calzolari, M.; Lelli, D.; Carrieri, M.; Urbanelli, S.; Pudar, D.; Bellini, R. Does Aedes albopictus (Diptera: Culicidae) play any role in Usutu virus transmission in Northern Italy? Experimental oral infection and field evidences. *Acta Trop.* **2017**, *172*, 192–196. [CrossRef] [PubMed]

113. Beaman, J.R.; Turell, M.J. Transmission of Venezuelan Equine Encephalomyelitis Virus by Strains of Aedes albopictus (Diptera: Culicidae) Collected from North and South America. *J. Med. Entomol.* **1991**, *28*, 161–164. [CrossRef] [PubMed]

114. Fernandez, Z.; Moncayo, A.C.; Carrara, A.S.; Forattini, O.P.; Weaver, A.S.C. Vector Competence of Rural and Urban Strains of Aedes (Stegomyia) albopictus (Diptera: Culicidae) from São Paulo State, Brazil for IC, ID, and IF Subtypes of Venezuelan Equine Encephalitis Virus. *J. Med. Entomol* **2003**, *40*, 522–527. [CrossRef] [PubMed]

115. Turell, M.J.; Beaman, J.R. Experimental Transmission of Venezuelan Equine Encephalomyelitis Virus by a strain of Aedes albopictus (Diptera: Culicidae) from New Orleans, Lousiana. *J. Med. Entomol.* **1992**, *29*, 802–807. [CrossRef] [PubMed]

116. Baqar, S.; Hayes, C.G.; Murphy, J.R.; Watts, D.M. Vertical transmission of West Nile virus by Culex and Aedes species mosquitoes. *Am. J. Trop. Med. Hyg.* **1993**, *48*, 757–762. [CrossRef]

117. Brustolin, M.; Talavera, S.; Santamaría, C.; Rivas, R.; Pujol, N.; Aranda, C.; Marquès, E.; Valle, M.; Verdún, M.; Pagès, N.; et al. Culex pipiens and Stegomyia albopicta (=Aedes albopictus) populations as vectors for lineage 1 and 2 West Nile virus in Europe. *Med. Vet. Entomol.* **2016**, *30*, 166–173. [CrossRef]

118. Cupp, E.W.; Hassan, H.K.; Yue, X.; Oldland, W.K.; Lilley, B.M.; Unnasch, T.R. West Nile Virus Infection in Mosquitoes in the Mid-South USA, 2002–2005. *J. Med. Entomol.* **2007**, *44*, 117–125. [CrossRef]

119. Farajollahi, A.; Kesavaraju, B.; Price, D.C.; Williams, G.M.; Healy, S.P.; Gaugler, R.; Nelder, M.P. Field efficacy of BG-Sentinel and industry-standard traps for Aedes albopictus (Diptera: Culicidae) and West Nile virus surveillance. *J. Med. Entomol.* **2009**, *46*, 919–925. [CrossRef]

120. Holick, J.; Kyle, A.; Ferraro, W.; Delaney, R.; Marta, I. Discovery of Aedes albopictus infected with west nile virus in Southeastern Pennsylvania. *J. Am. Mosq. Control Assoc. Inc.* **2002**, *18*, 131.

121. Philip, C.B.; Smadel, J.E. Transmission of West Nile Virus by Infected Aedes albopictus. *Proc. Soc. Exp. Biol. Med.* **1943**, *53*, 49–50. [CrossRef]

122. Sardelis, M.R.; Turell, M.J.; Guinn, M.L.O.; Andre, R.G.; Roberts, D.R. Vector Competence of Three North American Strains of Aedes Albopictus for West Nile Virus1. *J. Am. Mosq. Control Assoc.* **2002**, *18*, 284–289. [PubMed]

123. Tiawsirisup, S.; Platt, K.B.; Evans, R.B.; Rowley, W.A. A Comparision of West Nile Virus Transmission by Ochlerotatus trivittatus (COQ.), Culex pipiens (L.), and Aedes albopictus (Skuse). *Vector-Borne Zoonotic Dis.* **2005**, *5*, 40–47. [CrossRef] [PubMed]

124. Vanlandingham, D.L.; McGee, C.E.; Klinger, K.A.; Vessey, N.; Fredregillo, C.; Higgs, S. Short report: Relative susceptibilties of South Texas mosquitoes to infection with West Nile virus. *Am. J. Trop. Med. Hyg.* **2007**, *77*, 925–928. [CrossRef]

125. Fortuna, C.; Remoli, M.E.; Severini, F.; Di Luca, M.; Toma, L.; Fois, F.; Bucci, P.; Boccolini, D.; Romi, R.; Ciufolini, M.G. Evaluation of vector competence for West Nile virus in Italian Stegomyia albopicta (=Aedes albopictus) mosquitoes. *Med. Vet. Entomol.* **2015**, *29*, 430–433. [CrossRef]

126. Amraoui, F.; Ayed, W.B.; Madec, Y.; Faraj, C.; Himmi, O.; Btissam, A.; Sarih, M.; Failloux, A.B. Potential of Aedes albopictus to cause the emergence of arboviruses in Morocco. *PLoS Negl. Trop. Dis.* **2019**, *13*, e0006997. [CrossRef]

127. Couto-Lima, D.; Madec, Y.; Bersot, M.I.; Campos, S.S.; Motta, M.D.A.; Dos Santos, F.B.; Vazeille, M.; Da Costa Vasconcelos, P.F.; Lourenço-De-Oliveira, R.; Failloux, A.B. Potential risk of re-emergence of urban transmission of Yellow Fever virus in Brazil facilitated by competent Aedes populations. *Sci. Rep.* **2017**, *7*, 4848. [CrossRef]

128. Johnson, B.; Chambers, T.; Grabtree, M.; Filippis, A.; Vilarinhos, P.; Resende, M.; Marcoris, L.; Miller, B. Vector competence of Brazilian Aedes aegypti and Ae. abopictus for a Brazilian yellow fever virus isolate *. *Trans. R. Soc. Trop. Med. Hyg.* **2002**, *96*, 611–613. [CrossRef]

129. Lourenço de Oliveira, R.; Vazeille, M.; Maria, A.N.A.; Filippis, B.D.E.; Failloux, A. Large genetic differentiation and low variation in vector competence for dengue and yellow fever viruses of Aedes albopictus from Brazil, the United States, and the Cayman Islands. *Am. J. Trop. Med. Hyg.* **2003**, *69*, 105–114. [CrossRef]

130. Miller, B.R.; Mitchell, C.J.; Ballinger, M.E. Replication, tissue tropisms and transmission of yellow fever virus in Aedes albopictus. *Trans. R. Soc. Trop. Med. Hyg.* **1989**, *83*, 252–255. [CrossRef]

131. Ferreira-De-Lima, V.H.; Lima-Camara, T.N. Natural vertical transmission of dengue virus in Aedes aegypti and Aedes albopictus: A systematic review. *Parasites Vectors* **2018**, *11*, 77. [CrossRef] [PubMed]

132. Mitchell, C.J.; Haramis, L.D.; Karabatsos, N.; Smith, G.C.; Starwalt, V.J. Isolation of La Crosse, Cache Valley, and Potosi Viruses from Aedes Mosquitoes (Diptera: Culicidae) Collected at Used-Tire Sites in Illinois during 1994-1995. *J. Med. Entomol.* **1998**, *35*, 573–577. [CrossRef]

133. Mitchell, C.J.; Niebylski, M.L.; Smith, G.C.; Karabatsos, N.; Martin, D.; Mutebi, J.P.; Craig, G.B.J.; Mahler, M.J. Isolation of eastern equine encephalitis virus from Aedes albopictus in Florida. *Science* **1992**, *257*, 526–527. [CrossRef] [PubMed]

134. Westby, K.M.; Fritzen, C.; Paulsen, D.; Poindexter, S.; Moncayo, A.C. La Crosse Encephalitis Virus Infection in Field-Collected *Aedes albopictus*, *Aedes japonicus*, and *Aedes triseriatus* in Tennessee. *J. Am. Mosq. Control Assoc.* **2015**, *31*, 233–241. [CrossRef] [PubMed]

135. Gerhardt, R.R.; Gottfried, K.L.; Apperson, C.S.; Davis, B.S.; Erwin, P.C.; Smith, A.B.; Panella, N.A.; Powell, E.E.; Nasci, R.S. First isolation of La Crosse virus from naturally infected Aedes albopictus. *Emerg. Infect. Dis.* **2001**, *7*, 807–811. [CrossRef]

136. Harrison, B.A.; Mitchell, C.J.; Apperson, C.S.; Smith, G.C.; Karabatsos, N.; Engber, B.R.; Newton, N.H. Isolation of potosi virus from Aedes albopictus in North Carolina. *J. Am. Mosq. Control Assoc.* **1995**, *11*, 225–229.

137. Fortuna, C.; Remoli, M.E.; Di Luca, M.; Severini, F.; Toma, L.; Benedetti, E.; Bucci, P.; Montarsi, F.; Minelli, G.; Boccolini, D.; et al. Experimental studies on comparison of the vector competence of four Italian Culex pipiens populations for West Nile virus. *Parasit. Vectors* **2015**, *8*, 463. [CrossRef]

138. Fros, J.J.; Miesen, P.; Vogels, C.B.; Gaibani, P.; Sambri, V.; Martina, B.E.; Koenraadt, C.J.; van Rij, R.P.; Vlak, J.M.; Takken, W.; et al. Comparative Usutu and West Nile virus transmission potential by local Culex pipiens mosquitoes in north-western Europe. *One Heal.* **2015**, *1*, 31–36. [CrossRef]

139. Goddard, L.B.; Roth, A.E.; Reisen, W.K.; Scott, T.W.; States, U. Vector Competence of California Mosquitoes for. *Emerg. Infect. Dis.* **2002**, *8*, 1385–1391. [CrossRef]

140. Ebel, G.D.; Rochlin, I.; Longacker, J.; Kramer, L.D. Culex restuans (Diptera: Culicidae) relative abundance and vector competence for West Nile Virus. *J. Med. Entomol.* **2005**, *42*, 838–843. [CrossRef]

141. Kilpatrick, A.; Fonseca, D.M.; Ebel, G.D.; Reddy, M.R.; Kramer, L.D. Spatial and temporal variation in vector competence of Culex pipiens and Cx. restuans mosquitoes for West Nile virus. *Am. J. Trop. Med. Hyg.* **2010**, *83*, 607–613. [CrossRef] [PubMed]

142. Richards, S.L.; Mores, C.N.; Lord, C.C.; Tabachnick, W.J. Impact of Extrinsic Incubation Temperature and Virus Exposure on Vector Competence of *Culex pipiens quinquefasciatus* Say (Diptera: Culicidae) for West Nile Virus. *Vector-Borne Zoonotic Dis.* **2007**, *7*, 629–636. [CrossRef] [PubMed]

143. Richards, S.L.; Anderson, S.L.; Lord, C.C. Vector competence of Culex pipiens quinquefasciatus (Diptera: Culicidae) for West Nile virus isolates from Florida. *Trop. Med. Int. Heal.* **2014**, *19*, 610–617. [CrossRef] [PubMed]

144. Anderson, S.L.; Richards, S.L.; Tabachnick, W.J.; Smartt, C.T. Effects of West Nile Virus Dose and Extrinsic Incubation Temperature on Temporal Progression of Vector Competence in Culex pipiens quinquefasciatus. *J. Am. Mosq. Control Assoc.* **2010**, *26*, 103–107. [CrossRef] [PubMed]

145. Turell, M.J.; Guinn, M.L.O.; Dohm, D.J.; Jones, J.W. Vector Competence of North American Mosquitoes (Diptera: Culicidae) for West Nile Virus Vector Competence of North American Mosquitoes (Diptera: Culicidae) for West Nile Virus. *J. Med. Entomol.* **2001**, *38*, 130–134. [CrossRef] [PubMed]

146. Vaidyanathan, R.; Fleisher, A.E.; Minnick, S.L.; Simmons, K.A.; Scott, T.W. Nutritional Stress Affects Mosquito Survival and Vector Competence for West Nile Virus. *Vector-Borne Zoonotic Dis.* **2008**, *8*, 727–732. [CrossRef]

147. Paupy, C.; Ollomo, B.; Kamgang, B.; Moutailler, S.; Rousset, D.; Demanou, M.; Hervé, J.-P.; Leroy, E.; Simard, F. Comparative role of Aedes albopictus and Aedes aegypti in the emergence of Dengue and Chikungunya in central Africa. *Vector Borne Zoonotic Dis.* **2010**, *10*, 259–266. [CrossRef]

148. Vega-Rúa, A.; Lourenço-De-Oliveira, R.; Mousson, L.; Vazeille, M.; Fuchs, S.; Yébakima, A.; Gustave, J.; Girod, R.; Dusfour, I.; Leparc-Goffart, I.; et al. Chikungunya Virus Transmission Potential by Local Aedes Mosquitoes in the Americas and Europe. *PLoS Negl. Trop. Dis.* **2015**, *9*, e0003780. [CrossRef]

149. Vega-Rua, A.; Zouache, K.; Caro, V.; Diancourt, L.; Delaunay, P.; Grandadam, M.; Failloux, A.B. High Efficiency of Temperate Aedes albopictus to Transmit Chikungunya and Dengue Viruses in the Southeast of France. *PLoS ONE* **2013**, *8*, e59716. [CrossRef]

150. Honório, N.A.; Wiggins, K.; Câmara, D.C.P.; Eastmond, B.; Alto, B.W. Chikungunya virus vector competency of Brazilian and Florida mosquito vectors. *PLoS Negl. Trop. Dis.* **2018**, *12*, e0006521. [CrossRef]

151. Fortuna, C.; Toma, L.; Remoli, M.E.; Amendola, A.; Severini, F.; Boccolini, D.; Romi, R.; Venturi, G.; Rezza, G.; Di Luca, M. Vector competence of Aedes albopictus for the Indian Ocean Lineage (IOL) chikungunya viruses of the 2007 and 2017 outbreaks in italy: A comparison between strains with and without the E1:A226V mutation. *Eurosurveillance* **2018**, *23*, 1800246. [CrossRef] [PubMed]

152. Ngoagouni, C.; Kamgang, B.; Kazanji, M.; Paupy, C.; Nakouné, E. Potential of Aedes aegypti and Aedes albopictus populations in the Central African Republic to transmit enzootic chikungunya virus strains. *Parasites Vectors* **2017**, *10*, 164. [CrossRef] [PubMed]

153. Calvez, E.; Guillaumot, L.; Girault, D.; Richard, V.; O'Connor, O.; Paoaafaite, T.; Teurlai, M.; Pocquet, N.; Cao-Lormeau, V.M.; Dupont-Rouzeyrol, M. Dengue-1 virus and vector competence of Aedes aegypti (Diptera: Culicidae) populations from New Caledonia. *Parasites Vectors* **2017**, *10*, 381. [CrossRef] [PubMed]

154. Gonçalves, C.M.; Melo, F.F.; Bezerra, J.M.T.; Chaves, B.A.; Silva, B.M.; Silva, L.D.; Pessanha, J.E.M.; Arias, J.R.; Secundino, N.F.C.; Norris, D.E.; et al. Distinct variation in vector competence among nine field populations of Aedes aegypti from a Brazilian dengue-endemic risk city. *Parasites Vectors* **2014**, *7*, 320. [CrossRef]

155. Poole-Smith, B.K.; Hemme, R.R.; Delorey, M.; Felix, G.; Gonzalez, A.L.; Amador, M.; Hunsperger, E.A.; Barrera, R. Comparison of Vector Competence of Aedes mediovittatus and Aedes aegypti for Dengue Virus: Implications for Dengue Control in the Caribbean. *PLoS Negl. Trop. Dis.* **2015**, *9*, e0003462. [CrossRef]

156. Richards, S.L.; Anderson, S.L.; Alto, B.W. Vector Competence of Aedes aegypti and Aedes albopictus (Diptera: Culicidae) for Dengue Virus in the Florida Keys. *J. Med. Entomol.* **2012**, *49*, 942–946. [CrossRef]

157. Brustolin, M.; Santamaria, C.; Napp, S.; VerdÚn, M.; Rivas, R.; Pujol, N.; Talavera, S.; Busquets, N. Experimental study of the susceptibility of a European Aedes albopictus strain to dengue virus under a simulted Mediterranean temperature regime. *Med. Vet. Entomol.* **2018**, *32*, 393–398. [CrossRef]

158. Chepkorir, E.; Lutomiah, J.; Mutisya, J.; Mulwa, F.; Limbaso, K.; Orindi, B.; Ng'Ang'a, Z.; Sang, R. Vector competence of Aedes aegypti populations from Kilifi and Nairobi for dengue 2 virus and the influence of temperature. *Parasites Vectors* **2014**, *7*, 435. [CrossRef]

159. Diallo, M.; Ba, Y.; Faye, O.; Soumare, L.; Dia, I.; Sall, A.A. Vector competence of Aedes aegypti populations from Senegal for sylvatic and epidemic dengue 2 virus isolated in West Africa. *Trans. R. Soc. Trop. Med. Hyg.* **2008**, *102*, 493–498. [CrossRef]

160. Heitmann, A.; Jansen, S.; Lühken, R.; Leggewie, M.; Badusche, M.; Pluskota, B.; Becker, N.; Vapalahti, O.; Schmidt-Chanasit, J.; Tannich, E. Experimental transmission of zika virus by mosquitoes from central Europe. *Eurosurveillance* **2017**, *22*, 14–17. [CrossRef]

161. Liu, Z.; Zhou, T.; Lai, Z.; Zhang, Z.; Jia, Z.; Zhou, G.; Williams, T. Competence of Aedes aegypti, Aedes albopictus, and Culex quinquefasciatus mosquitoes as Zika Virus Vectors, China. *Emerg. Infect. Dis.* **2017**, *23*, 1085–1091. [CrossRef] [PubMed]

162. Di Luca, M.; Severini, F.; Toma, L.; Boccolini, D.; Romi, R.; Remoli, M.E.; Sabbatucci, M.; Rizzo, C.; Venturi, G.; Rezza, G. Experimental studies of susceptibility of Italian Aedes albopictus to Zika virus. *Euro Surveill.* **2016**, *21*, 18.

163. Hugo, L.E.; Stassen, L.; La, J.; Gosden, E.; Ekwudu, O.; Winterford, C.; Viennet, E.; Faddy, H.M.; Devine, G.J.; Frentiu, F.D. Vector competence of Australian Aedes aegypti and Aedes albopictus for an epidemic strain of Zika virus. *PLoS Negl. Trop. Dis.* **2019**, *13*, e0007281. [CrossRef] [PubMed]

164. González, M.A.; Pavan, M.G.; Fernandes, R.S.; Busquets, N.; David, M.R.; Lourenço-Oliveira, R.; García-Pérez, A.L.; MacIel-De-Freitas, R. Limited risk of Zika virus transmission by five Aedes albopictus populations from Spain. *Parasites Vectors* **2019**, *12*, 150. [CrossRef]

165. Bonica, M.B.; Goenaga, S.; Martin, M.L.; Feroci, M.; Luppo, V.; Muttis, E.; Fabbri, C.; Morales, M.A.; Enria, D.; Micieli, M.V.; et al. Vector competence of Aedes aegypti for different strains of Zika virus in Argentina. *PLoS Negl. Trop. Dis.* **2019**, *13*, e0007433. [CrossRef]

166. Main, B.J.; Nicholson, J.; Winokur, O.C.; Steiner, C.; Riemersma, K.K.; Stuart, J.; Takeshita, R.; Krasnec, M.; Barker, C.M.; Coffey, L.L. Vector competence of Aedes aegypti, Culex tarsalis, and Culex quinquefasciatus from California for Zika virus. *PLoS Negl. Trop. Dis.* **2018**, *12*, e0006524. [CrossRef]

167. Calvez, E.; O'Connor, O.; Pol, M.; Rousset, D.; Faye, O.; Richard, V.; Tarantola, A.; Dupont-Rouzeyrol, M. Differential transmission of Asian and African Zika virus lineages by Aedes aegypti from New Caledonia. *Emerg. Microbes Infect.* **2018**, *7*, 159. [CrossRef]

168. Chouin-Carneiro, T.; Vega-Rua, A.; Vazeille, M.; Yebakima, A.; Girod, R.; Goindin, D.; Dupont-Rouzeyrol, M.; Lourenço-de-Oliveira, R.; Failloux, A.B. Differential Susceptibilities of Aedes aegypti and Aedes albopictus from the Americas to Zika Virus. *PLoS Negl. Trop. Dis.* **2016**, *10*, e4543. [CrossRef]

169. Hery, L.; Boullis, A.; Delannay, C.; Vega-Rúa, A. Transmission potential of African, Asian and American Zika virus strains by Aedes aegypti and Culex quinquefasciatus from Guadeloupe (French West Indies). *Emerg. Microbes Infect.* **2019**, *8*, 699–706. [CrossRef]

170. Richard, V.; Paoaafaite, T.; Cao-Lormeau, V.M. Vector Competence of French Polynesian Aedes aegypti and Aedes polynesiensis for Zika Virus. *PLoS Negl. Trop. Dis.* **2016**, *10*, e0005024. [CrossRef]

171. Hall-Mendelin, S.; Pyke, A.T.; Moore, P.R.; Ritchie, S.A.; Moore, F.A.J.; Van Den Hurk, A.F. Characterization of a Western Pacific Zika Virus Strain in Australian Aedes aegypti. *Vector-Borne Zoonotic Dis.* **2018**, *18*, 317–322. [CrossRef] [PubMed]

172. Ellis, B.R.; Sang, R.C.; Horne, K.M.; Higgs, S.; Wesson, D.M. Yellow fever virus susceptibility of two mosquito vectors from Kenya, East Africa. *Trans. R. Soc. Trop. Med. Hyg.* **2012**, *106*, 387–389. [CrossRef] [PubMed]

173. Yen, P.S.; Amraoui, F.; Rúa, A.V.; Failloux, A.B. Aedes aegypti mosquitoes from guadeloupe (French west indies) are able to transmit yellow fever virus. *PLoS ONE* **2018**, *13*, e0204710. [CrossRef] [PubMed]

174. Van Den Hurk, A.F.; McElroy, K.; Pyke, A.T.; McGee, C.E.; Hall-Mendelin, S.; Day, A.; Ryan, P.A.; Ritchie, S.A.; Vanlandingham, D.L.; Higgs, S. Vector competence of Australian mosquitoes for yellow fever virus. *Am. J. Trop. Med. Hyg.* **2011**, *85*, 446–451. [CrossRef]

175. Jupp, P.G.; Kemp, A. Laboratory vector competence experiments with yellow fever virus and five South African mosquito species including Aedes aegypti. *Trans R Soc Trop Med Hyg* **2002**, *96*, 493–498. [CrossRef]

176. Focks, D.A. *Review of Entomological Sampling Methods and Indicators for Dengue Vectors*; World Health Organization: Geneva, Switzerland, 2004.

177. De Abreu, F.V.S.; Delatorre, E.; Dos Santos, A.A.C.; Ferreira-De-Brito, A.; De Castro, M.G.; Ribeiro, I.P.; Furtado, N.D.; Vargas, W.P.; Ribeiro, M.S.; Meneguete, P.; et al. Combination of surveillance tools reveals that yellow fever virus can remain in the same atlantic forest area at least for three transmission seasons. *Mem. Inst. Oswaldo Cruz* **2019**, *114*, e190076. [CrossRef]

178. Silver, J.B. *Mosquito Ecology: Field Sampling Methods*, 3rd ed.; Springer Netherlands: Dordrecht, The Netherlands, 2008.

179. De Abreu, F.V.S.; Ribeiro, I.P.; Ferreira-de-Brito, A.; dos Santos, A.A.C.; de Miranda, R.M.; Bonelly, I.; Neves, M.S.A.S.; Bersot, M.I.; dos Santos, T.P.; Gomes, M.Q.; et al. Haemagogus leucocelaenus and Haemagogus janthinomys are the primary vectors in the major yellow fever outbreak in Brazil, 2016–2018. *Emerg. Microbes Infect.* **2019**, *8*, 218–231. [CrossRef]

180. Kilpatrick, A.; Kramer, L.; Campbell, S.; Alleyne, O.; Dobson, A.; Daszak, P. West Nile Virus Risk Assessment and the Bridge Vector Paradigm. *Emerg. Infect. Dis.* **2005**, *11*, 425–429. [CrossRef]

181. Muñoz, J.; Ruiz, S.; Soriguer, R.; Alcaide, M.; Viana, D.S.; Roiz, D.; Vázquez, A.; Figuerola, J. Feeding patterns of potential West Nile virus vectors in south-west Spain. *PLoS ONE* **2012**, *7*, 3954. [CrossRef]

182. Failloux, A.; Vazeille, M.; Rodhain, F. Geographic genetic variation in populations of the dengue virus vector Aedes aegypti. *J. Mol. Evol.* **2002**, *55*, 653–663. [CrossRef]

183. Lambrechts, L.; Chevillon, C.; Albright, R.; Thaisomboonsuk, B.; Richardson, J.; Jarman, R.; Scott, T. Genetic specificity and potential for local adaptation between dengue viruses and mosquito vectors. *Biomed Cent. Evol. Biol.* **2009**, *9*, 160. [CrossRef] [PubMed]

184. Zouache, K.; Fontaine, A.; Vega-Rua, A.; Mousson, L.; Thiberge, J.M.; Lourenco-De-Oliveira, R.; Caro, V.; Lambrechts, L.; Failloux, A.B. Three-way interactions between mosquito population, viral strain and temperature underlying chikungunya virus transmission potential. *Proc. R. Soc. Bbiol. Sci.* **2014**, *281*, 20141078. [CrossRef] [PubMed]

185. Possas, C.; Lourenço de Oliveira, R.; Tauil, P.L.; Pinheiro, F.D.P.; Pissinatti, A.; Venancio da Cunha, R.; Freire, M.; Menezes Martins, R.; Homma, A. Yellow Fever outbreak in Brazil: The puzzle of rapid viral spread and challenges for immunization. *Mem. Inst. Oswaldo Cruz* **2018**, *113*, e180278. [CrossRef] [PubMed]

186. Amraoui, F.; Pain, A.; Piorkowski, G.; Vazeille, M.; Couto-Lima, D.; de Lamballerie, X.; Lourenço-de-Oliveira, R.; Failloux, A.B. Experimental Adaptation of the Yellow Fever Virus to the Mosquito Aedes albopictus and Potential risk of urban epidemics in Brazil, South America. *Sci. Rep.* **2018**, *8*, 14337. [CrossRef]

187. Tsetsarkin, K.A.; Weaver, S.C. Sequential adaptive mutations enhance efficient vector switching by chikungunya virus and its epidemic emergence. *Plos Pathog.* **2011**, *7*, e1002412. [CrossRef]

188. Forattini, O.P.; Marques, G.R.A.M.; Kakitani, I.; De Brito, M.; Sallum, M.A.M. Significado epidemiológico dos criadouros de Aedes albopictus em bromélias. *Rev. Saude Publica* **1998**, *32*, 186–188. [CrossRef]

189. Li, Y.; Kamara, F.; Zhou, G.; Puthiyakunnon, S.; Li, C.; Liu, Y.; Zhou, Y.; Yao, L.; Yan, G.; Chen, X.G. Urbanization Increases Aedes albopictus Larval Habitats and Accelerates Mosquito Development and Survivorship. *PLoS Negl. Trop. Dis.* **2014**, *8*, e3301. [CrossRef]

190. Hiscox, A.; Kaye, A.; Vongphayloth, K.; Banks, I.; Piffer, M.; Khammanithong, P.; Sananikhom, P.; Kaul, S.; Hill, N.; Lindsay, S.W.; et al. Risk factors for the presence of Aedes aegypti and Aedes albopictus in domestic water-holding containers in areas impacted by the Nam Theun 2 hydroelectric project, Laos. *Am. J. Trop. Med. Hyg.* **2013**, *88*, 1070–1078. [CrossRef]

191. Amerasinghe, F.P.; Alagoda, T.S.B. Mosquito oviposition in bamboo traps, with special reference to Aedes albopictus, Aedes novalbopictus and Armigeres subalbatus. *Insect Sci. Applic.* **1984**, *5*, 493–500. [CrossRef]

192. Brito, M.D.; Forattini, O.P. Produtividade de criadouros de Aedes aegypti no Vale do Paraíba, SP, Brasil. *Rev. Saúde Pública* **2004**, *38*, 209–215. [CrossRef]

193. Chan, K.L.; Ho, B.C.; Chan, Y.C. Aedes aegypti (L.) and Aedes albopictus (Skuse) in Singapore City. 2. Larval habitats. *Bull. World Health Organ.* **1971**, *44*, 629–633. [PubMed]

194. Delatte, H.; Toty, C.; Boyer, S.; Bouetard, A.; Bastien, F.; Fontenille, D. Evidence of Habitat Structuring Aedes albopictus Populations in Réunion Island. *PLoS Negl. Trop. Dis.* **2013**, *7*, e2111. [CrossRef] [PubMed]

195. Tuten, H.C.; Bridges, W.C.; Paul, K.S.; Adler, P.H. Blood-feeding ecology of mosquitoes in zoos. *Med. Vet. Entomol.* **2012**, *26*, 407–416. [CrossRef] [PubMed]

196. Hess, A.D.; Hayes, R.O.; Tempelis, C.H. The use of th forage ratio technique in mosquito host preference studies. *Mosq. News* **1968**, *28*, 386–389.

Review

Mayaro Virus Pathogenesis and Transmission Mechanisms

Cheikh Tidiane Diagne [1,*], Michèle Bengue [1], Valérie Choumet [2], Rodolphe Hamel [1], Julien Pompon [1] and Dorothée Missé [1,*]

[1] MIVEGEC, IRD, Univ. Montpellier, CNRS, 34394 Montpellier, France; michele.bengue@ird.fr (M.B.); rodolphe.hamel@ird.fr (R.H.); julien.pompon@ird.fr (J.P.)

[2] Unité Environnement Risques Infectieux Groupe Arbovirus, Institut Pasteur, 75724 Paris, France; valerie.choumet@pasteur.fr

* Correspondence: c.diagne@icloud.com (C.T.D.); dorothee.misse@ird.fr (D.M.)

Received: 10 July 2020; Accepted: 3 September 2020; Published: 8 September 2020

Abstract: Mayaro virus (MAYV), isolated for the first time in Trinidad and Tobago, has captured the attention of public health authorities worldwide following recent outbreaks in the Americas. It has a propensity to be exported outside its original geographical range, because of the vast distribution of its vectors. Moreover, most of the world population is immunologically naïve with respect to infection with MAYV which makes this virus a true threat. The recent invasion of several countries by *Aedes albopictus* underscores the risk of potential urban transmission of MAYV in both tropical and temperate regions. In humans, the clinical manifestations of MAYV disease range from mild fever, rash, and joint pain to arthralgia. In the absence of a licensed vaccine and clinically proven therapeutics against Mayaro fever, prevention focuses mainly on household mosquito control. However, as demonstrated for other arboviruses, mosquito control is rather inefficient for outbreak management and alternative approaches to contain the spread of MAYV are therefore necessary. Despite its strong epidemic potential, little is currently known about MAYV. This review addresses various aspects of MAYV, including its epidemiology, vector biology, mode of transmission, and clinical complications, as well as the latest developments in MAYV diagnosis.

Keywords: Mayaro; emerging arbovirus; alphavirus; Togaviridae; Aedes; vector competence

---

1. Introduction

Following the major Zika virus (ZIKV) outbreaks in 2016, special attention has been given to neglected arboviruses. Arboviral infections are likely to spread globally as a result of the expansion of mosquitoes which, in turn, is caused by increased trade and climate change [1–3]. Mayaro virus (MAYV) is a single-stranded RNA virus. It belongs to the Togaviridae family and was first detected in 1954 in Trinidad in the sera of forest workers [4–6]. Like other alphaviruses, MAYV can infect, replicate and disseminate in both vertebrate and invertebrate hosts [7–9]. In humans, the virus causes Mayaro fever, which is characterized by long-lasting arthralgia similar to that occurring in Dengue fever. Although mortality among infected people is low as yet, Mayaro fever may become a major public health problem, particularly in rural areas, with increasing prevalence in the Amazon region due to ecosystem changes [10]. Although several outbreaks of Mayaro fever have already been reported more than three decades ago in Northern Brazil [11], MAYV is now spreading rapidly to other regions in Latin America [12]. It is likely that the global burden of Mayaro fever is still largely underestimated because of the lack of adequate and accurate diagnostics. At the same time, the *Aedes* mosquito, which can spread MAYV and other viruses, is invading new habitats and regions at an alarming pace [13]. Perhaps the most significant threat of a potential MAYV epidemic comes from increased colonization of urbanized areas by *Aedes aegypti*. In order to better predict the capacity of the virus to spread and infect new areas, further studies are urgently needed. At present, it cannot be excluded that MAYV infections, like those of ZIKV and Chikungunya (CHIKV) viruses, will one day become a major health issue. At present, MAYV is still neglected and few studies have been carried out on its pathogenesis, the biology of its potential vectors or the dynamics of its transmission. We present here a current and comprehensive review of published scientific research and technical reports on MAYV, focusing on its vectors, in addition to the epidemiological and clinical aspects of Mayaro fever.

2. Distribution and Significance to Health

Mosquitoes are responsible for the transmission of a variety of deadly pathogens, including MAYV, that affect the lives and livelihoods of millions of people across the world each year. The global expansion of the two major urban vectors (*Ae. aegypti* and *Ae. albopictus*) and the increasing number of international travelers is certain to increase the risk of MAYV transmission [14]. There has been an underestimation of reported Mayaro fever cases reported since its first detection in 1954 which is not only due to its shared symptoms with Dengue and Chikungunya fever, but also to the occurrence of unreported asymptomatic cases [4,15,16]. Figure 1 illustrates the spatial distribution of the virus that was isolated from vectors, non-human primates (NHP) or humans. The presence of MAYV has been mainly reported in Latin America, with Brazil having the highest number of reported cases of Mayaro fever [4,6,10,17–33] and MAYV infections that were reported in Haiti indicate that the virus is also likely to circulate in the Caribbean [31,34]. A phylogenetic study carried out on whole-genome sequences of viruses isolated from Haitian patients suggests that the virus has been introduced into Haiti on three separate occasions [34]. Several imported cases of MAYV infections have also been reported from 1997 onwards in tourists returning from endemic areas of South and Central America, with several cases reported in Europe [14,35–39]. This scenario shows the importance of the potential role of air travel in the global spread of MAYV [14,35–38]. Due to a lack of information and reporting in Africa and Asia, it is as yet uncertain whether MAYV is present in these continents.

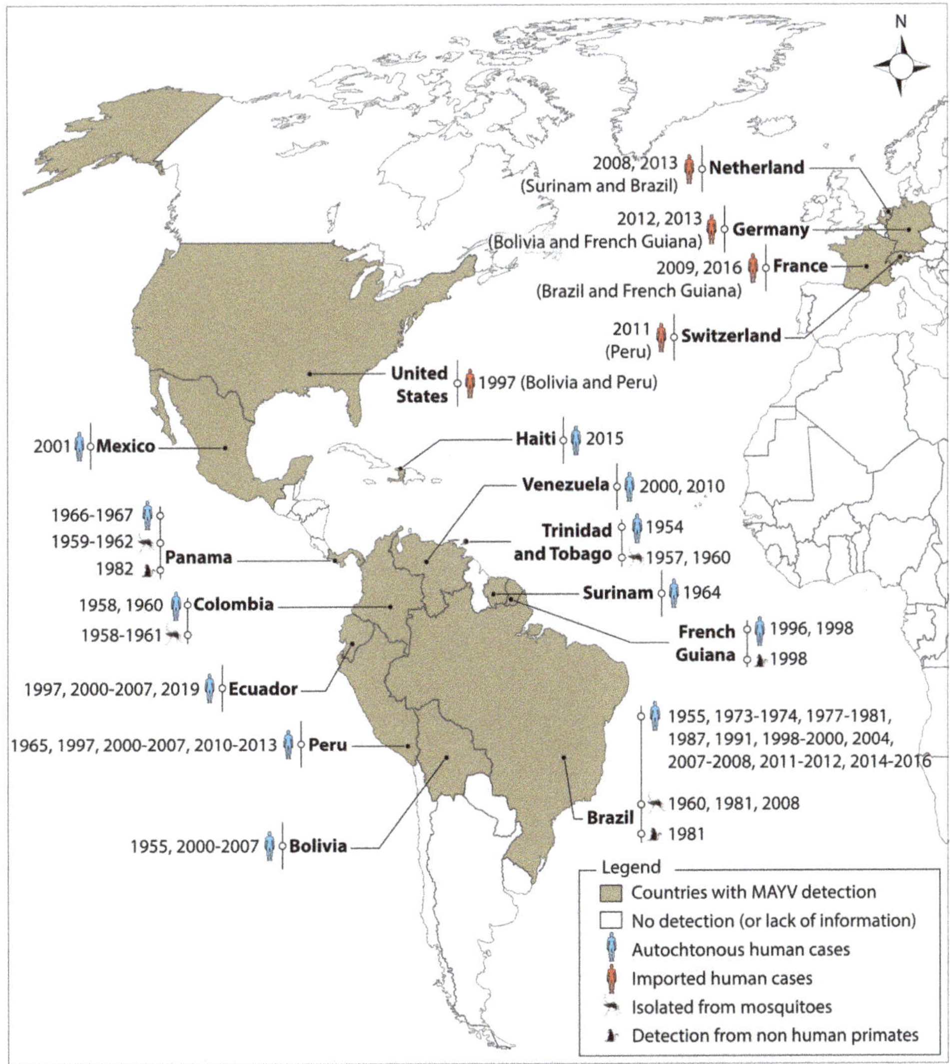

Figure 1. Global distribution of Mayaro virus (MAYV). The virus was detected in vectors, non-human primates and humans. The first imported cases were reported in the United States in 1997 and in Europe in 2008. The origin of imported cases is shown in parentheses. These reports do not exclude the possible presence of MAYV in other countries, notably Africa and Asia. MAYV is not necessarily present throughout the countries/territories shaded in this map. ArcGIS Desktop V 10.5 software was used to generate this map (ESRI 2019, Redlands, CA, USA: Environmental Systems Research Institute; https://desktop.arcgis.com/en/).

3. Clinical Manifestations and Pathogenesis of Mayaro Fever in Humans

Phylogenetically, MAYV is closely related to CHIKV and, like the latter, causes a debilitating flu-like illness in the infected host that is indistinguishable from Chikungunya fever. The main symptoms include chills, fever, gastrointestinal manifestations, eye pain, myalgia and arthralgia [18,29] (Figure 2). In particular, arthralgia can last for months to years, making the Mayaro fever even more debilitating than flavivirus infections, whose symptoms last only one to three weeks [4,11,15,40,41]. Acute symptoms generally last three to five days in most patients with Mayaro disease [15,25]. Arthralgia and myalgia account for 50–89% and 75% of infected patients, respectively [42]. Dizziness and itching are the other clinical manifestations of the disease [20,38]. Severe complications can occur due to MAYV infection, among which are myocarditis, hemorrhagic and neurological manifestations [17]. After a bite from an infected vector, MAYV spreads via the blood vessels into the body of the susceptible host (Figure 3). The virus replicates in white blood cells (e.g., monocytes, macrophages) and spreads to bones, muscles and joints via the main sites of replication, the spleen and the liver [15,43–48]. The severity of Mayaro fever is associated with the production of pro-inflammatory cytokines and mediators (MCP-1, IL-2, IL-9, IL-13, IL-7, VEGF, IL-17, and IP-10) both in humans and experimental mouse models [49–51]. Some of these cytokines have been shown to be associated with other pathologies involving bones and joints [47,52,53]. Mayaro fever can also induce oxidative stress (OS) in the liver of infected mice and in vitro infected HepG2 cells [54,55]. The induction of OS has also been observed in several other arbovirus infections, including Chikungunya [56–58]. The latter process is likely to play a role in the pathogenesis of MAYV, because of its important role in the initiation and control of the production of many soluble mediators such as reactive oxygen species (ROS), and is furthermore involved in apoptosis and inflammation [59]. Currently, little is known about the beneficial or detrimental effects of this process on the infected host. Many cell types have been shown to be implicated in the pathogenesis of MAYV. The role of macrophages, being targets for MAYV infection, in the development of arthritis has been demonstrated through the secretion of TNF-α and ROS [60]. In this respect, it is of interest to note that high serum levels of TNF-α have previously been reported in MAYV-infected patients and mice [50,61]. A recent study has also shown that in vitro infection of bone marrow-derived macrophages with MAYV results in the overexpression of key inflammasome proteins such as NLRP3, AIM2, ASC and caspase 1 [61]. Interestingly, the authors demonstrated that the induction of ROS was linked to the activation of the NLRP3 pathway. Because many studies have shown that ROS production can be associated with immune regulation [62], it would be of interest to investigate whether ROS secretion impacts the polarization of macrophages during MAYV infection. Results from studies using experimental mouse models and human cell lines have revealed that osteoblasts are susceptible to infection with alphaviruses, resulting in the secretion of MCP-1, IL-6, and IL-1β [43,63,64]. We have previously shown that primary human chondrocytes, osteoblasts and synoviocytes, which are the main cell types involved in arthralgia, are permissive to MAYV infection, indicating that these cells may be involved in the pathogenesis of the disease, notably as a result of the overexpression of arthritis-related genes [65]. Infection of the host by MAYV results in the sensitization of monocytes and the induction of their osteoclastogenic activity which leads, in turn, to bone erosion and cartilage damage. Taking into consideration the debilitating condition of MAYV-infected patients, which can last for months to years, further studies should be carried out to determine the mechanisms leading to severe arthritis as a result of MAYV infection. This also begs the question as to why some patients develop arthritis and others do not. It would also be interesting to develop studies leading to the identification of biomarkers or risk factors associated with severe forms of the disease. There is considerable variation in antibody production and persistence from one host to another. MAYV infection induces a transient production of immunoglobulin M (IgM) antibodies, indicating the occurrence of recent infection that generally lasts for at least three months after the onset of clinical symptoms [29,66,67] (Figure 2). In addition, the presence of virus-specific immunoglobulin G (IgG) serum antibodies that persists throughout the lifespan of an infected person is an indicator of prior infection with MAYV, in particular when present at increased levels [29,66,67]. Interestingly, a study by Santiago et al. shows that neutralizing antibodies alone are not sufficient to prevent the occurrence of chronic arthritis [50]. Earnest and collaborators generated a series of neutralizing broad-spectrum mouse antibodies by the use of recombinant MAYV E2 protein [68] and showed their therapeutic utility against MAYV, thereby underscoring the importance that should be given to IgG subclasses and their effector functions.

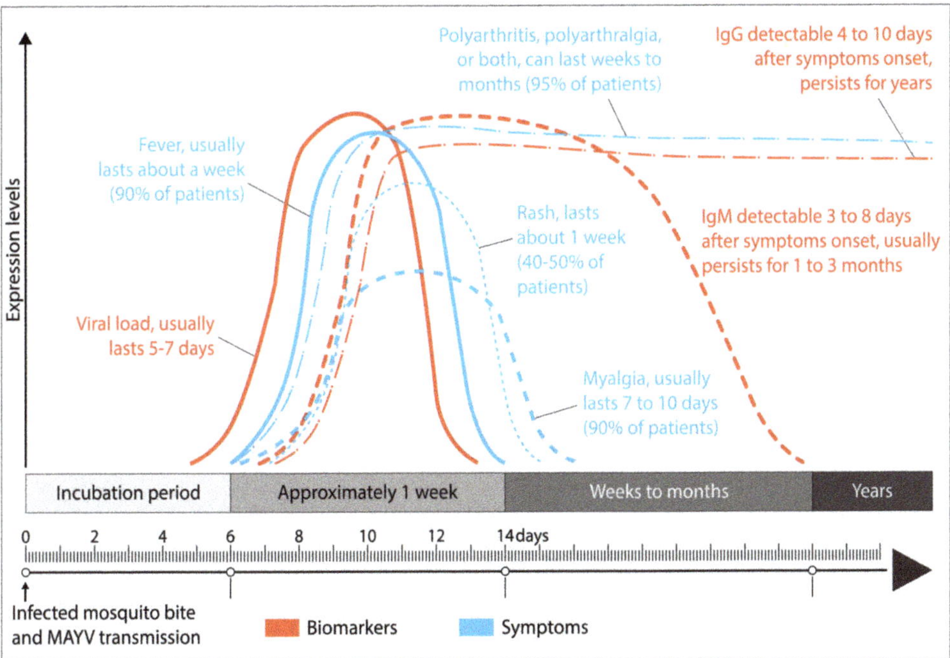

Figure 2. Pathogenesis of Mayaro fever. Time course of MAYV viremia and detection of immunoglobulin M/G (IgM/IgG) after the inoculation of MAYV into a susceptible host by an infected mosquito. Duration of clinical manifestations that may appear in an infected individual. The figure was designed using Adobe Creative Cloud apps (https://www.adobe.com/creativecloud.html).

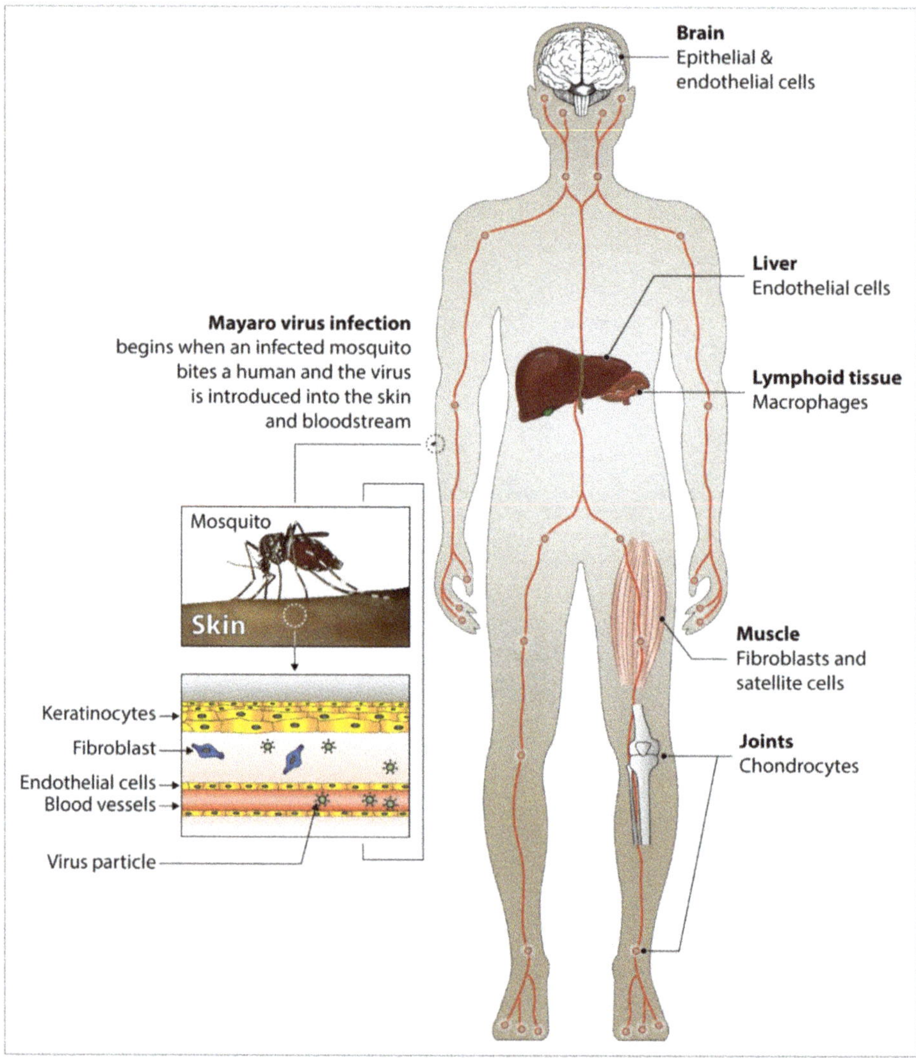

Figure 3. Diagram showing the probable dissemination of MAYV in humans, based on animal experiments and clinical data for similar alphaviruses. Transmission of MAYV occurs following its inoculation by an infected mosquito (*Haemagogus janthinomys*, *Aedes aegypti*, etc.). The virus then replicates in the skin (more precisely at the inoculation site of the virus by the competent vector) and propagates into the target tissues (muscle, the liver, joints, etc.) via the numerous blood vessels, followed by the recruitment of inflammatory cells in these tissues. Most of the target cells have not yet been identified for MAYV but the diagram shows an extrapolation based on other alphaviruses. The figure was designed using Adobe Creative Cloud apps (https://www.adobe.com/creativecloud.html).

4. Diagnosis

The symptoms caused by MAYV infection are often difficult to be recognized because of the limited diagnostic laboratory facilities in many of the endemic regions. In addition to the variety of symptoms, which are not very specific, diagnosis of MAYV infection is rendered difficult by the existence of coinfections and the lack of rapid tests [25,69]. In MAYV surveillance, the discriminating diagnosis must also consider possible concomitant CHIKV, DENV, and ZIKV infections due to the co-circulation these viruses [20,31,33]. Neutralization and reverse-transcription polymerase chain reaction (PCR) tests are now available in almost all laboratories and can help in the confirmation of MAYV infection [31,38]. In a cross-sectional study in Ecuador, the prevalence of MAYV was estimated by evaluating the presence of IgM and IgG antibodies by ELISA [70]. It is of note that when samples are collected after the acute phase of clinical signs (which then no longer represents the viremia phase, lasting two to three days for MAYV), testing based on isolation is impractical. Nevertheless, this technique, which is generally performed in vivo in mice or in vitro by inoculation into cell cultures of arthropods (C6/36, Aag2, etc.) or vertebrates (Vero, BHK-21, etc.), remains the most accurate method for virus detection [20]. It is generally performed in vivo in mice or in vitro by inoculation into the cell cultures of arthropods (C6/36, Aag2, etc.) or vertebrates (Vero, BHK-21, etc.). In some circumstances, the isolation method is not the most appropriate due to several factors, including the costs involved and the need for additional tests, which increase the time required to deliver results. Complement fixation and neutralization tests are among the serological methods that can also be used in certain contexts for the diagnosis of MAYV. The detection of circulating IgM antibodies has the advantage of being specific and also constitutes the fastest technique. Indeed, the production of IgM usually occurs about five days after infection with MAYV [11,67]. For MAYV diagnosis, possible cross-reactions must be taken into account, in particular those between MAYV-, CHIKV- and O'nyong-nyong virus-specific antibodies [71–74]. In recent years, modern, more sensitive and faster methods using molecular biology techniques (i.e., PCR) have been widely used for the diagnosis of several pathogens of medical interest [75]. Accurate diagnosis of MAYV in human communities can also be rendered difficult due to the high prevalence of asymptomatic cases [25,31].

5. Genomic Structure and Viral Life Cycle

The genome of MAYV, the causative agent of Mayaro fever, is a single-stranded RNA virus with positive polarity and a length of approximately 11.5 kilobases (kb). Its genomic structure forms two open reading frames, one coding for structural proteins and the other coding for non-structural proteins (Figure 4). The latter are expressed as a polyprotein that is cleaved both during and after translation into four non-structural proteins (nsP1, nsP2, nsP3, and nsP4), whereas the structural genes generate the six structural proteins (C, E1, E2, E3, 6K, and transframe) [71]. The viral particles are icosahedral, comprising a closed nucleocapsid in a compact envelope consisting of the host cell membrane sealed with E1-E2 complex [8,9,76]. The life cycle of MAYV starts with the fusion of the viral envelope with the plasma membrane of the target cells via specific receptors on their surface such as MXRA8, which has recently been revealed to be a receptor for several arthritogenic alphaviruses [7,77]. This fusion process, however, merits further study to determine the precise mechanisms involved. Next, the virus enters inside the cell by endocytosis and is subsequently found within the endosomal compartment. Once in the cytoplasm, the pH of the endosome decreases and the virus disassembles its basic structure and releases the genomic RNA which is translated into non-structural proteins that generate messenger RNA and various structural proteins. The latter are then transformed into a nucleocapsid and its glycoproteins, which, after combination with the plasma membrane, leads to the release of new viral particles outside the cell. The released particles then infect other cells and the cycle repeats itself [71].

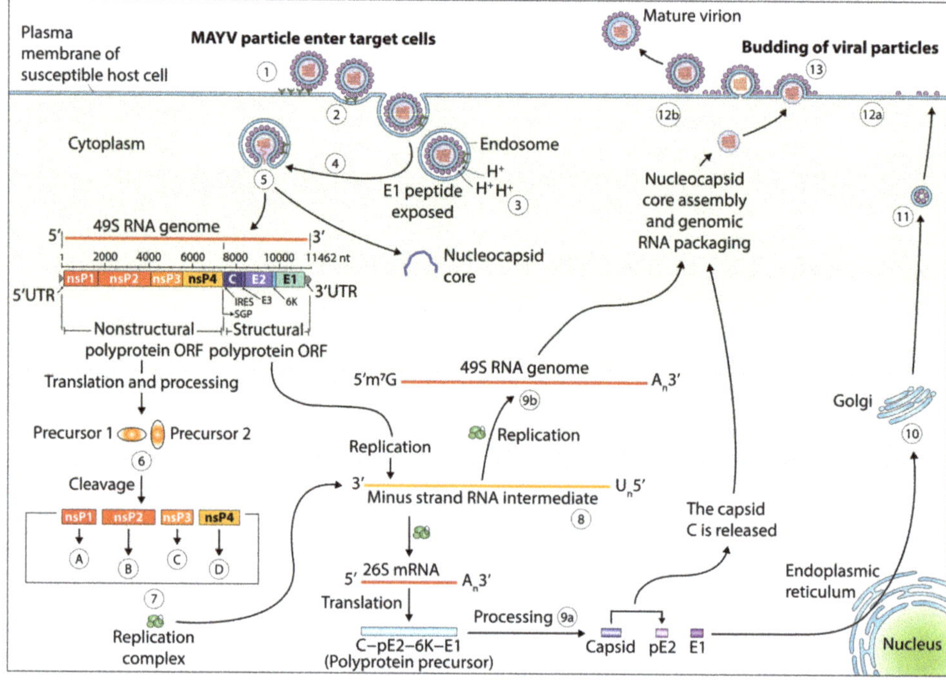

Figure 4. The replicative cycle and genomic structure of MAYV. E: envelope; nsP: nonstructural protein; UTR: untranslated region; nt: nucleotides. A: nsP1 is required in the formation of the viral RNA and the RNA capping processing; B: nsP2 presents several enzymes required in transcription; C: nsP3 is one element of the replication complex; D: nsP4 is known as the viral RNA polymerase. 1: endocytosis through a clathrin-coated pit or, alternatively, through a caveolin-coated pit; 2: start of virus internalization with receptors of the host cell; 3: the low pH obtained in the endosome leads to structural modifications in the envelope of the virus that reveal the E1. The latter mediates cell membrane-virus fusion; 4: endosomal cell membrane-virus fusion; 5: release of the capsid core and genome of the virus; 6: from the mRNA of the virus, a pair of nsP precursors are generated; 7: the replication complex is obtained from the gathering of nsP proteins. This complex thus allows the synthesis of a minus-strand RNA; 8: this RNA is used as the model to generate both genomic (49S) and subgenomic (26S) RNAs; 9a: processing of the polyprotein precursor by an autoproteolytic serine protease; 9b: genomic RNA involved in nucleocapsid core assembly and genomic RNA packaging; 10: processing and maturation of glycoproteins pE2 and E1; 11: in the Golgi, processed glycoproteins are associated. Then they are transported into the cell membrane; 12a: at the cell membrane, the pE2 is split into E2 and E3; 12b: association of the viral RNA with the capsid C and the recruitment of E1 allows viral assembly; 13: particles of MAYV associated with the core are released outside of the host cell through the membrane. The figure was designed using Adobe Creative Cloud apps (https://www.adobe.com/creativecloud.html).

6. Phylogeny

The complete nucleotide sequence of MAYV was determined in 2005 [78]. MAYV is part of the Semliki Forest Complex due to its antigenic composition and is phylogenetically close to other alphaviruses such as CHIKV, Sindbis, and Ross river viruses that induce an identical clinical illness in infected patients [4,15]. Its antigenic resemblance to CHIKV has been well described [72]. Nucleotide sequence homologies and recent phylogenetic analysis carried out on MAYV strains isolated from the Americas indicate the existence of three major phylogroups: disseminated (D), limited (L), and new

(N) [79,80]. D is the more widely disseminated group and contains strains identified in several countries in the Americas (Trinidad and Tobago, Peru, French Guiana, Venezuela, and Bolivia), while the phylogroup limited (L) is found only in a few countries (Haiti and Brazil) [31]. A minor genotype, new (N), which is limited to only one known sequence, was isolated in 2010 in Peru [81].

7. Transmission Cycle

In MAYV-endemic regions in Latin America, a viral transmission cycle in the sylvatic environment is observed. This cycle mainly involves the mosquito *Haemagogus janthinomys* and non-human primates (NHP), considered to be the primary hosts of the virus [82–84] (Figure 5). To a lesser extent, this cycle may involve not only other vertebrates or secondary hosts (rodents, reptiles, etc.), whose role is not clearly identified, but also other types of mosquitoes such as *Mansonia*, *Culex*, *Sabethes*, *Aedes*, and *Psorophora* which may play an accessory or occasional role in this transmission cycle [10,26,85–87]. Due to both the zoophagous nature of *Haemagogus* spp. and their limited biotope in rural areas close to forests, humans are only sporadically infected [88]. At present, no animal reservoir of MAYV has been formally documented in Africa, Asia, and Europe. In urban areas, domestic mosquitoes are potential vectors of MAYV that could transmit the virus to humans under appropriate conditions. The introduction of MAYV into urban habitats and their establishment a domestic cycle is increasingly likely. If this process becomes a reality, this situation could be a real burden, not only for endemic regions but also on a global scale, especially when possible mutations of the virus will favor more efficient dissemination by domestic vectors [25]. In addition to this classical mode of transmission, there is also vertical transmission (passage of the virus from an infected female to her offspring), a mechanism by which the vector contains the virus for long periods of time, thereby constituting a true reservoir. Vertical transmission was revealed by the isolation of MAYV from two pools of adult *Ae. aegypti* which were collected from the egg stage in Brazil in 2017 [89]. Further studies are needed to identify the importance of vertical transmission during epidemics and the molecular mechanism that underlies viral maintenance in the vectors during unfavorable periods [90]. It is, however, clear that the effectiveness of this mechanism depends on climate change, geographical location, arbovirus genotype, and vector population dynamics [91]. Although this mechanism has been extensively reported for DENV, CHIKV, ZIKV, and Yellow fever virus infections [92–95], very few studies have addressed its role in MAYV and other arboviruses [89]. Airborne transmission of MAYV has also been reported among laboratory personnel [96].

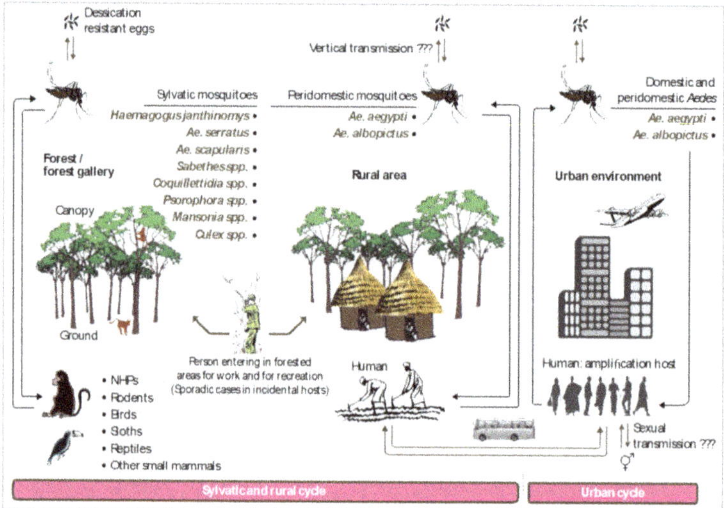

Figure 5. MAYV transmission cycles in endemic South and Central Americas. The list of arthropod and vertebrate species or groups mentioned in this figure may not be exhaustive due to the lack of data for certain regions. The figure was designed using Adobe Creative Cloud apps (https://www.adobe.com/creativecloud.html).

8. MAYV Vectors: Field and Laboratory Observations

To incriminate a mosquito species as a vector of MAYV, two conditions must be met, preferably in this order: 1) first, the virus must be detected in natura from specimens of the species and 2) experiences in controlled conditions must provide evidence of its capacity to transmit the virus. Several entomological surveillances across the world have provided the suspicious involvement of many species but have not given enough formal evidence for the incrimination of a particular species. Indeed, any pathogen is likely to be detected in blood ingested by an arthropod [97]. On the other hand, although transmission experiments are able to clarify potential implications, the number of species to be tested for vector competence is limited by certain technical restrictions and this approach is not always relevant due to the lack of contact between the arthropod and humans in natura [98]. Therefore, the synthesis of all observations obtained from these studies will enable us to define the true MAYV vectors implicated in the viral transmission cycles (Figure 5).

8.1. Evidence from the Field

MAYV isolates have been obtained from diverse mosquito genera (*Haemagogus, Culex, Mansonia, Sabethes, Psorophora, Aedes*) (see Table 1). *Haemagogus* spp., especially *Hg. janthinomys*, appears to be the primary vector [84]. In 1957, a strain of MAYV was isolated from a pool (49 specimens) of the mosquito *Mansonia venezuelensis*, collected in northeastern Trinidad. This isolation was the first one recorded from arthropods collected in natura [99]. Subsequently, isolates were frequently obtained from *Hg. janthinomys* and less regularly from other mosquito species [83,100,101]. In addition, many MAYV isolates were reported in 1978 in Brazil from *Hg. janthinomys* [84]. *Cx. quinquefasciatus* mosquitoes found naturally infected with MAYV were also reported in Brazil [28]; nevertheless, their ability to transmit the virus is still unproven [102].

Table 1. Mosquito species in which MAYV was detected in natura.

Country of Origin	Species	References
Brazil	*Aedes aegypti*	[28,89]
Panama	*Sabethes* spp.	[21]
Brazil	*Haemagogus janthinomys*	[20,83,84]
Colombia	*Aedes serratus*	[101]
Brazil	*Cx. quinquefasciatus*	[28]
Panama	*Cx. vomerifer*	[26]
Trinidad and Tobago	*Mansonia venezuelensis*	[99]
Colombia	*Psorophora ferox*	[101]
Colombia	*Psorophora albipes*	[101]

8.2. Vector Competence Studies

There are a few studies on vector competence for MAYV [26,98,103,104] and those carried out concern *Ae. aegypti* populations from Iquitos, Florida and French Polynesia, and *Ae. albopictus* from Brazil, Florida and Reunion [98,103–106]. These studies have shown that *Ae. albopictus* and *Ae. aegypti* are capable of transmitting MAYV. In addition, several other geographically divergent Anopheles species can transmit MAYV (Table 2) [98,103,104]. The latter observation suggests the possible involvement of some species of this genus in the amplification of MAYV in naïve areas.

Table 2. Vector species found experimentally competent.

Country of Origin	Species	References
Iquitos, Peru	*Ae. aegypti*	[98,102,106]
Bora-Bora, French Polynesia	*Ae. aegypti*	[105]
Florida, USA	*Ae. aegypti*	[103]
Brazil	*Ae. albopictus*	[104]
La Reunion	*Ae. albopictus*	[105]
Florida, USA	*Ae. albopictus*	[103]
Trinidad and Tobago	*Ae. scapularis*	[86]
Virginia, USA	*An. freeborni*	[102]
Maryland, USA	*An. gambiae*	[102]
Virginia, USA	*An. quadrimaculatus*	[102]
Maryland, USA	*An. stephensi*	[102]

9. Major Urban MAYV Vectors: Extension and Risk to Developed and Developing Countries

9.1. Aedes aegypti

The mosquito *Ae. aegypti* is a highly anthropophilic and extremely opportunistic species, with a nearly worldwide distribution. This is a domestic vector involved in arbovirus transmission at a global scale [1,107]. This vector has not yet been reported in European countries, except for some regions (Figure 6), although global warming could lead to their colonization in the near future. The delay of this phenomena will be determined particularly by global climate policies such as the Paris Agreement [108]. A few studies are have also dealt with climatic factors and mechanisms involved in the invasion and proliferation of this species [107,109–111].

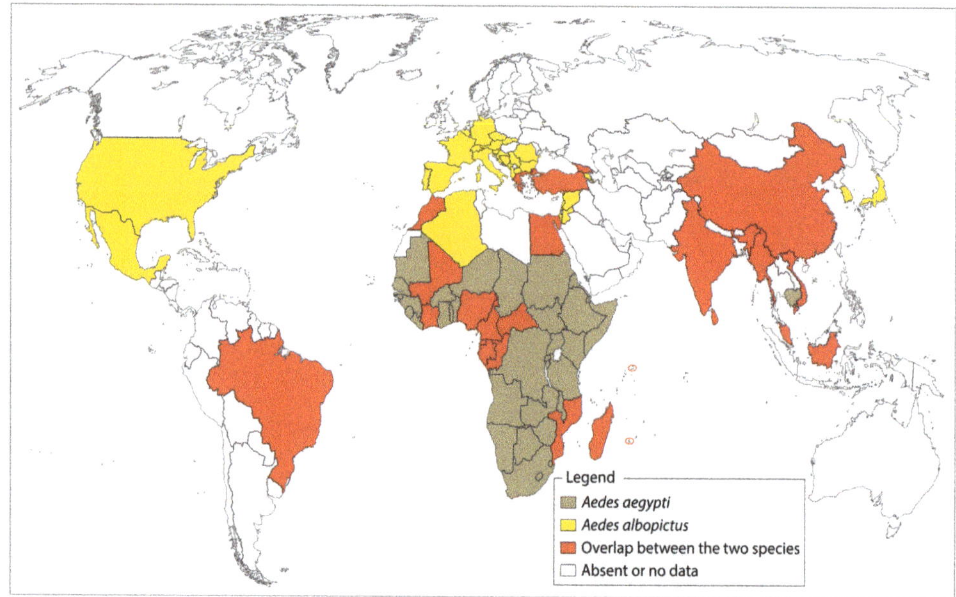

Figure 6. Worldwide distribution of *Ae. albopictus* and *Ae. aegypti*. The "tiger mosquito", *Ae. albopictus*, has currently conquered several temperate regions including France, Italy, and Spain as shown on this map. The species is indeed a competent vector for several alphaviruses and flaviviruses which raises concerns when considering the occurrence of imported cases reported in the USA and Europe. On the other hand, *Ae. aegypti* is present in almost all tropical regions across the globe. On many occasions, the World Health Organization and the Rockefeller foundation have tried unsuccessfully to eradicate it. These two main urban vectors are not necessarily present throughout the countries/territories shaded in this map. ArcGIS Desktop V 10.5 software was used to generate this map (ESRI 2019, Redlands, CA, USA: Environmental Systems Research Institute; https://desktop.arcgis.com/en/).

9.2. Aedes albopictus

Ae. albopictus, commonly known as the "tiger mosquito", is native to Southeast Asia. In recent years, the species has established itself in several regions across the world (Figure 6). In this process of invasion, scientists incriminate the transport of used tires which can harbor desiccation-resistant eggs. Following recent evidence from laboratory experiments on the ability of this species to transmit the virus, health officials expressed their concern and warned about the potential threat of a MAYV epidemic in areas already colonized by this species, as well as those threatened by its introduction.

10. Other Potential Vertebrate Hosts for MAYV

The presence of MAYV and/or anti-MAYV antibodies has been detected in vertebrate hosts in diverse localities in South America. In French Guiana, MAYV has been isolated from a wide variety of mammals, although their relevance in MAYV transmission is as yet unknown [112]. The virus has also been isolated from various other wild animals, including birds [113]. Anti-MAYV antibodies have been detected in marmosets (*Callithrix argentata*) as well.

11. MAYV Transmission in Nonhuman Primates

The rainforests in the Amazon region harbor diverse NHP species [114]. This diversity underlines the importance of identifying the level of involvement of each species in the bio-ecology of MAYV in order to acquire the necessary tools to better prepare or to avoid a worst-case scenario that could be generated by a possible amplification in human populations, starting from an enzootic cycle. As we are aware of these issues, it is important and urgent to carry out studies on the nature of NHPs present in this MAYV-endemic region. This lack of knowledge can be overcome by encouraging scientists who take an interest in this field and by setting up adequate infrastructure.

12. Emergence Mechanisms of Urban MAYV Transmission

Generally, the use of inappropriate vector control, coupled with the high density and the large distribution of urban *Aedes* spp., including *Ae. albopictus* and *Ae. aegypti*, is crucial in MAYV spreading [13]. Specifically, in low-income settings, it is mainly deforestation, unplanned policies concerning urbanization and infrastructure, as well as a lack of resources that benefit mosquito development and their contact with humans. Human movement may also increase the spread of MAYV, as shown for other similar vector-borne viruses [115–118]. Nevertheless, the incidence of disparities in transmission should be determined for vulnerable or minority communities most in need by improving access to high-quality telemedicine, increasing the use of mobile care and testing, in addition to reforming public health data practices, not only for MAYV, but also for other arboviruses as well.

13. Entomological Surveillance

At present, no surveillance system has been put in place for Mayaro fever and the first detections were obtained incidentally during investigations of febrile illnesses. During the epidemics of 1995 in Peru and 2008 in Brazil, results from surveillance efforts were based exclusively on reports provided by sentinel structures based on different techniques in use by the virology community. The occurrence of imported cases stresses the importance of global MAYV entomological surveillance. Such detection programs could provide an opportunity to bring together forces around the "one health" concept with an emphasis on vectors. The ambition, above all, would be to act more widely, more efficiently and more quickly on these issues in both the short and long term. Generally, entomological surveillance in a MAYV epidemic situation involves making an inventory of the mosquito species present at that time in the affected locality from the larval and adult stages, determining their abundance and, where appropriate, their distribution and potential involvement [15]. The latter can be provided by the results of attempts to isolate the virus from specimens collected in the field, including forests and human habitations. Until now, no efficient entomological surveillance of MAYV has been established. At present, the "human-landing catch" method is the preferred method for collecting those vectors likely to come into contact with humans, but unfortunately it is not appropriate in epidemic situations and is more expensive compared to other methods [119,120]. However, alternative methods (although they are less effective, but safer for the operator), through the use of different types of traps, can be used as well. In addition, surveillance of animal fauna (both wild and domestic) could provide additional details and thus allow us to refine the decision-making process to control MAYV. Taken together, the overall interpretation of the results from these surveys will provide a reliable estimation of the MAYV risk in the affected area.

14. Prevention and Control

In concrete terms, the prevention of Mayaro is contingent on the suppression of mosquito bites and, consequently, of human–vector contact. In the case of *Haemagogus* and *Aedes* vectors, whose peak activity coincides at the beginning or the end of the day, the use of insecticide-treated nets does not fully protect against these vectors. Because of the lack of a specific vaccine, the only effective way of managing the spread of the disease in endemic areas is vector control. This strategy, feasible in both domestic and peridomestic settings, focuses primarily on the suppression of mosquito breeding sites through the use of predators, larvicides, etc., which can be reinforced by global and multisectoral coordination across the fields of healthcare, community engagement and social communication [121]. The latter can be carried out among the local population and can take the form of awareness-raising actions about MAYV disease and individual protection measures against mosquito bites using different broadcasting channels and available media (TV, radio, social networks, etc.). Other vector control methods focusing on the elimination of adult stages by spraying with insecticides, particularly in epidemic situations, can also be implemented [84]. Unfortunately, the range of available molecules and formulations is limited, hence the importance of focusing on resistance levels in MAYV vector populations and understanding the underlying mechanisms. In the past, measures that were considered effective, based on the management of household water storage, were implemented in Brazil during the 1981 epidemic. Being among the less studied mosquito-borne diseases, MAYV is still neglected by health authorities and the general public across the world, specifically in non-endemic regions. Communities need to be aware of Mayaro fever epidemics, the damage the virus can cause to their health, and preventive measures, including cleaning their homes and cleaning water containers and the environment around their households. All necessary actions including social and behavior change should be taken in order to inspire community members to engage more strongly to prevent MAYV for a safer future. Surveillance, diagnosis of MAYV and vector control are the three main pillars in terms of prevention of the disease. In terms of vaccine research, efforts are being made and two promising vaccines are currently being developed and will hopefully soon be tested in clinical trials [122]. In order to respond to this crucial issue, the activities carried out in this field must be the culmination of a collective approach. At present, many questions about MAYV remain unresolved and require additional efforts by all multidisciplinary actors (virologists, entomologists, sociologists, statisticians, etc.) in order to find appropriate or alternative solutions.

15. Factors Associated with MAYV Emergence and Invasion

Nowadays, most outbreaks of MAYV are restricted to rural settings close to tropical forests, but a possible future urbanization of MAYV is to be expected, as suggested by the following observations [40,123]: (1) urbanization as a possible result of the virus being carried by travelers or infected birds [14,124]; (2) the strong homology between MAYV with CHIKV, an alphavirus that circulates in urban areas; (3) the frequent observation of MAYV cases near tropical cities in which *Ae. aegypti*, the principal urban vector, is endemic [15]; (4) results from laboratory studies showing that both *Ae. albopictus* [103–105] and *Ae. aegypti* [98,103,105] are able to transmit MAYV (Table 2). Features related to the emergence of MAYV include (1) anthropic disturbances in the ecosystem, notably caused by farming and agrobusiness activities, urbanization, a demographic boom, poor sanitary conditions, deforestation and human migration (Figure 7) [3,125–127], (2) the genetic mutation of the virus and its adaptation to other arthropod vectors [128], (3) the ability of these vectors to tolerate a lethal dose of major insecticides, as well as increased pathogen resistance to drugs [129] and, finally, (4) changing weather conditions that favor the invasion of mosquito vectors into new areas in which humans have not encountered the virus and have not yet acquired a protective immune response [2,130,131]. Ultimately, is it fair to state that changing climatic conditions are mainly responsible for the increased incidence and risk of viral spread to humans. These act in favor of the vectors that subsequently occupy hitherto unscathed areas and consequently allow the virus to emerge in these areas, as illustrated by the MAYV epidemic in Haiti in 2015. The positive correlation between the occurrence of epidemics and the peak density of vector populations during the wet season in tropical regions are further proof of the impact of climate on the transmission of MAYV [2,5,31,113].

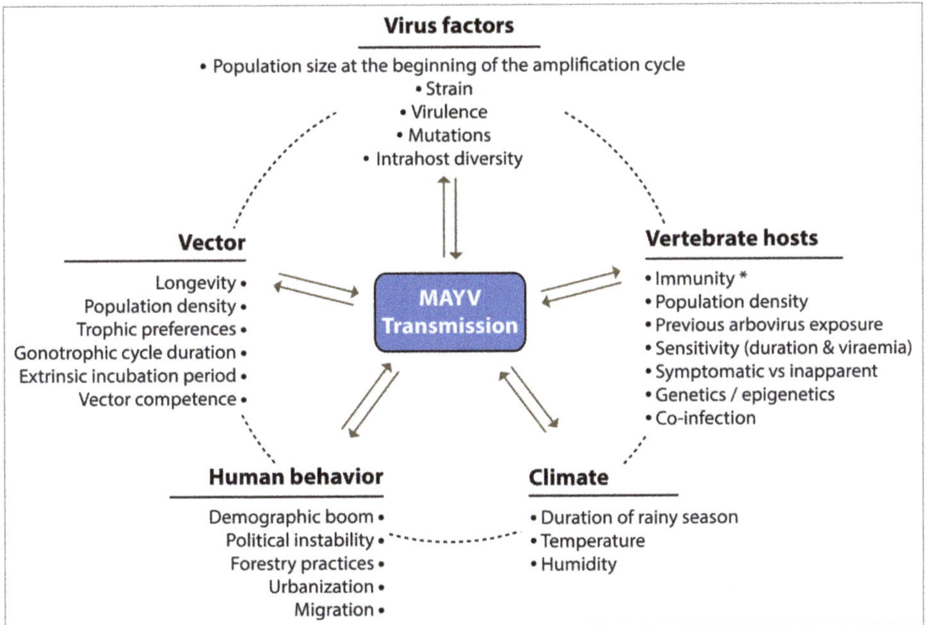

Figure 7. Schematic diagram showing the ecological factors that affect MAYV transmission. The dashed line and double arrows illustrate possible interactions between factors. * Population-level susceptibility. The figure was designed using Adobe Creative Cloud apps (https://www.adobe.com/creativecloud.html).

16. Future Trends

Like CHIKV and ZIKV, MAYV could grow significantly and become one of the most widespread arboviruses on the planet. This scenario could be triggered by an evolution of the virus, changes in the host–vector relationship, etc. To anticipate possible adverse trends in Mayaro fever, there is a need to implement improved strategies and strengthen existing diagnostic platforms. The lack of optimal sampling methods for mosquitoes remains one of the major gaps in MAYV surveillance. The "human-landing collection" method remains the only effective method for the correct estimation of human–vector relationships, but its use is being questioned and debated because of ethical considerations [119,120]. For the coming years, it will therefore be urgent to think about implementation of alternative methods that can provide more perfect sampling data. Given the complexity of vector-borne diseases, experiments on vector competence for MAYV should include, in addition to species of the genera *Aedes* and *Haemagogus*, both *An. gambiae* and *Cx. quinquefasciatus*, which are species of great importance and whose presence is strongly dependent on human habitation. Studies on the efficacy of antiviral drugs and vaccines are generally supported by animal models, but not all models are suitable for a particular scientific question. In the case of MAYV, it is necessary to define which animal model (NHP, mouse, hamster, etc.) is suitable for vaccine research. For this purpose, we can learn from past years of alphavirus research and refer to available and relevant studies. Brazil, French Guyana, Peru, Colombia and Panama are among the regions in the world most affected by Mayaro disease. To help fight MAYV, it is necessary to combine simple community-based experiments with scientific innovation. Mobile phones have become important for disease mapping and surveillance through the use of short messaging services in affected areas. Adapting user-friendly innovations to the living conditions of people in remote locations can change the dynamics of disease control. The development of innovative scientific products from research and development takes time, but easy access to those innovations is needed to combat vector-borne diseases. Public–private partnerships, as well as university and national programs should be promoted to develop innovative tools for prevention. Technology is and will continue to be undoubtedly changing the way we monitor health information systems and discover new innovative diagnostic or control strategies against arboviruses. From sample collection, to diagnostics, to data sharing, technology has firmly taken its place in our day-to-day research activities.

17. Conclusions

Variabilities in susceptibility and ability to disseminate and transmit MAYV have been reported depending on the extrinsic incubation conditions, the virus strains used and the mosquito species tested. These variations in the expression of oral receptivity have been repeatedly described in natural populations of vector mosquitoes. Several factors, both intrinsic and extrinsic to the vectors, may explain these variations. Among these, possible interactions between the virus and the microbiota of the mosquito's midgut seem to be particularly important, knowing that this symbiosis could provide important physiological functions to the host, including the synthesis of essential nutrients, resistance to infection and stimulation of the immune system. In many studies on the involvement of the immune system, bacteria have been used, but it was only very recently that their role in the immune response against virus infections was discovered. Differences in the expression of genes involved in the immune response are probably due to changes in the specific composition of the bacterial flora present in the midgut of the host mosquito and are important factors influencing the variation in transmission. To date, no data on potential vectors are available on this aspect. Therefore, for further analysis, studies should be carried out on the diversity and relative proportions of the different taxa of this midgut microbiota. Given that virus–host interactions are often accompanied by variations in viral proteins, resulting in the emergence of disease, it would be interesting to investigate the contribution of recombination phenomena to the natural evolution of MAYV strains and the epidemiological risk that recombinant strains may present. From a technical point of view, the identification and subsequent inactivation, by homologous recombination, of genes involved in the replication or adsorption of viral particles, for example, could significantly reduce transmission. It would also be helpful to improve our knowledge of the mechanisms underlying MAYV replication in mosquitos. In this respect, further studies providing valuable information on the basic ecology and spatiotemporal dynamics of vector abundance and their association with MAYV should be carried out. Transmission probability models could be used to estimate the natural periodicity of virus incidence cycles in each potential host. These predictive models will certainly make it possible to improve monitoring methods, thereby providing predictive capacity for the identification and early warning of emergence risks, as well as their control, by appropriate and accepted preventive methods. From the same perspective, two axes should be explored: (i) understanding how human contact with MAYV via vectors may increase with intensive use of forest resources; (ii) understanding how wildlife populations and communities, acting as reservoirs and potential vectors, respond to changes in the environment. These modifications will have an impact on wildlife populations, leading to the displacement of native forest species and the introduction of new ones capable of colonizing disturbed areas, resulting in the emergence of arboviruses, including MAYV, which were previously confined to the biotopes of natural reservoirs, thereby increasing the risk of contact with new mammalian and human hosts.

Author Contributions: C.T.D. and M.B. contributed to writing and editing the review. D.M. contributed to designing the tables, figures, and correcting the manuscript. J.P., V.C., and R.H. contributed mostly to the final form of the manuscript and its improvement. All authors contributed to the idea and formation of the manuscript and approved the submitted version. All authors have read and agreed to the published version of the manuscript.

Funding: This work was supported by the Méditerranée Infection Foundation and was publicly funded by the French National Research Agency (ANR-15-CE15-00029).

Conflicts of Interest: The authors declare that the research was conducted in the absence of any commercial or financial relationships that could be construed as potential conflicts of interest.

References

1. Semenza, J.C.; Suk, J.E. Vector-borne diseases and climate change: A European perspective. *FEMS Microbiol. Lett.* **2018**, *365*. [CrossRef] [PubMed]
2. Gould, E.A.; Higgs, S. Impact of climate change and other factors on emerging arbovirus diseases. *Trans. R. Soc. Trop. Med. Hyg.* **2009**, *103*, 109–121. [CrossRef]
3. Weaver, S.C.; Reisen, W.K. Present and future arboviral threats. *Antiviral Res.* **2010**, *85*, 328–345. [CrossRef]
4. Anderson, C.R.; Downs, W.G.; Wattley, G.H.; Ahin, N.W.; Reese, A.A. Mayaro Virus: A New Human Disease Agent. *Am. J. Trop. Med. Hyg.* **1957**, *6*, 1012–1016. [CrossRef]
5. LeDuc, J.W.; Pinheiro, F.P.; da Rosa, A.P.A.T. An Outbreak of Mayaro Virus Disease in Belterra, Brazil II. Epidemiology. *Am. J. Trop. Med. Hyg.* **1981**, *30*, 682–688. [CrossRef]
6. Schaeffer, M.; Gajdusek, D.C.; Lema, A.B.; Eichenwald, H. Epidemic Jungle Fevers Among Okinawan Colonists in the Bolivian Rain Forest. *Am. J. Trop. Med. Hyg.* **1959**, *8*, 372–396. [CrossRef] [PubMed]
7. Carvalho, C.A.M.; Silva, J.L.; Oliveira, A.C.; Gomes, A.M.O. On the entry of an emerging arbovirus into host cells: Mayaro virus takes the highway to the cytoplasm through fusion with early endosomes and caveolae-derived vesicles. *PeerJ* **2017**, *5*, e3245. [CrossRef] [PubMed]
8. Mezencio, J.M.S.; de Souza, W.; Fonseca, M.E.F.; Rebello, M.A. Ultrastructural study of Mayaro virus replication in BHK-21 cells. *Arch. Virol.* **1990**, *114*, 229–235. [CrossRef] [PubMed]
9. Mezencio, J.M.S.; de Souza, W.; Fonseca, M.E.F.; Rebello, M.A. Replication of Mayaro virus in Aedes albopictus cells: An electron microscopic study. *Arch. Virol.* **1989**, *104*, 299–308. [CrossRef]
10. Vasconcelos, P.F.; Travassos da Rosa, A.; Rodrigues, S.G.; Travassos da Rosa, E.S.; Dégallier, N.; Travassos da Rosa, J.F. Inadequate management of natural ecosystem in the Brazilian Amazon region results in the emergence and reemergence of arboviruses. *Cad. Saúde Pública* **2001**, *17*, S155–S164. [CrossRef]
11. Pinheiro, F.P.; Freitas, R.B.; da Rosa, J.F.T.; Gabbay, Y.B.; Mello, W.A.; LeDuc, J.W. An Outbreak of Mayaro Virus Disease in Belterra, Brazil I. Clinical and Virological Findings. *Am. J. Trop. Med. Hyg.* **1981**, *30*, 674–681. [CrossRef]
12. Izurieta, R.O.; DeLacure, D.A.; Izurieta, A.; Hoare, I.A.; Reina Ortiz, M. Mayaro virus: The jungle flu. *Virus Adapt. Treat.* **2018**, *10*, 9–17. [CrossRef]
13. Kraemer, M.U.; Sinka, M.E.; Duda, K.A.; Mylne, A.Q.; Shearer, F.M.; Barker, C.M.; Moore, C.G.; Carvalho, R.G.; Coelho, G.E.; Van Bortel, W.; et al. The global distribution of the arbovirus vectors Aedes aegypti and Ae. albopictus. *eLife* **2015**, *4*, e08347. [CrossRef]
14. Theilacker, C.; Held, J.; Allering, L.; Emmerich, P.; Schmidt-Chanasit, J.; Kern, W.V.; Panning, M. Prolonged polyarthralgia in a German traveller with Mayaro virus infection without inflammatory correlates. *BMC Infect. Dis.* **2013**, *13*, 369. [CrossRef]
15. Tesh, R.B.; Watts, D.M.; Russell, K.L.; Damodaran, C.; Calampa, C.; Cabezas, C.; Ramirez, G.; Vasquez, B.; Hayes, C.G.; Rossi, C.A.; et al. Mayaro Virus Disease: An Emerging Mosquito-Borne Zoonosis in Tropical South America. *Clin. Infect. Dis.* **1999**, *28*, 67–73. [CrossRef]
16. Phillips, P.A.; Lehmann, D.; Spooner, V.; Barker, J.; Tulloch, S.; Sungu, M.; Canil, K.A.; Pratt, R.D.; Lupiwa, T.; Alpers, M.P. Viruses associated with acute lower respiratory tract infections in children from the Eastern Highlands of Papua New Guinea (1983–1985). *Southeast Asian J. Trop. Med. Public Health* **1990**, *21*, 373–382.
17. Acosta-Ampudia, Y.; Monsalve, D.M.; Rodríguez, Y.; Pacheco, Y.; Anaya, J.-M.; Ramírez-Santana, C. Mayaro: An emerging viral threat? *Emerg. Microbes Infect.* **2018**, *7*, 1–11. [CrossRef]
18. Aguilar-Luis, M.A.; del Valle-Mendoza, J.; Silva-Caso, W.; Gil-Ramirez, T.; Levy-Blitchtein, S.; Bazán-Mayra, J.; Zavaleta-Gavidia, V.; Cornejo-Pacherres, D.; Palomares-Reyes, C.; del Valle, L.J. An emerging public health threat: Mayaro virus increases its distribution in Peru. *Int. J. Infect. Dis.* **2020**, *92*, 253–258. [CrossRef]
19. Pego, P.N.; Gomes, L.P.; Provance, D.W., Jr.; De Simone, S.G. Mayaro Virus Disease. *J. Hum. Virol. Retrovirol.* **2014**. [CrossRef]
20. Azevedo, R.S.S.; Silva, E.V.P.; Carvalho, V.L.; Rodrigues, S.G.; Neto, J.P.N.; Monteiro, H.A.O.; Peixoto, V.S.; Chiang, J.O.; Nunes, M.R.T.; Vasconcelos, P.F.C. Mayaro Fever Virus, Brazilian Amazon. *Emerg. Infect. Dis.* **2009**, *15*, 1830–1832. [CrossRef] [PubMed]
21. Galindo, P.; Srihongse, S.; Rodaniche, E.D.; Grayson, M.A. An Ecological Survey for Arboviruses in Almirante, Panama, 1959–1962. *Am. J. Trop. Med. Hyg.* **1966**, *15*, 385–400. [CrossRef] [PubMed]

22. Pinheiro, F.; Travassos da Rosa, A.; Freitas, R.; Travassos da Rosa, J.; Vasconcelos, P. Aspectos clínico-epidemiológicos das arboviroses. *Inst. Evandro Chagas* **1986**, *50*, 375–408.

23. Seymour, C.; Peralta, P.H.; Montgomery, G.G. Serologic Evidence of Natural Togavirus Infections in Panamanian Sloths and Other Vertebrates. *Am. J. Trop. Med. Hyg.* **1983**, *32*, 854–861. [CrossRef]

24. de Oliveira Mota, M.T.; Ribeiro, M.R.; Vedovello, D.; Nogueira, M.L. Mayaro virus: A neglected arbovirus of the Americas. *Future Virol.* **2015**, *10*, 1109–1122. [CrossRef]

25. Mackay, I.M.; Arden, K.E. Mayaro virus: A forest virus primed for a trip to the city? *Microbes Infect.* **2016**, *18*, 724–734. [CrossRef]

26. Galindo, P.; Srihongse, S. Transmission of Arboviruses to Hamsters by the Bite of Naturally Infected Culex (Melanoconion) Mosquitoes. *Am. J. Trop. Med. Hyg.* **1967**, *16*, 525–530. [CrossRef]

27. Muñoz, M.; Navarro, J.C. Mayaro: A re-emerging Arbovirus in Venezuela and Latin America. *Biomédica* **2012**, *32*, 286–302. [CrossRef]

28. Serra, O.P.; Cardoso, B.F.; Ribeiro, A.L.M.; dos Santos, F.A.L.; Slhessarenko, R.D. Mayaro virus and dengue virus 1 and 4 natural infection in culicids from Cuiabá, state of Mato Grosso, Brazil. *Mem. Inst. Oswaldo Cruz* **2016**, *111*, 20–29. [CrossRef]

29. Mourão, M.P.G.; de Souza Bastos, M.; de Figueiredo, R.P.; Gimaque, J.B.L.; dos Santos Galusso, E.; Kramer, V.M.; de Oliveira, C.M.C.; Naveca, F.G.; Figueiredo, L.T.M. Mayaro Fever in the City of Manaus, Brazil, 2007–2008. *Vector-Borne Zoonotic Dis.* **2011**, *12*, 42–46. [CrossRef]

30. Halsey, E.S.; Siles, C.; Guevara, C.; Vilcarromero, S.; Jhonston, E.J.; Ramal, C.; Aguilar, P.V.; Ampuero, J.S. Mayaro Virus Infection, Amazon Basin Region, Peru, 2010–2013. *Emerg. Infect. Dis.* **2013**, *19*, 1839–1842. [CrossRef]

31. Lednicky, J.; De Rochars, V.M.B.; Elbadry, M.; Loeb, J.; Telisma, T.; Chavannes, S.; Anilis, G.; Cella, E.; Ciccozzi, M.; Okech, B.; et al. Mayaro Virus in Child with Acute Febrile Illness, Haiti, 2015. *Emerg. Infect. Dis.* **2016**, *22*, 2000–2002. [CrossRef] [PubMed]

32. Diaz, L.A.; Spinsanti, L.I.; Almiron, W.R.; Contigiani, M.S. UNA virus: First report of human infection in Argentina. *Rev. Inst. Med. Trop. São Paulo* **2003**, *45*, 109–110. [CrossRef]

33. Forshey, B.M.; Guevara, C.; Laguna-Torres, V.A.; Cespedes, M.; Vargas, J.; Gianella, A.; Vallejo, E.; Madrid, C.; Aguayo, N.; Gotuzzo, E. Arboviral etiologies of acute febrile illnesses in Western South America, 2000–2007. *PLoS Negl. Trop. Dis.* **2010**, *4*, e787. [CrossRef] [PubMed]

34. Blohm, G.; Elbadry, M.A.; Mavian, C.; Stephenson, C.; Loeb, J.; White, S.; Telisma, T.; Chavannes, S.; Beau De Rochar, V.M.; Salemi, M.; et al. Mayaro as a Caribbean traveler: Evidence for multiple introductions and transmission of the virus into Haiti. *Int. J. Infect. Dis.* **2019**, *87*, 151–153. [CrossRef] [PubMed]

35. Hassing, R.-J.; Leparc-Goffart, I.; Blank, S.N.; Thevarayan, S.; Tolou, H.; van Doornum, G.; van Genderen, P.J. Imported Mayaro virus infection in the Netherlands. *J. Infect.* **2010**, *61*, 343–345. [CrossRef] [PubMed]

36. Gautret, P.; Cramer, J.P.; Field, V.; Caumes, E.; Jensenius, M.; Gkrania-Klotsas, E.; de Vries, P.J.; Grobusch, M.P.; Lopez-Velez, R.; Castelli, F.; et al. Infectious diseases among travellers and migrants in Europe, EuroTravNet 2010. *Eurosurveillance* **2012**, *17*, 20205. [CrossRef]

37. Receveur, M.C.; Grandadam, M.; Pistone, T.; Malvy, D. Infection with Mayaro virus in a French traveller returning from the Amazon region, Brazil, January, 2010. *Eurosurveillance* **2010**, *15*, 19563. [CrossRef] [PubMed]

38. Neumayr, A.; Gabriel, M.; Fritz, J.; Günther, S.; Hatz, C.; Schmidt-Chanasit, J.; Blum, J. Mayaro Virus Infection in Traveler Returning from Amazon Basin, Northern Peru. *Emerg. Infect. Dis.* **2012**, *18*, 695–696. [CrossRef]

39. Llagonne-Barets, M.; Icard, V.; Leparc-Goffart, I.; Prat, C.; Perpoint, T.; André, P.; Ramière, C. A case of Mayaro virus infection imported from French Guiana. *J. Clin. Virol.* **2016**, *77*, 66–68. [CrossRef]

40. Coimbra, T.L.M.; Santos, C.L.S.; Suzuki, A.; Petrella, S.M.C.; Bisordi, I.; Nagamori, A.H.; Marti, A.T.; Santos, R.N.; Fialho, D.M.; Lavigne, S.; et al. Mayaro virus: Imported cases of human infection in São Paulo State, Brazil. *Rev. Inst. Med. Trop. São Paulo* **2007**, *49*, 221–224. [CrossRef]

41. Talarmin, A.; Chandler, L.J.; Kazanji, M.; de Thoisy, B.; Debon, P.; Lelarge, J.; Labeau, B.; Bourreau, E.; Vié, J.C.; Shope, R.E.; et al. Mayaro virus fever in French Guiana: Isolation, identification, and seroprevalence. *Am. J. Trop. Med. Hyg.* **1998**, *59*, 452–456. [CrossRef] [PubMed]

42. Arenívar, C.; Rodríguez, Y.; Rodríguez-Morales, A.J.; Anaya, J.-M. Osteoarticular manifestations of Mayaro virus infection. *Curr. Opin. Rheumatol.* **2019**, *31*, 512–516. [CrossRef] [PubMed]

43. Dupuis-Maguiraga, L.; Noret, M.; Brun, S.; Grand, R.L.; Gras, G.; Roques, P. Chikungunya Disease: Infection-Associated Markers from the Acute to the Chronic Phase of Arbovirus-Induced Arthralgia. *PLoS Negl. Trop. Dis.* **2012**, *6*, e1446. [CrossRef] [PubMed]

44. Chow, A.; Her, Z.; Ong, E.K.S.; Chen, J.; Dimatatac, F.; Kwek, D.J.C.; Barkham, T.; Yang, H.; Rénia, L.; Leo, Y.-S.; et al. Persistent Arthralgia Induced by Chikungunya Virus Infection is Associated with Interleukin-6 and Granulocyte Macrophage Colony-Stimulating Factor. *J. Infect. Dis.* **2011**, *203*, 149–157. [CrossRef]

45. Labadie, K.; Larcher, T.; Joubert, C.; Mannioui, A.; Delache, B.; Brochard, P.; Guigand, L.; Dubreil, L.; Lebon, P.; Verrier, B.; et al. Chikungunya disease in nonhuman primates involves long-term viral persistence in macrophages. *J. Clin. Invest.* **2010**, *120*, 894–906. [CrossRef]

46. Noret, M.; Herrero, L.; Rulli, N.; Rolph, M.; Smith, P.N.; Li, R.W.; Roques, P.; Gras, G.; Mahalingam, S. Interleukin 6, RANKL, and Osteoprotegerin Expression by Chikungunya Virus-Infected Human Osteoblasts. *J. Infect. Dis.* **2012**, *206*, 455–457. [CrossRef]

47. Morrison, T.E.; Oko, L.; Montgomery, S.A.; Whitmore, A.C.; Lotstein, A.R.; Gunn, B.M.; Elmore, S.A.; Heise, M.T. A Mouse Model of Chikungunya Virus–Induced Musculoskeletal Inflammatory Disease: Evidence of Arthritis, Tenosynovitis, Myositis, and Persistence. *Am. J. Pathol.* **2011**, *178*, 32–40. [CrossRef]

48. Gardner, J.; Anraku, I.; Le, T.T.; Larcher, T.; Major, L.; Roques, P.; Schroder, W.A.; Higgs, S.; Suhrbier, A. Chikungunya Virus Arthritis in Adult Wild-Type Mice. *J. Virol.* **2010**, *84*, 8021–8032. [CrossRef]

49. Tappe, D.; Pérez-Girón, J.V.; Just-Nübling, G.; Schuster, G.; Gómez-Medina, S.; Günther, S.; Muñoz-Fontela, C.; Schmidt-Chanasit, J. Sustained Elevated Cytokine Levels during Recovery Phase of Mayaro Virus Infection. *Emerg. Infect. Dis.* **2016**, *22*, 750–752. [CrossRef]

50. Santiago, F.W.; Halsey, E.S.; Siles, C.; Vilcarromero, S.; Guevara, C.; Silvas, J.A.; Ramal, C.; Ampuero, J.S.; Aguilar, P.V. Long-Term Arthralgia after Mayaro Virus Infection Correlates with Sustained Pro-inflammatory Cytokine Response. *PLoS Negl. Trop. Dis.* **2015**, *9*, e0004104. [CrossRef]

51. Santos, F.M.; Dias, R.S.; de Oliveira, M.D.; Costa, I.C.T.A.; de Souza Fernandes, L.; Pessoa, C.R.; da Matta, S.L.P.; Costa, V.V.; Souza, D.G.; da Silva, C.C.; et al. Animal model of arthritis and myositis induced by the Mayaro virus. *PLoS Negl. Trop. Dis.* **2019**, *13*, e0007375. [CrossRef] [PubMed]

52. Churchman, S.M.; Ponchel, F. Interleukin-7 in rheumatoid arthritis. *Rheumatology* **2008**, *47*, 753–759. [CrossRef] [PubMed]

53. Yoo, S.-A.; Yoon, H.-J.; Kim, H.-S.; Chae, C.-B.; Falco, S.D.; Cho, C.-S.; Kim, W.-U. Role of placenta growth factor and its receptor flt-1 in rheumatoid inflammation: A link between angiogenesis and inflammation. *Arthritis Rheum.* **2009**, *60*, 345–354. [CrossRef] [PubMed]

54. Camini, F.C.; da Silva Caetano, C.C.; Almeida, L.T.; da Costa Guerra, J.F.; de Mello Silva, B.; de Queiroz Silva, S.; de Magalhães, J.C.; de Brito Magalhães, C.L. Oxidative stress in Mayaro virus infection. *Virus Res.* **2017**, *236*, 1–8. [CrossRef] [PubMed]

55. da Silva Caetano, C.C.; Camini, F.C.; Almeida, L.T.; Ferraz, A.C.; da Silva, T.F.; Lima, R.L.S.; de Freitas Carvalho, M.M.; de Freitas Castro, T.; Carneiro, C.M.; de Mello Silva, B.; et al. Mayaro Virus Induction of Oxidative Stress is Associated With Liver Pathology in a Non-Lethal Mouse Model. *Sci. Rep.* **2019**, *9*, 15289. [CrossRef] [PubMed]

56. Dhanwani, R.; Khan, M.; Alam, S.I.; Rao, P.V.L.; Parida, M. Differential proteome analysis of Chikungunya virus-infected new-born mice tissues reveal implication of stress, inflammatory and apoptotic pathways in disease pathogenesis. *Proteomics* **2011**, *11*, 1936–1951. [CrossRef]

57. Hosakote, Y.M.; Jantzi, P.D.; Esham, D.L.; Spratt, H.; Kurosky, A.; Casola, A.; Garofalo, R.P. Viral-mediated Inhibition of Antioxidant Enzymes Contributes to the Pathogenesis of Severe Respiratory Syncytial Virus Bronchiolitis. *Am. J. Respir. Crit. Care Med.* **2011**, *183*, 1550–1560. [CrossRef]

58. Kayesh, M.E.H.; Ezzikouri, S.; Sanada, T.; Chi, H.; Hayashi, Y.; Rebbani, K.; Kitab, B.; Matsuu, A.; Miyoshi, N.; Hishima, T.; et al. Oxidative Stress and Immune Responses During Hepatitis C Virus Infection in Tupaia belangeri. *Sci. Rep.* **2017**, *7*, 9848. [CrossRef]

59. Gil, L.; Martínez, G.; Tápanes, R.; Castro, O.; González, D.; Bernardo, L.; Vázquez, S.; Kourí, G.; Guzmán, M.G. Oxidative Stress in Adult Dengue Patients. *Am. J. Trop. Med. Hyg.* **2004**, *71*, 652–657. [CrossRef]

60. Cavalheiro, M.G.; Costa, L.S.D.; Campos, H.S.; Alves, L.S.; Assunção-Miranda, I.; Poian, A.T.D.; Cavalheiro, M.G.; Costa, L.S.D.; Campos, H.S.; Alves, L.S.; et al. Macrophages as target cells for Mayaro virus infection: Involvement of reactive oxygen species in the inflammatory response during virus replication. *An. Acad. Bras. Ciênc.* **2016**, *88*, 1485–1499. [CrossRef]

61. de Castro-Jorge, L.A.; de Carvalho, R.V.H.; Klein, T.M.; Hiroki, C.H.; Lopes, A.H.; Guimarães, R.M.; Fumagalli, M.J.; Floriano, V.G.; Agostinho, M.R.; Slhessarenko, R.D.; et al. The NLRP3 inflammasome is involved with the pathogenesis of Mayaro virus. *PLOS Pathog.* **2019**, *15*, e1007934. [CrossRef]

62. Yang, Z.; Min, Z.; Yu, B. Reactive oxygen species and immune regulation. *Int. Rev. Immunol.* **2020**, 1–7. [CrossRef] [PubMed]

63. Chen, W.; Foo, S.-S.; Rulli, N.E.; Taylor, A.; Sheng, K.-C.; Herrero, L.J.; Herring, B.L.; Lidbury, B.A.; Li, R.W.; Walsh, N.C.; et al. Arthritogenic alphaviral infection perturbs osteoblast function and triggers pathologic bone loss. *Proc. Natl. Acad. Sci. USA* **2014**, *111*, 6040–6045. [CrossRef] [PubMed]

64. Rulli, N.E.; Rolph, M.S.; Srikiatkhachorn, A.; Anantapreecha, S.; Guglielmotti, A.; Mahalingam, S. Protection from Arthritis and Myositis in a Mouse Model of Acute Chikungunya Virus Disease by Bindarit, an Inhibitor of Monocyte Chemotactic Protein-1 Synthesis. *J. Infect. Dis.* **2011**, *204*, 1026–1030. [CrossRef] [PubMed]

65. Bengue, M.; Ferraris, P.; Baronti, C.; Diagne, C.T.; Talignani, L.; Wichit, S.; Liegeois, F.; Bisbal, C.; Nougairède, A.; Missé, D. Mayaro Virus Infects Human Chondrocytes and Induces the Expression of Arthritis-Related Genes Associated with Joint Degradation. *Viruses* **2019**, *11*, 797. [CrossRef] [PubMed]

66. Torres, J.R.; Russell, K.L.; Vasquez, C.; Barrera, R.; Tesh, R.B.; Salas, R.; Watts, D.M. Family Cluster of Mayaro Fever, Venezuela. *Emerg. Infect. Dis.* **2004**, *10*, 1304–1306. [CrossRef]

67. Figueiredo, L.T.M.; Nogueira, R.M.R.; Cavalcanti, S.M.B.; Schatzmayr, H.; da Rosa, A.T.; Figueiredo, L.T.M.; Nogueira, R.M.R.; Cavalcanti, S.M.B.; Schatzmayr, H.; da Rosa, A.T. Study of two different enzyme immunoassays for the detection of Mayaro virus antibodies. *Mem. Inst. Oswaldo Cruz* **1989**, *84*, 303–307. [CrossRef]

68. Earnest, J.T.; Basore, K.; Roy, V.; Bailey, A.L.; Wang, D.; Alter, G.; Fremont, D.H.; Diamond, M.S. Neutralizing antibodies against Mayaro virus require Fc effector functions for protective activity. *J. Exp. Med.* **2019**, *216*, 2282–2301. [CrossRef]

69. da Rosa, A.P.T.; Vasconcelos, P.F.; da Rosa, J.F.T. *An Overview of Arbovirology in Brazil and Neighbouring Countries*; Instituto Evandro Chagas: Ananindeua, Brazil, 1998; ISBN 85-86784-01-X.

70. Izurieta, R.O.; Macaluso, M.; Watts, D.M.; Tesh, R.B.; Guerra, B.; Cruz, L.M.; Galwankar, S.; Vermund, S.H. Hunting in the Rainforest and Mayaro Virus Infection: An emerging Alphavirus in Ecuador. *J. Glob. Infect. Dis.* **2011**, *3*, 317–323. [CrossRef]

71. Strauss, J.H.; Strauss, E.G. The alphaviruses: Gene expression, replication, and evolution. *Microbiol. Rev.* **1994**, *58*, 491–562. [CrossRef]

72. Karabatsos, N. Antigenic Relationships of Group a Arboviruses by Plaque Reduction Neutralization Testing. *Am. J. Trop. Med. Hyg.* **1975**, *24*, 527–532. [CrossRef] [PubMed]

73. Fischer, C.; Bozza, F.; Merino, X.J.M.; Pedroso, C.; de Oliveira Filho, E.F.; Moreira-Soto, A.; Schwalb, A.; de Lamballerie, X.; Netto, E.M.; Bozza, P.T.; et al. Robustness of Serologic Investigations for Chikungunya and Mayaro Viruses following Coemergence. *mSphere* **2020**, *5*. [CrossRef] [PubMed]

74. Calisher, C.H.; el-Kafrawi, A.O.; Mahmud, M.I.A.-D.; da Rosa, A.P.T.; Bartz, C.R.; Brummer-Korvenkontio, M.; Haksohusodo, S.; Suharyono, W. Complex-specific immunoglobulin M antibody patterns in humans infected with alphaviruses. *J. Clin. Microbiol.* **1986**, *23*, 155–159. [CrossRef] [PubMed]

75. Lambert, A.J.; Martin, D.A.; Lanciotti, R.S. Detection of North American Eastern and Western Equine Encephalitis Viruses by Nucleic Acid Amplification Assays. *J. Clin. Microbiol.* **2003**, *41*, 379–385. [CrossRef] [PubMed]

76. Firth, A.E.; Chung, B.Y.; Fleeton, M.N.; Atkins, J.F. Discovery of frameshifting in Alphavirus 6K resolves a 20-year enigma. *Virol. J.* **2008**, *5*, 108. [CrossRef]

77. Zhang, R.; Kim, A.S.; Fox, J.M.; Nair, S.; Basore, K.; Klimstra, W.B.; Rimkunas, R.; Fong, R.H.; Lin, H.; Poddar, S.; et al. Mxra8 is a receptor for multiple arthritogenic alphaviruses. *Nature* **2018**, *557*, 570–574. [CrossRef]

78. Lavergne, A.; de Thoisy, B.; Lacoste, V.; Pascalis, H.; Pouliquen, J.-F.; Mercier, V.; Tolou, H.; Dussart, P.; Morvan, J.; Talarmin, A.; et al. Mayaro virus: Complete nucleotide sequence and phylogenetic relationships with other alphaviruses. *Virus Res.* **2006**, *117*, 283–290. [CrossRef]

79. Powers, A.M.; Aguilar, P.V.; Chandler, L.J.; Brault, A.C.; Meakins, T.A.; Watts, D.; Russell, K.L.; Olson, J.; Vasconcelos, P.F.C.; Rosa, A.T.D.; et al. Genetic Relationships among Mayaro and Una Viruses Suggest Distinct Patterns of Transmission. *Am. J. Trop. Med. Hyg.* **2006**, *75*, 461–469. [CrossRef]

80. Auguste, A.J.; Adams, A.P.; Arrigo, N.C.; Martinez, R.; da Rosa, A.P.A.T.; Adesiyun, A.A.; Chadee, D.D.; Tesh, R.B.; Carrington, C.V.F.; Weaver, S.C. Isolation and Characterization of Sylvatic Mosquito-Borne Viruses in Trinidad: Enzootic Transmission and a New Potential Vector of Mucambo Virus. *Am. J. Trop. Med. Hyg.* **2010**, *83*, 1262–1265. [CrossRef]

81. Auguste, A.J.; Liria, J.; Forrester, N.L.; Giambalvo, D.; Moncada, M.; Long, K.C.; Morón, D.; de Manzione, N.; Tesh, R.B.; Halsey, E.S.; et al. Evolutionary and Ecological Characterization of Mayaro Virus Strains Isolated during an Outbreak, Venezuela, 2010. *Emerg. Infect. Dis.* **2015**, *21*, 1742–1750. [CrossRef]

82. Mota, M.T.O.; Vedovello, D.; Estofolete, C.; Malossi, C.D.; Araújo, J.P.; Nogueira, M.L. Complete Genome Sequence of Mayaro Virus Imported from the Amazon Basin to São Paulo State, Brazil. *Genome Announc.* **2015**, *3*. [CrossRef] [PubMed]

83. Pinheiro, F.P.; LeDuc, J.W. Mayaro virus disease. *Arboviruses Epidemiol. Ecol.* **1998**, *3*, 137–150.

84. Hoch, A.L.; Peterson, N.E.; LeDuc, J.W.; Pinheiro, F.P. An Outbreak of Mayaro Virus Disease in Belterra, Brazil III. Entomological and Ecological Studies. *Am. J. Trop. Med. Hyg.* **1981**, *30*, 689–698. [CrossRef] [PubMed]

85. Pauvolid-Corrêa, A.; Juliano, R.S.; Campos, Z.; Velez, J.; Nogueira, R.M.R.; Komar, N.; Pauvolid-Corrêa, A.; Juliano, R.S.; Campos, Z.; Velez, J.; et al. Neutralising antibodies for Mayaro virus in Pantanal, Brazil. *Mem. Inst. Oswaldo Cruz* **2015**, *110*, 125–133. [CrossRef] [PubMed]

86. Aitken, T.H.G.; Anderson, C.R. Virus Transmission Studies with Trinidadian Mosquitoes. *Am. J. Trop. Med. Hyg.* **1959**, *8*, 41–45. [CrossRef]

87. Monath, T.P. Dengue: The risk to developed and developing countries. *Proc. Natl. Acad. Sci. USA* **1994**, *91*, 2395–2400. [CrossRef]

88. Abad-Franch, F.; Grimmer, G.H.; de Paula, V.S.; Figueiredo, L.T.M.; Braga, W.S.M.; Luz, S.L.B. Mayaro Virus Infection in Amazonia: A Multimodel Inference Approach to Risk Factor Assessment. *PLoS Negl. Trop. Dis.* **2012**, *6*, e1846. [CrossRef]

89. Maia, L.M.S.; Bezerra, M.C.F.; Costa, M.C.S.; Souza, E.M.; Oliveira, M.E.B.; Ribeiro, A.L.M.; Miyazaki, R.D.; Slhessarenko, R.D. Natural vertical infection by dengue virus serotype 4, Zika virus and Mayaro virus in Aedes (Stegomyia) aegypti and Aedes (Stegomyia) albopictus. *Med. Vet. Entomol.* **2019**, *33*, 437–442. [CrossRef]

90. Gubler, D.J. The Global Emergence/Resurgence of Arboviral Diseases as Public Health Problems. *Arch. Med. Res.* **2002**, *33*, 330–342. [CrossRef]

91. Bosio, C.F.; Thomas, R.E.; Grimstad, P.R.; Rai, K.S. Variation in the Efficiency of Vertical Transmission of Dengue-1 Virus by Strains of Aedes albopictus (Diptera: Culicidae). *J. Med. Entomol.* **1992**, *29*, 985–989. [CrossRef]

92. Diallo, D.; Dia, I.; Diagne, C.T.; Gaye, A.; Diallo, M. Chapter 4—Emergences of Chikungunya and Zika in Africa. In *Chikungunya and Zika Viruses*; Higgs, S., Vanlandingham, D.L., Powers, A.M., Eds.; Academic Press: Cambridge, MA, USA, 2018; pp. 87–133, ISBN 978-0-12-811865-8.

93. de Toni Aquino da Cruz, L.C.; Serra, O.P.; Leal-Santos, F.A.; Ribeiro, A.L.M.; Slhessarenko, R.D.; dos Santos, M.A. Natural transovarial transmission of dengue virus 4 in Aedes aegypti from Cuiabá, State of Mato Grosso, Brazil. *Rev. Soc. Bras. Med. Trop.* **2015**, *48*, 18–25. [CrossRef] [PubMed]

94. Delatte, H.; Paupy, C.; Dehecq, J.S.; Thiria, J.; Failloux, A.B.; Fontenille, D. Aedes albopictus, vector of chikungunya and dengue viruses in Reunion Island: Biology and control. *Parasite Paris Fr.* **2008**, *15*, 3–13. [CrossRef]

95. Mondet, B.; Vasconcelos, P.F.C.; Travassos da Rosa, A.P.A.; Travassos da Rosa, E.S.; Rodrigues, S.G.; Travassos da Rosa, J.F.S.; Bicout, D.J. Isolation of Yellow Fever Virus from Nulliparous Haemagogus (Haemagogus) janthinomys in Eastern Amazonia. *Vector-Borne Zoonotic Dis.* **2002**, *2*, 47–50. [CrossRef] [PubMed]

96. Junt, T.; Heraud, J.M.; Lelarge, J.; Labeau, B.; Talarmin, A. Determination of natural versus laboratory human infection with Mayaro virus by molecular analysis. *Epidemiol. Infect.* **1999**, *123*, 511–513. [CrossRef] [PubMed]

97. Gutiérrez-Bugallo, G.; Piedra, L.A.; Rodriguez, M.; Bisset, J.A.; Lourenço-de-Oliveira, R.; Weaver, S.C.; Vasilakis, N.; Vega-Rúa, A. Vector-borne transmission and evolution of Zika virus. *Nat. Ecol. Evol.* **2019**, *3*, 561–569. [CrossRef] [PubMed]

98. Long, K.C.; Ziegler, S.A.; Thangamani, S.; Hausser, N.L.; Kochel, T.J.; Higgs, S.; Tesh, R.B. Experimental Transmission of Mayaro Virus by Aedes aegypti. *Am. J. Trop. Med. Hyg.* **2011**, *85*, 750–757. [CrossRef] [PubMed]

99. Aitken, T.H.G.; Downs, W.G.; Anderson, C.R.; Spence, L.; Casals, J. Mayaro Virus Isolated from a Trinidadian Mosquito, Mansonia venezuelensis. *Science* **1960**, *131*, 986. [CrossRef]

100. Karabatsos, N. *International Catalogue of arthrOpod-Borne Viruses*; American Society of Tropical Medicine and Hygiene for The Subcommittee on Information Exchange of the American Committee on Arthropod-borne Viruses: San Antonio, TX, USA, 1985; Volume 3.

101. Groot, H.; Morales, A.; Vidales, H. Virus Isolations from Forest Mosquitoes in San Vicente de Chucuri, Colombia. *Am. J. Trop. Med. Hyg.* **1961**, *10*, 397–402. [CrossRef]

102. Brustolin, M.; Pujhari, S.; Henderson, C.A.; Rasgon, J.L. Anopheles mosquitoes may drive invasion and transmission of Mayaro virus across geographically diverse regions. *PLoS Negl. Trop. Dis.* **2018**, *12*, e0006895. [CrossRef]

103. Wiggins, K.; Eastmond, B.; Alto, B.W. Transmission potential of Mayaro virus in Florida Aedes aegypti and Aedes albopictus mosquitoes. *Med. Vet. Entomol.* **2018**, *32*, 436–442. [CrossRef]

104. Smith, G.C.; Francy, D.B. Laboratory studies of a Brazilian strain of Aedes albopictus as a potential vector of Mayaro and Oropouche viruses. *J. Am. Mosq. Control Assoc.* **1991**, *7*, 89–93. [PubMed]

105. Diop, F.; Alout, H.; Diagne, C.T.; Bengue, M.; Baronti, C.; Hamel, R.; Talignani, L.; Liegeois, F.; Pompon, J.; Vargas, R.E.M.; et al. Differential Susceptibility and Innate Immune Response of Aedes aegypti and Aedes albopictus to the Haitian Strain of the Mayaro Virus. *Viruses* **2019**, *11*, 924. [CrossRef] [PubMed]

106. Hotez, P.J.; Murray, K.O. Dengue, West Nile virus, chikungunya, Zika—And now Mayaro? *PLoS Negl. Trop. Dis.* **2017**, *11*, e0005462. [CrossRef]

107. Bhatt, S.; Gething, P.W.; Brady, O.J.; Messina, J.P.; Farlow, A.W.; Moyes, C.L.; Drake, J.M.; Brownstein, J.S.; Hoen, A.G.; Sankoh, O.; et al. The global distribution and burden of dengue. *Nature* **2013**, *496*, 504–507. [CrossRef] [PubMed]

108. Kraemer, M.U.; Reiner, R.C.; Brady, O.J.; Messina, J.P.; Gilbert, M.; Pigott, D.M.; Yi, D.; Johnson, K.; Earl, L.; Marczak, L.B. Past and future spread of the arbovirus vectors Aedes aegypti and Aedes albopictus. *Nat. Microbiol.* **2019**, *4*, 854–863. [CrossRef]

109. Khormi, H.M.; Kumar, L. Climate change and the potential global distribution of Aedes aegypti: Spatial modelling using geographical information system and CLIMEX. *Geospatial Health* **2014**, *8*, 405–415. [CrossRef]

110. Caminade, C.; Medlock, J.M.; Ducheyne, E.; McIntyre, K.M.; Leach, S.; Baylis, M.; Morse, A.P. Suitability of European climate for the Asian tiger mosquito Aedes albopictus: Recent trends and future scenarios. *J. R. Soc. Interface* **2012**, *9*, 2708–2717. [CrossRef]

111. Medlock, J.M.; Avenell, D.; Barrass, I.; Leach, S. Analysis of the potential for survival and seasonal activity of Aedes albopictus (Diptera: Culicidae) in the United Kingdom. *J. Vector Ecol.* **2006**, *31*, 292–304. [CrossRef]

112. de Thoisy, B.; Gardon, J.; Salas, R.A.; Morvan, J.; Kazanji, M. Mayaro Virus in Wild Mammals, French Guiana. *Emerg. Infect. Dis.* **2003**, *9*, 1326–1329. [CrossRef]

113. Figueiredo, L.T.M. Emergent arboviruses in Brazil. *Rev. Soc. Bras. Med. Trop.* **2007**, *40*, 224–229. [CrossRef]

114. Rylands, A.B.; Mittermeier, R.A. The Diversity of the New World Primates (Platyrrhini): An Annotated Taxonomy. In *South American Primates: Comparative Perspectives in the Study of Behavior, Ecology, and Conservation*; Garber, P.A., Estrada, A., Bicca-Marques, J.C., Heymann, E.W., Strier, K.B., Eds.; Developments in Primatology: Progress and Prospects; Springer: New York, NY, USA, 2009; pp. 23–54. ISBN 978-0-387-78705-3.

115. Ali, S.; Gugliemini, O.; Harber, S.; Harrison, A.; Houle, L.; Ivory, J.; Kersten, S.; Khan, R.; Kim, J.; LeBoa, C.; et al. Environmental and Social Change Drive the Explosive Emergence of Zika Virus in the Americas. *PLoS Negl. Trop. Dis.* **2017**, *11*, e0005135. [CrossRef] [PubMed]

116. Bogoch, I.I.; Brady, O.J.; Kraemer, M.U.G.; German, M.; Creatore, M.I.; Brent, S.; Watts, A.G.; Hay, S.I.; Kulkarni, M.A.; Brownstein, J.S.; et al. Potential for Zika virus introduction and transmission in resource-limited countries in Africa and the Asia-Pacific region: A modelling study. *Lancet Infect. Dis.* **2016**, *16*, 1237–1245. [CrossRef]

117. Stoddard, S.T.; Forshey, B.M.; Morrison, A.C.; Paz-Soldan, V.A.; Vazquez-Prokopec, G.M.; Astete, H.; Reiner, R.C.; Vilcarromero, S.; Elder, J.P.; Halsey, E.S. House-to-house human movement drives dengue virus transmission. *Proc. Natl. Acad. Sci. USA* **2013**, *110*, 994–999. [CrossRef] [PubMed]

118. Harrington, L.C.; Scott, T.W.; Lerdthusnee, K.; Coleman, R.C.; Costero, A.; Clark, G.G.; Jones, J.J.; Kitthawee, S.; Kittayapong, P.; Sithiprasasna, R.; et al. Dispersal of the Dengue Vector Aedes aegypti within and between Rural Communities. *Am. J. Trop. Med. Hyg.* **2005**, *72*, 209–220. [CrossRef] [PubMed]

119. Higgs, S.; Vanlandingham, D. Chikungunya Virus and Its Mosquito Vectors. *Vector-Borne Zoonotic Dis.* **2015**, *15*, 231–240. [CrossRef]

120. Achee, N.L.; Gould, F.; Perkins, T.A.; Reiner, R.C., Jr.; Morrison, A.C.; Ritchie, S.A.; Gubler, D.J.; Teyssou, R.; Scott, T.W. A Critical Assessment of Vector Control for Dengue Prevention. *PLoS Negl. Trop. Dis.* **2015**, *9*, e0003655. [CrossRef]

121. Izurieta, R.O.; Macaluso, M.; Watts, D.M.; Tesh, R.B.; Guerra, B.; Cruz, L.M.; Galwankar, S.; Vermund, S.H. Assessing Yellow Fever Risk in the Ecuadorian Amazon. *J. Glob. Infect. Dis.* **2009**, *1*, 7–13. [CrossRef]

122. Esposito, D.L.A.; da Fonseca, B.A.L.; Esposito, D.L.A.; da Fonseca, B.A.L. Will Mayaro virus be responsible for the next outbreak of an arthropod-borne virus in Brazil? *Braz. J. Infect. Dis.* **2017**, *21*, 540–544. [CrossRef]

123. de Figueiredo, M.L.G.; Figueiredo, L.T.M. Emerging alphaviruses in the Americas: Chikungunya and Mayaro. *Rev. Soc. Bras. Med. Trop.* **2014**, *47*, 677–683. [CrossRef]

124. Calisher, C.H.; Gutierrez, E.; Maness, K.; Lord, R.D. Isolation of Mayaro virus from a migrating bird captured in Louisiana in 1967. *Bull. Pan Am. Health Organ. PAHO* **1974**, *8*, 3.

125. Weaver, S.C. Urbanization and geographic expansion of zoonotic arboviral diseases: Mechanisms and potential strategies for prevention. *Trends Microbiol.* **2013**, *21*, 360–363. [CrossRef] [PubMed]

126. Halstead, S.B. Travelling arboviruses: A historical perspective. *Travel Med. Infect. Dis.* **2019**, *31*, 101471. [CrossRef] [PubMed]

127. Cleton, N.; Koopmans, M.; Reimerink, J.; Godeke, G.-J.; Reusken, C. Come fly with me: Review of clinically important arboviruses for global travelers. *J. Clin. Virol.* **2012**, *55*, 191–203. [CrossRef] [PubMed]

128. Mavian, C.; Rife, B.D.; Dollar, J.J.; Cella, E.; Ciccozzi, M.; Prosperi, M.C.F.; Lednicky, J.; Morris, J.G.; Capua, I.; Salemi, M. Emergence of recombinant Mayaro virus strains from the Amazon basin. *Sci. Rep.* **2017**, *7*, 8718. [CrossRef] [PubMed]

129. Beaty, B.J.; Black, W.; Eisen, L.; Flores, A.E.; García-Rejón, J.E.; Loroño-Pino, M.; Saavedra-Rodriguez, K. The intensifying storm: Domestication of Aedes aegypti, urbanization of arboviruses, and emerging insecticide resistance. In *Global Health Impacts of Vector-Borne Diseases: Workshop Summary*; Alison Mack, R., Forum on Microbial Threats, Board on Global Health, Health and Medicine Division, National Academies of Sciences, Engineering, and Medicine, Eds.; National Academies Press: Washington, DC, USA, 2016; pp. 111–142.

130. Liu, B.; Gao, X.; Ma, J.; Jiao, Z.; Xiao, J.; Hayat, M.A.; Wang, H. Modeling the present and future distribution of arbovirus vectors Aedes aegypti and Aedes albopictus under climate change scenarios in Mainland China. *Sci. Total Environ.* **2019**, *664*, 203–214. [CrossRef] [PubMed]

131. Kamal, M.; Kenawy, M.A.; Rady, M.H.; Khaled, A.S.; Samy, A.M. Mapping the global potential distributions of two arboviral vectors Aedes aegypti and Ae. albopictus under changing climate. *PLoS ONE* **2018**, *13*, e0210122. [CrossRef]

Article

Highly Efficient Vertical Transmission for Zika Virus in *Aedes aegypti* after Long Extrinsic Incubation Time

Menchie Manuel [1], Dorothée Missé [2] and Julien Pompon [1,2,*

[1] Department of Emerging Infectious Diseases, Duke-NUS Medical School, Singapore 169857, Singapore; menchie.manuel@duke-nus.edu.sg
[2] CNRS, IRD, MIVEGEC, Univ. Montpellier, 34394 Montpellier, France; dorothee.misse@ird.fr
*   Correspondence: Julien.pompon@ird.fr

Received: 21 April 2020; Accepted: 9 May 2020; Published: 11 May 2020

Abstract: While the Zika virus (ZIKV) 2014–2017 pandemic has subsided, there remains active transmission. Apart from horizontal transmission to humans, the main vector *Aedes aegypti* can transmit the virus vertically from mother to offspring. Large variation in vertical transmission (VT) efficiency between studies indicates the influence of parameters, which remain to be characterized. To determine the roles of extrinsic incubation time and gonotrophic cycle, we deployed an experimental design that quantifies ZIKV in individual progeny and larvae. We observed an early infection of ovaries that exponentially progressed. We quantified VT rate, filial infection rate, and viral load per infected larvae at 10 days post oral infection (d.p.i.) on the second gonotrophic cycle and at 17 d.p.i. on the second and third gonotrophic cycle. As compared to previous reports that studied pooled samples, we detected a relatively high VT efficiency from 1.79% at 10 d.p.i. and second gonotrophic cycle to 66% at 17 d.p.i. and second gonotrophic cycle. At 17 d.p.i., viral load largely varied and averaged around 800 genomic RNA (gRNA) copies. Longer incubation time and fewer gonotrophic cycles promoted VT. These results shed light on the mechanism of VT, how environmental conditions favor VT, and whether VT can maintain ZIKV circulation.

Keywords: Zika virus; *Aedes aegypti*; vertical transmission

1. Introduction

Zika virus (ZIKV) is a mosquito-transmitted *Flavivirus* that recently emerged as a pandemic virus [1]. From 2014–2017, ZIKV infected more than 1.5 million people mostly in Polynesia and Latin and South America. Although Zika symptoms are mild (flu-like) to absent, it can have life-debilitating and life-threatening consequences, such as microcephaly for prenatally infected newborns and Guillain–Barre syndrome in adults [2]. The wide distribution of and health risks associated with the infection prompted the World Health Organization (WHO) to declare a public health emergency of international concern in 2016. However, after 2018, records of Zika cases sharply dropped in the Americas and Polynesia [3] and the emergency status was lifted. The collapse of the epidemic was likely caused by herd immunity, given that almost 60% of the population had immunity against ZIKV in the Americas and Polynesia [4,5]. Nonetheless, Zika cases are still reported in Asia and suspected in Africa [3], and autochthonous transmission in France occurred at the end of 2019 [6]. These reports indicate that ZIKV circulation persists. A better understanding of ZIKV ongoing circulation will help prevent future outbreaks. Identification of the mode of transmission to target will help curb residual circulation.

While ZIKV is transmitted horizontally from mosquito to human, it is also vertically transmitted from mother to offspring in mosquitoes as for many pathogenic flaviviruses [7,8]. Moreover, ZIKV can be transmitted between humans during sexual intercourse and by maternal–fetal infection [9].

Aedes aegypti is the most likely horizontal vector of ZIKV, although *Aedes albopictus* is also competent to a lower extent [9–11]. Evidence of mosquito vertical transmission (VT) comes from laboratory studies that evaluate the ability of the virus to be passed onto the next generation and from detection of the virus in immature stages in the field. In controlled conditions, following oral infection, which is more relevant than intrathoracic inoculation, ZIKV VT has repeatedly been shown for *Ae. aegypti* [12–14]. ZIKV was also repeatedly detected in eggs and larvae collected in urban and forested environments [15–18]. However, a few other studies report an absence of VT in the laboratory or lack of detection in immature stages in the field [19,20]. Altogether, there is enough evidence to support the existence of VT for ZIKV in *Ae. aegypti*. However, VT efficiency varies with several parameters including mosquito and virus genetics, temperature, gonotrophic cycle, and extrinsic incubation period [7]. Variation in these parameters may explain the lack of VT experimental observation and absence of detected VT in the field. To understand how ZIKV circulation persists and especially how horizontal and vertical transmissions are balanced, it is important to determine the conditions that promote ZIKV VT.

In arthropods, two mechanisms can result in VT [21]. Infection of eggs when fertilized or oviposited can occur in the oviduct and is called trans-ovum transmission. Alternatively, infection of the female germinal tissues results in infected progeny and is called transovarial transmission. Both mechanisms have been proposed for arbovirus VT in *Ae. aegypti* [22,23]. Following uptake of an infectious blood meal, the viruses first infect the mosquito midgut and then disseminate to other tissues including muscles, legs, wings, trachea, and salivary glands [24]. ZIKV salivary gland infection can occur as early as three days post infection, leading to virus secretion in saliva and horizontal transmission to another host [10]. The whole mosquito body can virtually be infected by arboviruses at late incubation time. There is evidence of virus infection of ovaries and oviduct [25], supporting both trans-ovum and transovarial mechanisms. Importantly, the impact of gonotrophic cycle (i.e., the reproductive cycle between two consecutive egg laying events, encompassing search for a host, blood feeding, oogenesis, and oviposition) and extrinsic incubation period (i.e., duration between infection onset in mosquitoes and the transmission, vertically in this study) on VT should vary depending on the mechanism. Trans-ovum transmission should be independent of the gonotrophic cycle as it is related to a tissue that is not altered by the oogenesis cycle, whereas transovarial transmission should be affected by oocyte development and release. The combined and separate impacts of gonotrophic cycle and incubation period for ZIKV VT in *Ae. aegypti* remain to be fully characterized.

In this study, we aimed at improving our understanding of ZIKV circulation by determining the impacts of gonotrophic cycle and extrinsic incubation period on VT in *Ae. aegypti*. To this end, we developed an experimental design that singled out progeny and larvae, allowing us to precisely determine: (i) oviposition rate, calculated as the proportion of blood-fed females that laid eggs; (ii) VT rate, defined as the proportion of infected females that transmit the virus to their progeny (at least one individual); (iii) filial infection rate, calculated as the proportion of infected larvae per progeny; and (iv) viral load per infected larvae, as determined by copy number of ZIKV genomic RNA (gRNA) [26]. The results shed light on the VT mechanism and help understand the balance between horizontal or vertical transmissions. By identifying the factors that influence ZIKV circulation in mosquitoes, our study will provide information to understand ZIKV persistence and inform vector control strategies.

2. Results

2.1. Infection of Ovaries Increases with Extrinsic Incubation Time

Transovarial VT requires infection of mosquito female germinal tissues [21]. To quantify ovary infection, we collected ovary pairs from female *Ae. aegypti* at 3, 7, 10, and 14 days post oral infection (d.p.i.) with ZIKV and quantified gRNA copies by RT-qPCR. The different time points covered infection initiation up to complete infection of the whole mosquito body [10]. To orally infect mosquitoes, we used an infectious titer of 10^5 particles forming unit (pfu) / mL, which is epidemiologically relevant as it corresponds to blood titers quantified in patients in Polynesia and the Americas [27,28]. We observed infection of ovaries as early as 3 d.p.i. with 80% of ovary pairs infected (Figure 1). Infection progressed to 100% of ovary pairs at 7, 10, and 14 d.p.i. Similarly, copies of ZIKV gRNA per infected ovary pairs gradually increased from an average of about 20,000 at 3 d.p.i. to an average of about 1.8×10^7 at 14 d.p.i. (Figure 1). ZIKV gRNA copies were significantly lower at 3 d.p.i. than at 7 and 10 d.p.i., while gRNA level at 14 d.p.i. was significantly higher than at all earlier time points. The results support the early infection of *Ae. aegypti* ovaries following oral infection and indicate that virus load increases with extrinsic incubation period.

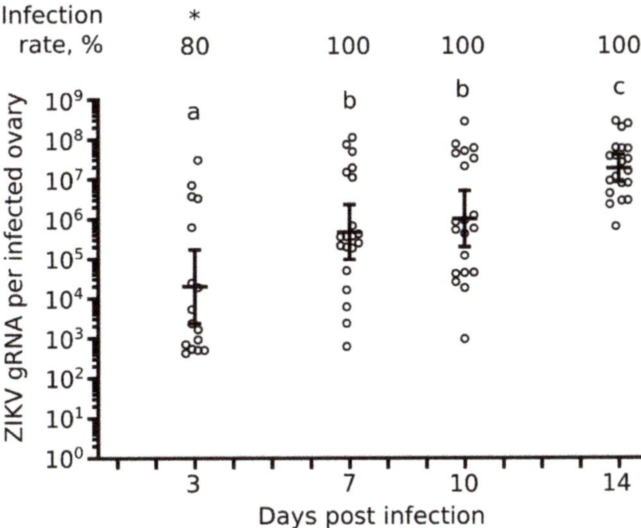

Figure 1. Kinetics of ovary infection. Zika virus (ZIKV) was quantified in single pairs of ovaries at 3, 7, 10, and 14 days post oral infection. Bars represent geometric means ± 95% CI. Each circle represents a pair of ovaries from one individual mosquito. N = 20 pair of ovaries per condition. * *p*-value < 0.05 from other conditions as calculated by Z-test. Different letters indicate significant difference (*p*-value < 0.05) as calculated by Tukey's test.

2.2. Experimental Design to Test the Impacts of Gonotrophic Cycle and Extrinsic Incubation Time on VT

We designed an experiment to test the separate and interacting impacts of gonotrophic cycle and extrinsic incubation time on VT (Figure 2). Female mosquitoes were orally fed with 10^5 pfu/mL of ZIKV as above, resulting in 100% infected mosquitoes (Figure 1). To minimize variation due to mosquito batch, the same batch of ZIKV-infected mosquitoes was divided into two cages. For each cage, we separated individual blood-fed mosquitoes in tubes prefilled with water and monitored oviposition at 3 d.p.i. The eggs from the first gonotrophic cycle were discarded as it has been shown that VT does not occur or is minimal in the first gonotrophic cycle for other arboviruses [29,30]. Mosquitoes that had oviposited and, thus, gone through their first gonotrophic cycle, were regrouped in their respective original cages.

In one of the "sister" cages, mosquitoes were offered a non-infectious blood meal at 7 d.p.i., whereas the other cage was maintained on sugar feeding. To obtain a precise picture of VT, we separated individual blood-fed mosquitoes in tubes prefilled with water and collected eggs at 10 d.p.i. Each progeny was reared separately until fourth instar larvae. ZIKV was quantified in 10 single larvae randomly selected from each progeny. By separately analyzing blood-fed mosquitoes and larvae, we were able to record ovipositing females, proportion of infected larvae in a sub-sample of each progeny, and to quantify the viral load per individual infected larva.

Mosquitoes that oviposited a second time had undergone a second gonotrophic cycle and were regrouped in the same cage. This cage and the one maintained on sugar feeding were then offered a non-infectious blood meal at 14 d.p.i. As previously, blood-fed mosquitoes from each cage were isolated in water-containing tubes. We collected the eggs at 17 d.p.i. and reared them until fourth instar larvae to quantify ZIKV in 10 single larvae per progeny. This experimental design quantified VT at 10 d.p.i. on the second gonotrophic cycle and at 17 d.p.i. on the second and third gonotrophic cycle.

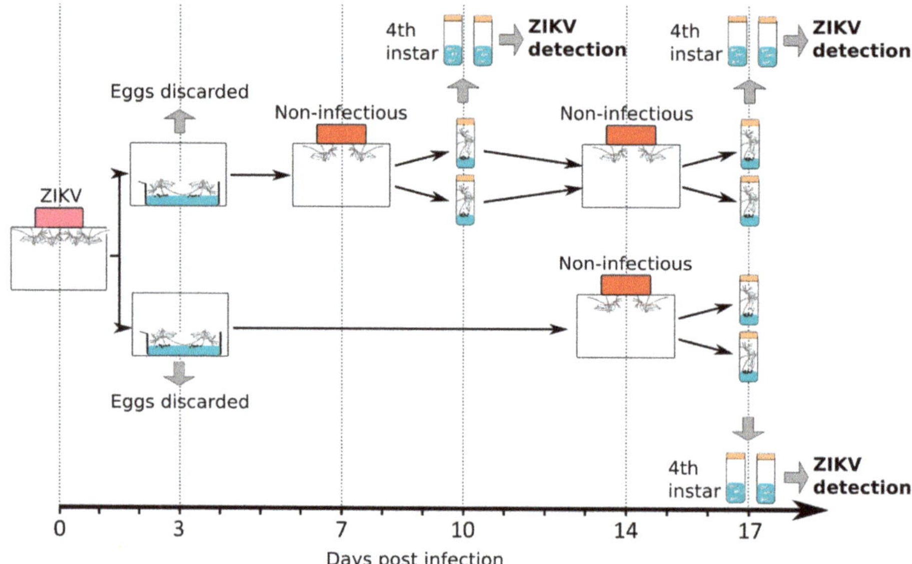

Figure 2. Experimental design to test the impact of gonotrophic cycle and time post infection on vertical transmission.

2.3. Oviposition Rate Moderately Increases with Incubation Time and Decreases with Additional Gonotrophic Cycle

Oviposition rate influences VT efficiency as absence of eggs prevents VT. We observed the highest oviposition rate for the second gonotrophic cycle at 17 d.p.i. with 79% of blood-fed mosquitoes laying eggs (Figure 3A). At the same extrinsic incubation time (17 d.p.i.), mosquitoes that had undergone an additional gonotrophic cycle (third) had a significantly lower oviposition rate at 54%. A shorter extrinsic incubation period at the second gonotrophic cycle reduced oviposition rate as compared to 17 d.p.i., although not significantly (p-value = 0.06, as indicated by Z-test). Our results indicate that oviposition rate is maximized by longer extrinsic incubation time and fewer gonotrophic cycles.

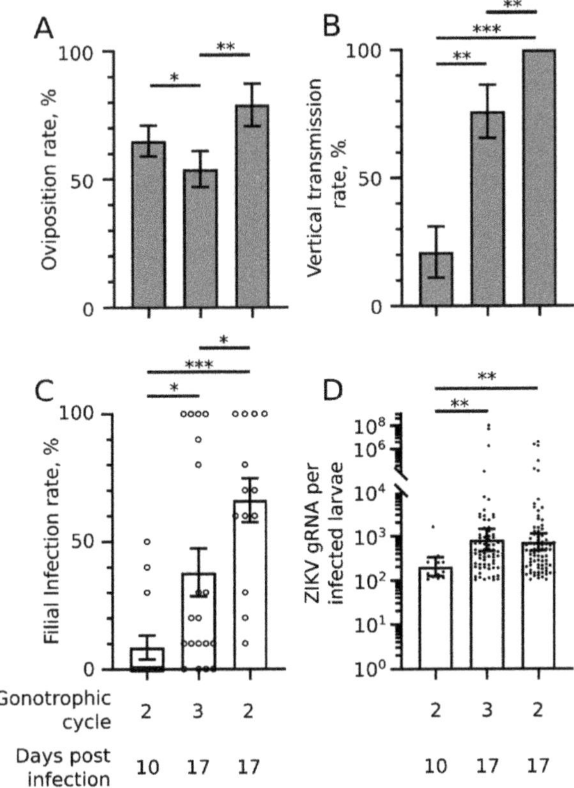

Figure 3. Impact of gonotrophic cycle and time post infection on vertical transmission. ZIKV was quantified in single fourth instar larvae on the second or third gonotrophic cycle at 10 or 17 days post oral infection. (**A**) Oviposition rate per condition. The number of mosquito females that laid eggs was divided by the total number of females that were blood fed. N blood-fed mosquitoes >24. (**B**) Vertical transmission rate. The number of progeny with at least one infected larva was divided by the total number of tested progeny. N progeny >14. (**A,B**) Bars indicate arithmetic means ± s.e. (**C**) Filial infection rate. Each circle represents the infection rate calculated from 10 single larvae randomly selected from all larvae from one infected mother (one progeny). Bars indicate arithmetic means ± s.e.m. (**D**) ZIKV genomic RNA (gRNA) copies per infected larvae. Each point represents one larva. Bars indicate geometric means ± 95% CI. (**C,D**) Larvae for each condition were collected from at least 14 progeny. * p-value < 0.05; **, p-value < 0.01; ***, p-value < 0.001, as calculated by Z-test (**A,B**) and Tukey's test (**C,D**). Number of samples analyzed in each condition are detailed in Supplementary Table S1.

2.4. VT Rate Drastically Increases with Incubation Time and Moderately Decreases with Gonotrophic Cycle

In each tested progeny, we quantified ZIKV in 10 randomly selected larvae. Although this experimental design may underestimate the VT rate by missing infected larvae in progeny with a low VT rate, the analysis of multiple progeny subsamples should provide an accurate observation. Strikingly, we observed that 100% of progeny were infected at 17 d.p.i. on the second gonotrophic cycle (Figure 3B). VT rate was significantly reduced to 76% when mosquitoes had undergone an additional gonotrophic cycle at 17 d.p.i. A larger reduction was caused by a shorter incubation time at the second gonotrophic cycle with only 21% of progeny infected. These results indicate that higher VT rate is mostly due to longer extrinsic incubation time, while processes associated with the gonotrophic cycle lower the VT rate.

2.5. Filial Infection Rate Increases with Incubation Time and Decreases with Gonotrophic Cycles

Calculation of filial infection rate was based on ZIKV detection in 10 single larvae randomly selected per progeny. Others quantify viruses in pools of larvae and calculate the minimum infection rate (MIR) based on the number of individuals per infected pool [7]. Both methods have pros and cons. However, our strategy also allowed us to calculate a complementary parameter: the exact viral load per infected larva (see below).

Similarly to the VT rate, we observed the highest filial infection rate for the second gonotrophic cycle at 17 d.p.i. with 66% of infected larvae per progeny (Figure 3C). At 17 d.p.i., the filial infection rate was significantly reduced by an additional gonotrophic cycle (38%), while a shorter incubation lowered the infection rate even more at the second gonotrophic cycle to 8.5%. These results indicate that, similarly to the VT rate, a higher filial infection rate results from longer incubation time and fewer gonotrophic cycles.

2.6. Viral Load per Infected Larva Increases with Incubation Time Only

We quantified the copy number of ZIKV gRNA per infected larva as an indication of viral load. Viral load was similar for larvae produced at 17 d.p.i. either from the second or third gonotrophic cycle (Figure 3D). The gRNA geometric mean was 816 for the second and 723 for the third gonotrophic cycle with values ranging from 103 to 10^8 gRNA. A shorter incubation time (10 d.p.i.) significantly reduced the larval viral load at the second gonotrophic cycle to a geometric mean of 203 ranging from 106 to 1610 gRNA. The larval viral load most probably reflects the number of viruses that initially infect the eggs. These results indicate that the number of viruses transferred from mother to eggs increases solely with extrinsic incubation time.

3. Discussion

Most importantly, our results confirm the ability of ZIKV to be vertically transmitted up to fourth instar larvae in *Ae. aegypti* and reveal a surprisingly high VT efficiency. Our experimental design, which separated individual ovipositing mosquitoes and individual offspring from each progeny, provided a precise evaluation of VT including VT rate, filial infection rate [26], and the exact viral load per infected larva. VT efficiency was then calculated by multiplying the VT rate by the filial infection rate. With regards to the VT rate, we observed that 100% of progeny contained at least one infected larva in the most optimal condition (second gonotrophic cycle at 17 d.p.i.), whereas in the least optimal condition (second gonotrophic cycle at 10 d.p.i.), VT rate was 21%. Although we did not quantify VT at the first gonotrophic cycle at 3 d.p.i., the observed impact of incubation time suggests that VT rate would have been minimal. We found only one other study that separated progeny to assess dengue virus (DENV) VT rate in another *Aedes* species; the authors found a similar range of VT rates, from 12.5% to 94.7% [31]. In most other studies, larvae from different progeny were analyzed as a pool [12–14], preventing the calculation of VT rate. Our results provide the first evaluation of ZIKV VT rate in controlled conditions and reveal that the VT rate can be extremely high, while confirming that VT can largely vary. With regards to the filial infection rate, we reported 8.5–66% in the worst (second gonotrophic cycle at 10 d.p.i.) and best (second gonotrophic cycle at 17 d.p.i.) conditions, respectively. Usual pooled analyses underestimate the filial infection rate because it is calculated by making the conservative assumption that only one larva is infected per pool. For ZIKV and *Ae. aegypti*, three studies calculated a filial infection rate of 1.7–.3% by studying pools of 20 individuals [13], 2.13% by studying pools of 30 [12], and 5.5% by studying pools of 10 [14]. Actual filial infection rates for these three studies could be as high as 66%, 64%, and 55%, respectively, if all pooled individuals were infected. These numbers are definitely overestimates but are within the range of our results when calculated from a single larva from the same progeny. Another study quantified vertically transmitted

ZIKV in single salivary glands and found a filial infection rate of 17% when eggs were collected at 5 d.p.i. [12], which is in line with the 21% we calculated in larvae from eggs collected at 10 d.p.i. Higher filial infection rate in salivary glands may be obtained when eggs are collected later, as observed for larvae from eggs we collected at 17 d.p.i. With regards to the viral load per infected larva, the best conditions (17 d.p.i. for either second or third gonotrophic cycle) averaged a bit lower than 1000 ZIKV gRNA per infected larva with a large variation among individual larvae. This is also the first time vertically transmitted viral load was calculated for individual larvae. The lowest infection levels we reported at the fourth larval instar may not persist through the mature phase and explain the lower filial infection rate observed in the mature than in the immature phase [7]. However, average and higher viral loads should be enough to infect the whole body of adults, including salivary glands. Vertically transmitted ZIKV infects salivary glands, enabling horizontal transmission to humans [14]. Overall, our precise results obtained by analyzing individual progeny and individual larvae indicate that VT efficiency in optimal conditions can be high (VT efficiency = VT rate × filial infection rate = 100% × 66% = 66%).

Several lines of evidence suggest that ZIKV is vertically transmitted by transovarial infection. Infection of germinal cells is a prerequisite for transovarial infection. We detected ZIKV in ovaries as early as 3 d.p.i., as previously reported [12]. We quantified an exponential increase in viral loads until 14 d.p.i., suggesting active replication in ovaries. In support of germ cell infection by flaviviruses, particles of dengue virus (DENV), another mosquito-borne *Flavivirus*, are present in the germarium and oocytes of *Ae. aegypti* [25]. As compared to trans-ovum infection, which occurs through the oviduct, transovarial infection occurs in the germinal cells [21] and, thus, should be affected by oogenesis. We observed that an additional gonotrophic cycle reduces VT and filial infection rates. A similar impact for the gonotrophic cycle in *Ae. aegypti* was previously observed for DENV [32] and another mosquito-borne *Flavivirus*, yellow fever virus [33]. Eventually, during transovarial infection, longer incubation time should provide more time for the virus to infect more germinal cells. We observed that incubation time increased VT for ZIKV, as for DENV [25]. However, for the same incubation time, an additional gonotrophic cycle reduced the egg infection rate (as evaluated by VT rate and filial infection rate) but not infection intensity (as determined by viral load in individual larvae). This suggests that infection occurs during a stage of oogenesis that is renewed by subsequent cycles. In other words, oviposition shortens the actual incubation of the infected stage. The viral load received depends on the infection intensity in ovaries, which is identical for the same incubation time. During oogenesis, oocytes mature, are fertilized, and deposited, and another round of oocyte production is initiated from the germline cells. We thus propose that ZIKV infection occurs during oocyte maturation, although this hypothesis requires further testing. Strongly supporting transovarial infection of flaviviruses, VT for DENV was reported after egg surface sterilization [25]. Transovarial transmission should be more efficient than trans-ovum transmission [34] and could explain the high VT efficiency we observed.

Our results shed light on the balance between VT in mosquitoes and horizontal transmission to humans. We revealed that VT is higher for a longer extrinsic incubation time and lower number of gonotrophic cycles. These same conditions also favor oviposition rate, further increasing VT efficiency. Horizontal transmission occurs through blood feeding, which then stimulates oocyte maturation and search for an oviposition site [35]. The relationship between blood feeding and oviposition is central to balance horizontal transmission and VT; frequent blood feedings will favor horizontal transmission at the expense of VT and vis versa. *Aedes aegypti* is highly anthropophilic [36] and the limited availability of humans can reduce blood feeding. Alternatively, oviposition induces host seeking for blood feeding, which, when acquired, will initiate a subsequent gonotrophic cycle [37]. The alternation of gonotrophic cycles can then be driven by the availability of oviposition sites. Arid weather, for instance, should reduce *Ae. aegypti* preferred oviposition sites in artificial containers [38], delaying oviposition, and favoring VT. Mathematical models predict that in endemic settings, DENV would not persist by VT unless efficiency was higher than 20–30% [39,40]. Based on available data from pooled analyses, the authors of the model concluded that VT for DENV has a minimal impact on epidemiology. The model was based on *Ae. aegypti* biting habits and should partially apply to ZIKV. Our results showing VT higher than 30% for ZIKV suggest that VT can maintain the virus, at least when conditions limit availability of human hosts or oviposition sites.

There is evidence that arboviruses can adapt to VT in mosquitoes. Laboratory studies showed that DENV can persist through several generations by VT [41–43]. DENV acquired by VT in *Ae. albopictus* females was then more efficiently vertically transmitted than DENV acquired by blood feeding [41]. It remains to be tested whether VT-adapted viruses conserve the ability to infect humans. Intriguingly, a meta-analysis found that VT efficiency was higher when flaviviruses were isolated from an arid climate [7], which, based on our results, corresponds to the optimal conditions for ZIKV VT. However, any adaptation to VT would have to be balanced with the potential fitness cost incurred to mosquito, as observed for slower growth of ZIKV vertically infected larvae [14]. The immune response in ovaries should also be taken into account. In our study, we used a low-passage ZIKV collected from humans (horizontally transmitted) in Brazil, which should not be adapted to VT.

In conclusion, we deployed an original experimental design that allowed us to precisely calculate VT in *Ae. aegypti* for ZIKV and reveal VT efficiency higher than that quantified from pooled analysis. We further show that incubation time increases whereas successive gonotrophic cycle decreases VT, informing about the conditions that favor VT. Altogether, while horizontal transmission to humans remains the most epidemiologically relevant transmission mode, our study indicates that ZIKV VT may maintain the virus when conditions are not amenable for horizontal transmission. Consequently, it is advisable to maintain vector control when horizontal transmission (evidenced by outbreaks) is low in order to curb residual vertical transmission.

4. Material and Methods

4.1. Mosquitoes

An *Ae. aegypti* mosquito colony was established from a thousand eggs collected in Singapore in 2010 and maintained in the laboratory. Eggs were hatched in tap water, larvae were fed a mix of fish food (TetraMin fish flakes, Melle, Germany) and liver powder (MP Biomedicals, Illkirch, France), and adults were held in rearing cages (Bioquip, Rancho Dominguez, CA, USA) supplemented with 10% sucrose and water. Adults were fed pig's blood (Primary Industries Pte Ltd., Singapore) twice weekly to produce eggs for maintaining the colony. The insectary was held at 28 °C with 50% humidity on a 12:12 h dark:light cycle.

4.2. Zika Virus

The strain BeH815744 was collected in the Paraiba state (northeast region), Brazil, in 2015 from a febrile non-pregnant woman with rashes. The virus was given from Pr. da Costa Vasconcelos, propagated in C6/36 (ATCC CRL-1660) cells and used after three passages. The virus stocks were titrated thrice by plaque assay in BHK-21 cells as previously described [10].

4.3. Oral Infection

Three- to five-day-old female mosquitoes were sugar deprived for 24 h before offering a blood meal containing a 40% volume of washed erythrocytes from serum pathogen free (SPF) pig's blood (SingHealth, Singapore), 5% of 10 mM ATP (Thermo Scientific, Waltham, MA, USA), 5% human serum (Sigma, St Louis, MO, USA) and 50% volume of virus in RPMI (Gibco, Waltham, MA, USA). The final blood viral titer was 10^5 pfu/mL and was confirmed by a BHK-based plaque assay on blood samples collected before and after feeding. Mosquitoes were exposed to the artificial blood meal for one hour using a Hemotek membrane feeder system (Discovery Workshops, Accrington, UK) with a porcine intestine membrane. Fully engorged females were selected and maintained with free access to a 10% sugar solution and water in an incubation chamber with conditions similar to those used for insect rearing.

4.4. Virus Quantification

Single ovary pairs or single larvae were homogenized in 350 µL of TRK lysis buffer (E.Z.N.A. Total RNA kit I (OMEGA Bio-Tek, Norcross, GA, USA)) using a bead Mill homogenizer (FastPrep-24, MP Biomedicals, Illkirch, France). Total RNA was extracted using the E.Z.N.A. Total RNA kit I (OMEGA Bio-Tek) and following the manufacturer's instruction with 30 µL of DEPC-treated water for filter elution. ZIKV genomic RNA (gRNA) copies were quantified with a one-step RT-qPCR with SensiFAST SYBR No-ROX one-step kit (BioLine, London, UK). Primers, targeting the conserved region in the envelope, were: 5′- AGGACAGGCCTTGACTTTTC -3′ and 5′- TGTTCCAGTGTGGAGTTC -3′, as previously used [10]. The 10 µL reaction mix contained 400 nM of forward and reverse primers, and 3 µL of RNA extract. Quantification was conducted on a CFX96 Touch Real-Time PCR Detection System (Bio-Rad, Hercule, CA, USA). The thermal profile was 45 °C for 10 min, 95 °C for 1 min, and 40 cycles of 95 °C for 5 sec and 60 °C for 20 sec.

An absolute standard curve was generated by amplifying fragments containing the qPCR targets using the qPCR forward primers tagged with a T7 promoter and the qPCR reverse primer. The fragment was reverse transcribed using a MegaScript T7 transcription kit (Ambion, Austin, TX, USA) and purified using the E.Z.N.A. Total RNA kit I. The total amount of RNA was quantified using a Nanodrop (Thermo Scientific) to estimate copy number based on an averaged base pair weight of 649 g/mole. Ten times serial dilutions were made and used to generate an absolute standard equation (Supplementary Figure S1). In each subsequent RT-qPCR plate with samples, we quantified four standard aliquot dilutions to adjust for threshold variation between plates.

4.5. Experimental Design to Test the Impact of Gonotrophic Cycle and Extrinsic Incubation Period

About 200 24 h starved 2- to 5-day-old female mosquitoes were offered a ZIKV infectious blood meal. Fully engorged mosquitoes were selected, and equally divided in two halves that were placed in separate 30 cm cubic cages (BioQuip, Rancho Dominguez, CA, USA). One day post-infectious blood feeding (d.p.i.), blood-fed mosquitoes were singled out in cylindrical vials of 2 cm diameter and 10 cm height. Vials were prefilled 1 cm high with distilled water and closed with a cotton plug. These eggs corresponding to the first gonotrophic cycle were collected at 3 d.p.i. and discarded. Only mosquitoes that oviposited were regrouped in their original cages. At 7 d.p.i., after 18 h of starvation, only one of the two cages was provided a non-infectious blood meal containing full SPF blood (SingHealth) for one hour using the Hemotek membrane feeder system with a porcine intestine membrane. Fully engorged mosquitoes were selected and kept in the cage, whereas non-fed mosquitoes were discarded. The other cage was identically starved but provided 10% sugar solution instead of a blood meal. At 8 d.p.i., mosquitoes that were fed a non-infectious blood meal were singled out as above in the cylindrical vials prefilled with distilled water. At 10 d.p.i., mosquitoes that oviposited were collected and kept in the same 30 cm cage. Eggs from each vial were reared in separated pans until the fourth instar larvae. At 14 d.p.i., after 18 h of starvation, the two cages were provided a non-infectious blood meal as above. Fully engorged females were selected and kept in their respective cages, whereas non-fed mosquitoes were discarded. At 15 d.p.i., mosquitoes from both cages were singled out in the vials prefilled with water. At 17 d.p.i., mosquitoes that oviposited were recorded and eggs were reared in separate pans until the fourth instar larvae. For each tested condition, we quantified ZIKV in 10 larvae randomly selected from one progeny. We analyzed 13–17 progeny per condition. During the experiment, mosquitoes were provided with 10% sugar solution when they were not offered a blood meal. Starvation consisted of removing the sugar solution. The experiment was repeated twice with different mosquito batches.

4.6. Statistics

Infection rate in the ovaries was calculated by dividing the number of infected ovaries over the total number of ovaries that were tested. VT rate was defined as the percentage of infected females that transmitted the virus to their progeny [21] and was calculated by dividing the number of progeny with at least one infected larva to the total number of females that oviposited. Filial infection rate was calculated by dividing the number of larvae with detectable levels of ZIKV gRNA out of 10 tested randomly selected larvae within progeny. Oviposition rate was calculated by dividing the number of female mosquitoes that oviposited over the total number of mosquitoes that were singled out in the vials prefilled with water. Differences in infection rates per progeny and ZIKV gRNA copies were evaluated using Tukey's multiple comparison tests. After evaluating distribution normality with D'Agostino omnibus K^2 test, ZIKV gRNA copies were log-transformed before statistical analysis. Differences infection rates in ovaries and oviposition rates were evaluated using Z-test. All tests were performed with Prism v8.0.2 (GraphPad, San Diego, CA, USA), except for Z-tests that were conducted with Systat v13.1 (Systat Inc., San Jose, CA, USA).

Supplementary Materials: The following are available online at http://www.mdpi.com/2076-0817/9/5/366/s1, Figure S1: Standard curve for absolute quantification of ZIKV gRNA. Table S1. Number of mosquitoes analyzed in the VT assay presented in Figure 3.

Author Contributions: Conceptualization, J.P.; Methodology, M.M. and J.P.; Formal Analysis, J.P.; Investigation, M.M.; Writing-Original Draft Preparation, J.P.; Writing-Review & Editing, M.M., D.M. and J.P.; Visualization, J.P.; Supervision, J.P.; Project Administration, J.P.; Funding Acquisition, J.P. All authors have read and agreed to the published version of the manuscript.

Funding: This research was funded by the National Medical Research Council of Singapore (NMRC/ZRRF/0007/2017) and the Ministry of Education, Singapore (MOE2015-T3-1-003). Support was also received through a Zika response allocation from the Duke-NUS Signature Research Program funded by the Agency for Science, Technology, and Research (A*Star), Singapore, and the Ministry of Health, Singapore.

Acknowledgments: We thank Mariano Garcia-Blanco for his support.

Conflicts of Interest: The authors declare not conflict of interest.

References

1. Hamel, R.; Liégeois, F.; Wichit, S.; Pompon, J.; Diop, F.; Talignani, L.; Thomas, F.; Desprès, P.; Yssel, H.; Missé, D. Zika virus: Epidemiology, clinical features and host-virus interaction. *Microbes Infect.* **2016**, *18*, 441–449. [CrossRef]

2. Cauchemez, S.; Besnard, M.; Bompard, P.; Dub, T.; Guillemette-Artur, P.; Eyrolle-Guignot, D.; Salje, H.; Van Kerkhove, M.D.; Abadie, V.; Garel, C.; et al. Association between Zika virus and microcephaly in French Polynesia, 2013–2015: A retrospective study. *Lancet Lond. Engl.* **2016**, *387*, 2125–2132. [CrossRef]

3. European Center for Disease Prevention and Control. *Zika Virus Transmission Worldwide—9 April 2019*; ECDC: Stockholm, Sweden, 2019.

4. Aubry, M.; Teissier, A.; Huart, M.; Merceron, S.; Vanhomwegen, J.; Roche, C.; Vial, A.-L.; Teururai, S.; Sicard, S.; Paulous, S.; et al. Zika Virus Seroprevalence, French Polynesia, 2014–2015. *Emerg Infect Dis.* **2017**, *23*, 669–672. [CrossRef] [PubMed]

5. Netto, E.M.; Moreira-Soto, A.; Pedroso, C.; Höser, C.; Funk, S.; Kucharski, A.J.; Rockstroh, A.; Kümmerer, B.M.; Sampaio, G.S.; Luz, E.; et al. High Zika Virus Seroprevalence in Salvador, Northeastern Brazil Limits the Potential for Further Outbreaks. *mBio* **2017**, *8*. [CrossRef] [PubMed]

6. WHO | Zika Virus Disease—France. Available online: http://www.who.int/csr/don/01-november-2019-zika-virus-disease-france/en/ (accessed on 7 April 2020).

7. Lequime, S.; Paul, R.E.; Lambrechts, L. Determinants of Arbovirus Vertical Transmission in Mosquitoes. *PLoS Pathog.* **2016**, *12*, e1005548. [CrossRef] [PubMed]

8. Lequime, S.; Lambrechts, L. Vertical transmission of arboviruses in mosquitoes: A historical perspective. *Infect. Genet. Evol.* **2014**, *28*, 681–690. [CrossRef]

9. Gutiérrez-Bugallo, G.; Piedra, L.A.; Rodriguez, M.; Bisset, J.A.; Lourenço-de-Oliveira, R.; Weaver, S.C.; Vasilakis, N.; Vega-Rúa, A. Vector-borne transmission and evolution of Zika virus. *Nat. Ecol. Evol.* **2019**, *3*, 561–569. [CrossRef]

10. Pompon, J.; Morales-Vargas, R.; Manuel, M.; Huat Tan, C.; Vial, T.; Hao Tan, J.; Sessions, O.M.; da Costa Vasconcelos, P.; Ng, L.C.; Missé, D. A Zika virus from America is more efficiently transmitted than an Asian virus by Aedes aegypti mosquitoes from Asia. *Sci. Rep.* **2017**, *7*, 1215. [CrossRef]

11. Zimler, R.A.; Alto, B.W. Florida Aedes aegypti (Diptera: Culicidae) and Aedes albopictus Vector Competency for Zika Virus. *J. Med. Entomol.* **2019**, *56*, 341–346. [CrossRef]

12. Li, C.-X.; Guo, X.-X.; Deng, Y.-Q.; Xing, D.; Sun, A.-J.; Liu, Q.-M.; Wu, Q.; Dong, Y.; Zhang, Y.-M.; Zhang, H.-D.; et al. Vector competence and transovarial transmission of two Aedes aegypti strains to Zika virus. *Emerg. Microbes Infect.* **2017**, *6*, e23. [CrossRef]

13. Phumee, A.; Chompoosri, J.; Intayot, P.; Boonserm, R.; Boonyasuppayakorn, S.; Buathong, R.; Thavara, U.; Tawatsin, A.; Joyjinda, Y.; Wacharapluesadee, S.; et al. Vertical transmission of Zika virus in Culex quinquefasciatus Say and Aedes aegypti (L.) mosquitoes. *Sci. Rep.* **2019**, *9*, 1–9. [CrossRef] [PubMed]

14. Chaves, B.A.; Junior, A.B.V.; Silveira, K.R.D.; Paz A da, C.; Vaz, E.B.d.C.; Araujo, R.G.P.; Rodrigues, N.B.; Campolina, T.B.; Orfano A da, S.; Nacif-Pimenta, R.; et al. Vertical Transmission of Zika Virus (Flaviviridae, Flavivirus) in Amazonian Aedes aegypti (Diptera: Culicidae) Delays Egg Hatching and Larval Development of Progeny. *J. Med. Entomol.* **2019**. [CrossRef] [PubMed]

15. Maniero, V.C.; Rangel, P.S.C.; Coelho, L.M.C.; Silva, C.S.B.; Aguiar, R.S.; Lamas, C.C.; Cardozo, S.V. Identification of Zika virus in immature phases of Aedes aegypti and Aedes albopictus: A surveillance strategy for outbreak anticipation. *Braz. J. Med. Biol. Res. Rev. Bras. Pesqui. Medicas E Biol.* **2019**, *52*, e8339. [CrossRef] [PubMed]

16. Maia, L.M.S.; Bezerra, M.C.F.; Costa, M.C.S.; Souza, E.M.; Oliveira, M.E.B.; Ribeiro, A.L.M.; Miyazaki, R.D.; Slhessarenko, R.D. Natural vertical infection by dengue virus serotype 4, Zika virus and Mayaro virus in Aedes (Stegomyia) aegypti and Aedes (Stegomyia) albopictus. *Med. Vet. Entomol.* **2019**. [CrossRef] [PubMed]

17. Costa, C.F.d.; Silva, A.V.d.; Nascimento, V.A.d.; Souza, V.C.d.; Monteiro, D.C.d.S.; Terrazas, W.C.M.; Passos, R.A.d.; Nascimento, S.; Lima, J.B.P.; Naveca, F.G. Evidence of vertical transmission of Zika virus in field-collected eggs of Aedes aegypti in the Brazilian Amazon. *PLoS Negl. Trop. Dis.* **2018**, *12*, e0006594. [CrossRef]

18. Smartt, C.T.; Stenn, T.M.S.; Chen, T.-Y.; Teixeira, M.G.; Queiroz, E.P.; Souza Dos Santos, L.; Queiroz, G.A.N.; Ribeiro Souza, K.; Kalabric Silva, L.; Shin, D.; et al. Evidence of Zika Virus RNA Fragments in Aedes albopictus (Diptera: Culicidae) Field-Collected Eggs From Camaçari, Bahia, Brazil. *J. Med. Entomol.* **2017**, *54*, 1085–1087. [CrossRef]

19. Hernández-Triana, L.M.; Barrero, E.; Delacour-Estrella, S.; Ruiz-Arrondo, I.; Lucientes, J.; Fernández de Marco, M.d.M.; Thorne, L.; Lumley, S.; Johnson, N.; Mansfield, K.L.; et al. Evidence for infection but not transmission of Zika virus by Aedes albopictus (Diptera: Culicidae) from Spain. *Parasit. Vectors* **2019**, *12*, 204. [CrossRef]

20. Felix, G.E.; Barrera, R.; Vazquez, J.; Ryff, K.R.; Munoz-Jordan, J.L.; Matias, K.Y.; Hemme, R.R. Entomological Investigation of Aedes aegypti In Neighborhoods With Confirmed Human Arbovirus Infection In Puerto Rico. *J. Am. Mosq. Control Assoc.* **2018**, *34*, 233–236. [CrossRef]

21. Clements, A.N. *Biology of Mosquitoes: Development Nutrition And Reproduction By A. N. Clements*; CABI: London, UK, 1992.

22. Rosen, L. Further observations on the mechanism of vertical transmission of flaviviruses by Aedes mosquitoes. *Am. J. Trop. Med. Hyg.* **1988**, *39*, 123–126. [CrossRef]

23. Higgs, S. Natural cycles of vector-borne pathogens. In *Biology of Disease Vectors*; Marquardt, W.C., Ed.; Elsevier: Amsterdam, The Netherlands, 2004.

24. Salazar, M.I.; Richardson, J.; Sanchez-Vargas, I.; Olson, K.; Beaty, B. Dengue virus type 2: Replication and tropisms in orally infected Aedes aegypti mosquitoes. *BMC Microbiol.* **2007**, *7*, 9. [CrossRef]

25. Sánchez-Vargas, I.; Harrington, L.C.; Doty, J.B.; Black 4th, W.C.; Olson, K.E. Demonstration of efficient vertical and venereal transmission of dengue virus type-2 in a genetically diverse laboratory strain of Aedes aegypti. *PLoS Negl. Trop. Dis.* **2018**, *12*, e0006754. [CrossRef] [PubMed]

26. Turell, M.J. Horizontal and vertical transmission of viruses by insect and tick vectors. In *The arboviruses: Epidemiology and Ecology*; Monath, T.P., Ed.; CRC: Boca Raton, FL, USA, 1988.

27. Waggoner, J.J.; Gresh, L.; Vargas, M.J.; Ballesteros, G.; Tellez, Y.; Soda, K.J.; Sahoo, M.K.; Nuñez, A.; Balmaseda, A.; Harris, E.; et al. Viremia and Clinical Presentation in Nicaraguan Patients Infected With Zika Virus, Chikungunya Virus, and Dengue Virus. *Clin. Infect. Dis. Off. Publ. Infect. Dis. Soc. Am.* **2016**, *63*, 1584–1590. [CrossRef] [PubMed]

28. Lanciotti, R.S.; Kosoy, O.L.; Laven, J.J.; Velez, J.O.; Lambert, A.J.; Johnson, A.J.; Stanfield, S.M.; Duffy, M.R. Genetic and serologic properties of Zika virus associated with an epidemic, Yap state, Micronesia, 2007. *Emerg. Infect. Dis. J.* **2008**, *14*, 1232. [CrossRef] [PubMed]

29. Diallo, M.; Fontenille, D.; Thonnon, J. Vertical transmission of the yellow fever virus by Aedes aegypti (Diptera, Culicidae): Dynamics of infection in F1 adult progeny of orally infected females. *Am. J. Trop. Med. Hyg.* **2000**, *62*, 151–156. [CrossRef] [PubMed]

30. Anderson, J.F.; Main, A.J.; Delroux, K.; Fikrig, E. Extrinsic Incubation Periods for Horizontal and Vertical Transmission of West Nile Virus by Culex pipiens pipiens (Diptera: Culicidae). *J. Med. Entomol.* **2008**, *43*, 445–451. [CrossRef]

31. Freier, J.E.; Rosen, L. Vertical transmission of dengue viruses by Aedes mediovittatus. *Am. J. Trop. Med. Hyg.* **1988**, *39*, 218–222. [CrossRef] [PubMed]

32. Rosen, L.; Shroyer, D.A.; Tesh, R.B.; Freier, J.E.; Lien, J.C. Transovarial transmission of dengue viruses by mosquitoes: Aedes albopictus and Aedes aegypti. *Am. J. Trop. Med. Hyg.* **1983**, *32*, 1108–1119. [CrossRef]

33. Beaty, B.J.; Tesh, R.B.; Aitken, T.H. Transovarial transmission of yellow fever virus in Stegomyia mosquitoes. *Am. J. Trop. Med. Hyg.* **1980**, *29*, 125–132. [CrossRef]

34. Tesh, R.B. Transovarial transmission of arboviruses in their invertebrate Vectors. In *Current Topics in Vector Research*; Harris, K.F., Ed.; Praeger: New York, NY, USA, 1984.

35. Day, J.F. Mosquito Oviposition Behavior and Vector Control. *Insects* **2016**, *7*, 65. [CrossRef]

36. McBride, C.S.; Baier, F.; Omondi, A.B.; Spitzer, S.A.; Lutomiah, J.; Sang, R.; Ignell, R.; Vosshall, L.B. Evolution of mosquito preference for humans linked to an odorant receptor. *Nature* **2014**, *515*, 222–227. [CrossRef]

37. Chadee, D.D. Studies on the post-oviposition blood-feeding behaviour of Aedes aegypti (L.) (Diptera: Culicidae) in the laboratory. *Pathog. Glob. Health* **2012**, *106*, 413–417. [CrossRef] [PubMed]

38. Wong, J.; Stoddard, S.T.; Astete, H.; Morrison, A.C.; Scott, T.W. Oviposition Site Selection by the Dengue Vector Aedes aegypti and Its Implications for Dengue Control. *PLoS Negl. Trop. Dis.* **2011**, *5*, e1015. [CrossRef] [PubMed]

39. Grunnill, M.; Boots, M. How important is vertical transmission of Dengue viruses by mosquitoes (Diptera: Culicidae)? *J. Med. Entomol.* **2016**. [CrossRef] [PubMed]

40. Adams, B.; Boots, M. How important is vertical transmission in mosquitoes for the persistence of dengue? Insights from a mathematical model. *Epidemics* **2010**, *2*, 1–10. [CrossRef] [PubMed]

41. Shroyer, D.A. Vertical maintenance of dengue-1 virus in sequential generations of Aedes albopictus. *J. Am. Mosq. Control Assoc.* **1990**, *6*, 312–314. [PubMed]

42. Joshi, V.; Mourya, D.T.; Sharma, R.C. Persistence of dengue-3 virus through transovarial transmission passage in successive generations of Aedes aegypti mosquitoes. *Am. J. Trop. Med. Hyg.* **2002**, *67*, 158–161. [CrossRef]

43. Rohani, A.; Zamree, I.; Joseph, R.; Lee, H. Persitency of transovarial dengue virus in Aedes aegypti (Linn.). *Southeast Asian J. Trop. Med. Public Health* **2008**, *39*, 4.

 *pathogens*

Article

Venereal Transmission of Vesicular Stomatitis Virus by *Culicoides sonorensis* Midges

Paula Rozo-Lopez [1], **Berlin Londono-Renteria** [1,*] and **Barbara S. Drolet** [2,*]

1   Department of Entomology, Vector Biology Laboratory, Kansas State University, Manhattan, KS 66506, USA; paularozo@ksu.edu

2   United States Department of Agriculture, Agricultural Research Service, Arthropod-Borne Animal Diseases Research Unit, Manhattan, KS 66502, USA

*   Correspondence: blondono@ksu.edu (B.L.-R.); barbara.drolet@usda.gov (B.S.D.)

Received: 31 March 2020; Accepted: 22 April 2020; Published: 24 April 2020

Abstract: *Culicoides sonorensis* biting midges are well-known agricultural pests and transmission vectors of arboviruses such as vesicular stomatitis virus (VSV). The epidemiology of VSV is complex and encompasses a broad range of vertebrate hosts, multiple routes of transmission, and diverse vector species. In temperate regions, viruses can overwinter in the absence of infected animals through unknown mechanisms, to reoccur the next year. Non-conventional routes for VSV vector transmission may help explain viral maintenance in midge populations during inter-epidemic periods and times of adverse conditions for bite transmission. In this study, we examined whether VSV could be transmitted venereally between male and female midges. Our results showed that VSV-infected females could venereally transmit virus to uninfected naïve males at a rate as high as 76.3% (RT-qPCR), 31.6% (virus isolation) during the third gonotrophic cycle. Additionally, VSV-infected males could venereally transmit virus to uninfected naïve females at a rate as high as 76.6% (RT-qPCR), 49.2% (virus isolation). Immunofluorescent staining of micro-dissected reproductive organs, immunochemical staining of midge histological sections, examination of internal reproductive organ morphology, and observations of mating behaviors were used to determine relevant anatomical sites for virus location and to hypothesize the potential mechanism for VSV transmission in *C. sonorensis* midges through copulation.

Keywords: vesicular stomatitis virus; *Culicoides* midges; non-conventional transmission; venereal transmission; reproductive anatomy; mating behavior

1. Introduction

Vesicular stomatitis virus (VSV) (Rhabdoviridae: *Vesiculovirus*) is a single-stranded, negative-sense, RNA pathogen responsible for vesicular stomatitis (VS) disease in cattle, horses, and swine [1]. VSV causes annual outbreaks in enzootic regions from northern South America to southern Mexico, infecting a large percentage of susceptible species [1]. In the U.S., VSV re-emerges sporadically with incursions originating from these southern enzootic regions moving northward into southwestern states at approximately 3 to 10-year intervals [1–3]. Epizootic viruses can overwinter, in an as yet identified natural reservoir, resulting in a second-year outbreak of the same viral genotype [3,4]. The epizootiology of VS is complex and comprises a wide variety of variables from a broad vertebrate host range, with variation in clinical outcome due to host species and site of initial infection, to the rapid transmission within animal herds by direct contact and fomites [1,5]. Furthermore, there is a diversity of suspected and potential transmission vector species acting as both mechanical and biological vectors throughout temperate and tropical ecosystems [6]. During VSV outbreaks in the U.S., *Culicoides* biting midges (Diptera: Ceratopogonidae) and *Simulium* black flies (Diptera: Simuliidae) have important roles in the initial introduction of VSV into animal herds and contribute to outbreak spread in the absence of animal movement [1–3,7]. Specifically, *Culicoides sonorensis* is one of the most common midge species associated with livestock agriculture [8,9] and a known biological transmission vector of VSV [10–15].

Transmission of VSV via *Culicoides* female bites is dependent upon available viremic hosts or infected hosts exhibiting skin-associated vesicular lesions containing large amounts of virus [16]. Blood feeding midges may acquire virus from blood [6], vesicular lesions, or from feeding on intact skin contaminated by vesicular fluid or virus-laden saliva [17,18]. However, the resulting pantropic systemic infection of *C. sonorensis* midges following oral ingestion of VSV [10], suggests that the interrelationships between the virus and vector may not be restricted to a bloodmeal-midgut-salivary gland-bloodmeal transmission route [6]. VSV infection and replication in reproductive tissues indicate that non-conventional routes of transmission might also occur. Specifically, VSV replication has been shown to occur in the ovarial epithelium and within the developing oocytes, suggesting that transovarial transmission might be possible [10]. Likewise, VSV infection of other relevant reproductive tissues and the rectal ampulla [10] suggests potential scenarios for trans-ovum transmission and transmission during sexual contact. Previously, VSV infection in *Culicoides* males has not been of interest because males were not believed to be involved in the transmission of viruses [19]. Since only females feed on blood, studies have been confined to the role females play in transmission and virus maintenance. However, in recent years, it has been suggested that males of some vector species might have a synergistic involvement in arbovirus transmission [20,21].

Therefore, determining the role of males, specifically the role venereal transmission (VNT) plays, in VSV maintenance in *Culicoides* populations, could lead to a more comprehensive understanding of 1) potential virus persistence in nature during interepidemic periods; 2) the overwintering of some viral genotypes leading to multi-year outbreaks; and 3) vector transmission dynamics during outbreaks. Herein, we report the first evidence for venereal transmission of any arbovirus in *Culicoides* spp. biting midges, and the first evidence for venereal transmission of VSV in any vector species. Additionally, we detail the mating behavior and morphological descriptions of *C. sonorensis* female and male reproductive anatomy with localization of VSV to provide insights into the potential mechanism of VNT.

2. Results

2.1. Venereal Transmission from Orally Infected C. sonorensis Females to Naïve Males

Colonized *C. sonorensis* midges typically survive an average of 14–21 days depending on the number and types of manipulations to which they are subject. If provided blood meals and allowed to cohabitate with males, female survival rates are adequate to analyze three gonotrophic cycles. To determine rates of venereal transmission from infected females to age-matched naïve males, four mating experiments, two tested by RT-qPCR and two tested by virus isolation (VI), were conducted through three sequential bloodmeal-induced gonotrophic cycles (GC). Naïve adult males were made available for copulation by cohabiting with VSV-fed females from 0 to 4 days post-feeding (dpf) (1 GC), 4 to 8 dpf (2 GC), and 8 to 12 dpf (3 GC) (Figure 1A). All 88 surviving males collected at the end of 1 GC tested negative for viral RNA (Table 1). During the second GC, 15.2% and 20% of the surviving males paired with orally infected females tested positive for viral RNA by RT-qPCR and for infectious virus by cytopathic effects (CPE) in cell culture, respectively. Cycle threshold (Ct) values of males paired with 2 GC females ranged from 34.7 to 32.9 (10^1 to 10^2 genome equivalents) (Figure 2A). At the end of 3 GC, 31.6% of the males were CPE positive and 76.3% RT-qPCR positive (Table 1) with Ct values ranging from 34.7 to 31.1 (10^1 to10^2 genome equivalents) (Figure 2A). Additionally, orally infected females (10 to 20 from each of two mating experiment) were collected at the end of each bloodmeal-induced gonotrophic cycle to be tested by RT-qPCR (Figure 2B). Ct values for females collected at the end of 1 GC ranged from 35.4 to 30.3, 35.6 to 29.3 for 2 GC, and 34.9 to 30.8 for 3 GC.

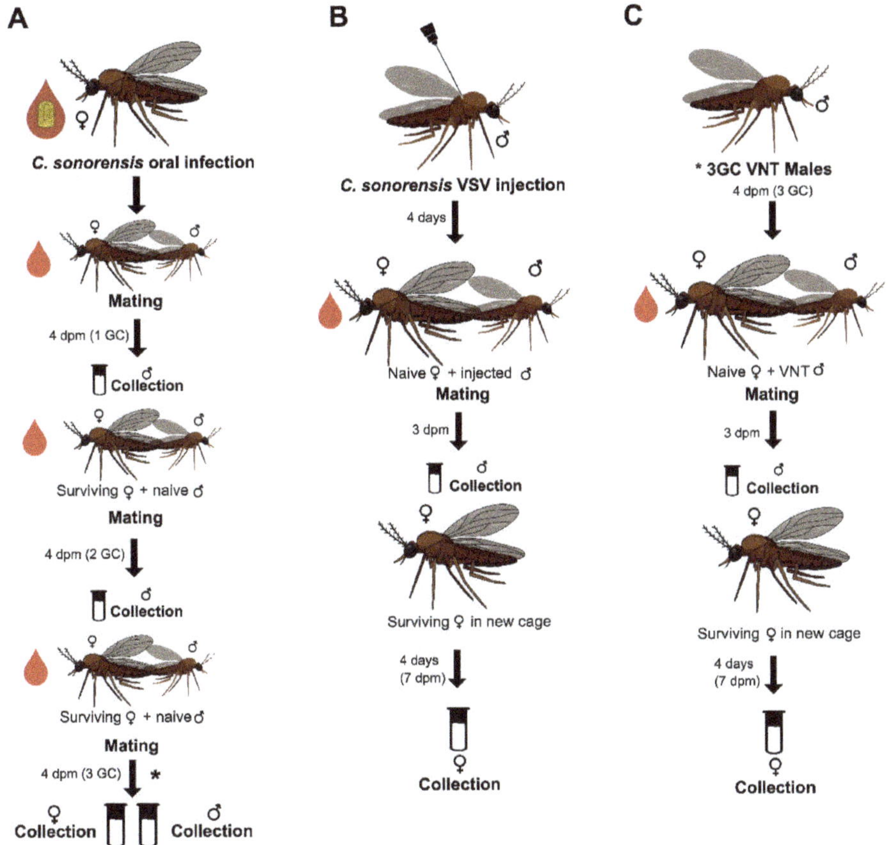

Figure 1. Experimental design to test venereal transmission of vesicular stomatitis virus (VSV) in *Culicoides sonorensis* midges. (**A**) Venereal transmission from VSV-fed females to naïve males. (**B**) Venereal transmission from intrathoracically VSV-injected males to naïve females. (**C**) Venereal transmission from 3rd gonotrophic cycle (GC)-mated males * (obtained in experiment (**A**)) to naïve females.

Table 1. Venereal transmission rates from orally infected females to naïve non-infected males during three gonotrophic cycles (GC) as detected by whole-body RT-qPCR or CPE.

GC	Initial Number of Midges	Surviving ♂ (%) [1]	VSV RT-qPCR+ ♂ (%) [2]	VSV CPE+ ♂ (%) [3]
1	240 ♀ and 120 ♂	88/120 (73.3%)	0	ND
2	199 ♀ and 100 ♂	59/100 (59%)	9/59 (15.2%)	ND
2	104 ♀ and 52 ♂	30/52 (57.7%)	ND	6/30 (20%)
3	131 ♀ and 65 ♂	38/65 (58.5%)	29/38 (76.3%)	ND
3	64 ♀ and 32 ♂	19/32 (59.4%)	ND	6/19 (31.6%)

[1] Surviving males were sampled 4 days after initial cohabitation, which was determined as the end of each blood meal-induced gonotrophic cycle (GC). [2] Vesicular stomatitis virus (VSV)-positive males detected by RT-qPCR. [3] VSV-positive males detected by cytopathic effect (CPE). ND, not determined.

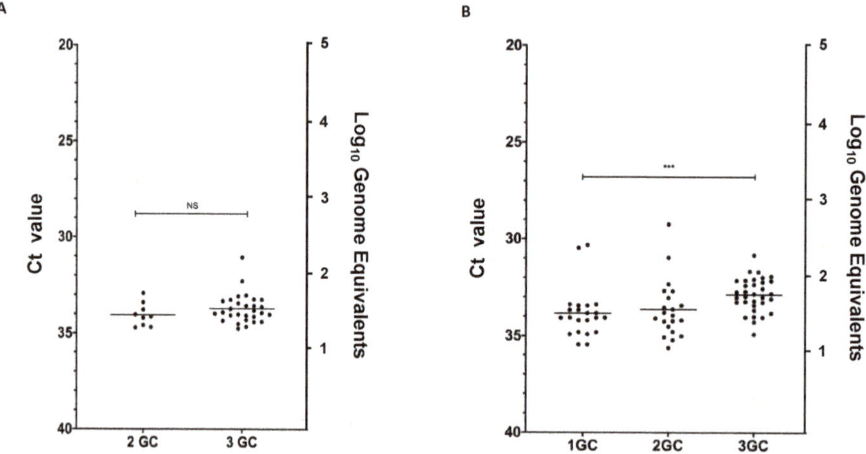

Figure 2. RT-qPCR cycle threshold (Ct) values (left Y-axis) and Log10 viral genome equivalents (right Y-axis) for individual VSV-positive (Ct ≤ 36) *C. sonorensis* midges. Non-parametric tests were used to compare distributions of Ct values between gonotrophic cycles. (**A**) *C. sonorensis* males infected with VSV following cohabitation with orally infected females at the end of the second (2 GC) and third (3 GC) gonotrophic cycles (*p* value = 0.196; NS, not significant). (**B**) *C. sonorensis* females orally infected with VSV tested at the end of each bloodmeal-induced gonotrophic cycle (1 GC, 2 GC, 3 GC) (*p*-value = 0.0002; *** *p* < 0.001).

2.2. Venereal Transmission of VSV from Intrathoracically Injected C. sonorensis Males to Naïve Females

Mating experiments were conducted to determine whether males 4 days post-injection (dpi) can venereally transmit VSV to age-matched naïve adult females, as tested by RT-qPCR and CPE (Table 2, Figure 1B). Of females surviving 7 days after the first exposure to infected males (11–14 days post emergence for both males and females), 49.2% were CPE positive (Table 2). To determine if the virus acquired by venereal transmission could disseminate into the salivary glands of females, we separately tested bodies (n = 77) and then tested heads with glands from RT-qPCR-positive bodies (n = 59). The bodies of 76.6% of the females tested positive for viral RNA (Table 2) with Ct values ranging from 34.8 to 22 (10^1 to 10^5 genome equivalents) (Figure 3). The heads of 11.63% of the females tested positive for viral RNA (Table 2) with Ct values ranging from 34.8 to 32.3 (10^1 to 10^2 genome equivalents) (Figure 3). Additionally, a sub-sample of the inoculated males was tested at 4 dpi (when introduced into the mating cages) and at 7 dpi (when removed from mating cages) for the presence of VSV RNA (N = 5) and infectious virus (N = 5). All tested males were positive, with titers ranging from 1.35×10^5 to 2.8×10^5 PFU/mL by plaque assay.

Table 2. Venereal transmission from VSV-injected *Culicoides* males (4 dpi) to age-matched naïve females as detected by RT-qPCR of individual bodies and heads and CPE of whole midges.

Initial Number of Midges	Surviving ♀(%) [1]	VSV RT-qPCR+ ♀Bodies (%)	VSV RT-qPCR+ ♀Heads (%)	VSV CPE+ Whole ♀(%)
58 ♂ and 116 ♀	77/116 (66.4%)	59/77 (76.6%)	6/59 (10.2%)	ND
39 ♂ and 78 ♀	61/78 (78.2%)	ND	ND	30/61 (49.2%)

[1] Surviving females were sampled 7 days after the initial cohabitation.

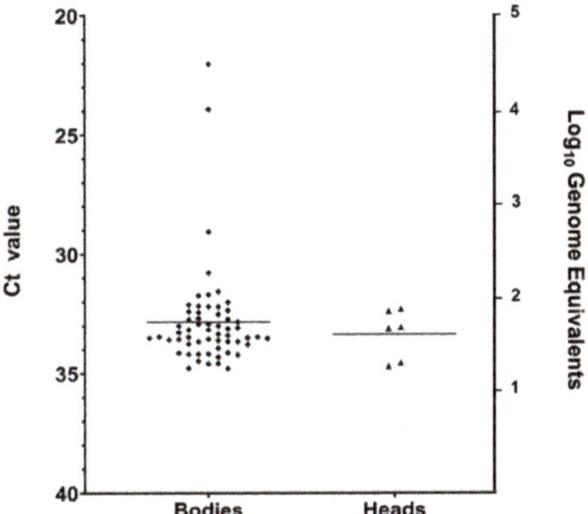

Figure 3. RT-qPCR cycle threshold (Ct) values (left Y-axis) and Log10 of viral genome equivalents (right Y-axis) for positive (Ct ≤ 36) *C. sonorensis* female bodies and heads infected with VSV following cohabitation with males infected by microinjection.

2.3. Venereal Transmission of VSV from Venereally Infected C. sonorensis Males to Naïve Females

Mating experiments were conducted to test venereal transmission of VSV from venereally infected males to younger naïve females (Figure 1C). Males used in these experiments had cohabitated with orally infected females during the third blood meal-induced gonotrophic cycle (3 GC). Younger females were used instead of age-matched in order to increase the chance of female survival at 7 days post mating. Of the 84 surviving females at 7 days post mating, 9.5% tested positive by RT-qPCR (Table 3) with Ct values ranging from 36.3 to 33.2 (10^1 to 10^2 genome equivalents) (Figure 4).

Table 3. Venereal transmission of VSV from venereally infected *Culicoides* males (4 days post-mating) to naïve females as detected by whole midge RT-qPCR.

Initial Number of Midges	Surviving ♀(%) [1]	VSV RT-qPCR+ Whole ♀(%)
100 ♂ and 200 ♀	84/200 (42%)	8/84 (9.5%)

[1] Surviving females were sampled 7 days after the initial cohabitation.

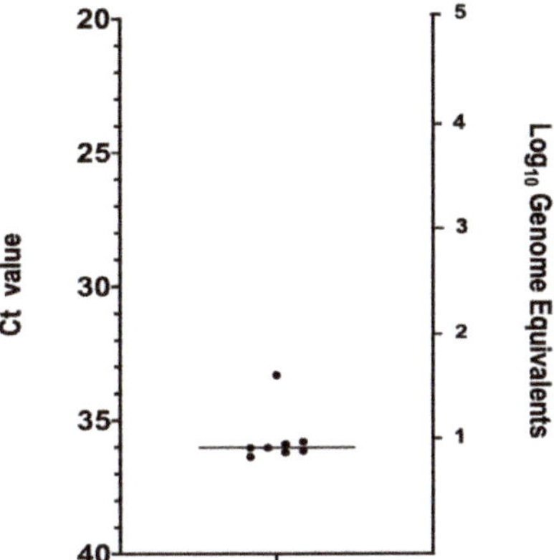

Figure 4. RT-qPCR cycle threshold (Ct) values (left Y-axis) and Log10 of viral genome equivalents (right Y-axis) for positive (Ct ≤ 36) *C. sonorensis* female bodies infected with VSV following cohabitation with venereally infected males.

2.4. VSV Infection in Reproductive Tracts

We conducted three trials using VSV-immunofluorescent staining of reproductive organs of intrathoracically inoculated males and females to establish tissue tropism for VSV in the reproductive tract of midges at 4 dpi. Initially, virgin females were used; however, the underdeveloped ovary morphology did not allow visualization of precise locations for VSV. Consequently, to add clarity to the virus location within developing oocytes, mated and blood-fed females were used in all subsequent trials.

The VSV-positive fluorescent puncta in males (Figure 5B,E, Table 4) indicated viral infection of the epithelial layer at the base of the testes (91.7%), in the outer epithelial surface of the accessory gland (58.3%), and throughout tissues in the hindgut and the rectal region (100%). In contrast, vas deferens, ejaculatory duct, and terminalia did not show positive fluorescence.

The intensity of positive fluorescence puncta in whole reproductive organs of females (Figure 5G, Table 4) was detected in the tracheal branches located in the space between ovarian sheaths (91.7%) and throughout tissues in the hindgut and the rectal region (100%). There were no apparent positive puncta in the spermatheca and spermathecal gland. Due to the inability to get intact whole reproductive tracts with undamaged ducts, we examined sagittal sections from 19 sequentially sampled virus fed females (9–13 dpf, time corresponding to 3 GC) used to first describe the temporal and spatial progression of VSV infection in *Culicoides* [10]. The positive staining (Figure 6B,C,E,F, Table 5) was detected in the ovaries (Ov) (36.8%), oviduct (Od) (16.8%), spermathecal duct (Sd) (26.3%), gonotreme (Go) and gonopore (Gp) (57.8%), and throughout tissues in the hindgut (Hg) and the rectal region (R).

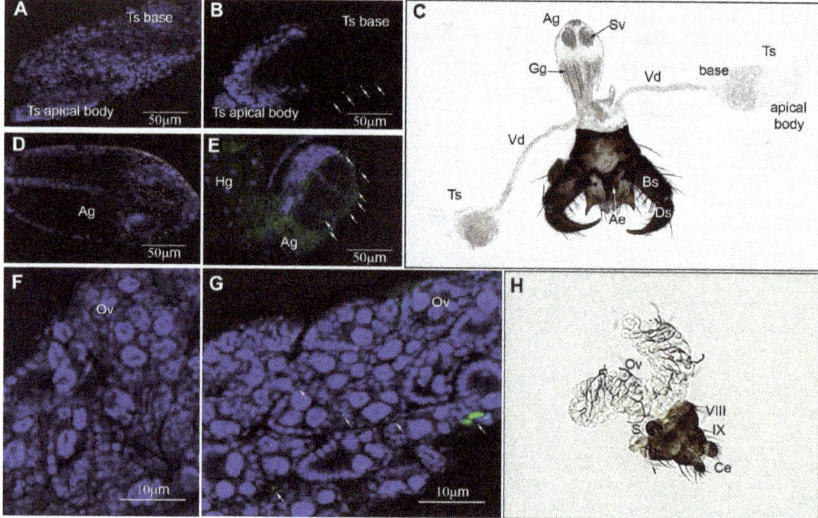

Figure 5. Immunofluorescent VSV-staining of intrathoracically infected *C. sonorensis*. (**A**) Testis (Ts) and (**D**) accessory gland (Ag) dissected from non-infected (negative control) males. (**B**) Testis (Ts) and (**E**) accessory gland (Ag) dissected from males 4 days post inoculation. Arrows denote VSV-positive staining (FITC-green puncta) in the epithelial layer of the testis base and outer epithelial layer of the Ag. Cellular nuclei were stained with DAPI (blue). (**C**) Male reproductive anatomy (brightfield 200×). Abbreviations: Ae: aedeagus, Ag: accessory gland, Bs: basistyle, Ds: dististyle, Gg: glutinous gland, Sv: seminal vesicle, Ts: testis, Vd: vas deferens. (**F**) Ovaries (Ov) dissected from non-infected (negative control) females. (**G**) Ovaries (Ov) dissected from females 4 days post inoculation. Arrows denote VSV-positive staining (FITC-green puncta) with DAPI nuclear stain (blue). (**H**) Female reproductive anatomy (brightfield 200×). Abbreviations: VIII: 8th abdominal segment, IX: 9th abdominal segment, Ce: cerci, Ov: ovary, S: spermatheca.

Table 4. Organs with positive fluorescent puncta staining of intrathoracically inoculated *C. sonorensis* midges 4 dpi.

Males		Females	
Testes	**Accessory Glands**	**Ovaries**	**Spermatheca**
15/17 (88.2%)	7/17 (41.2%)	11/12 (91.7%)	0/12 (0%)

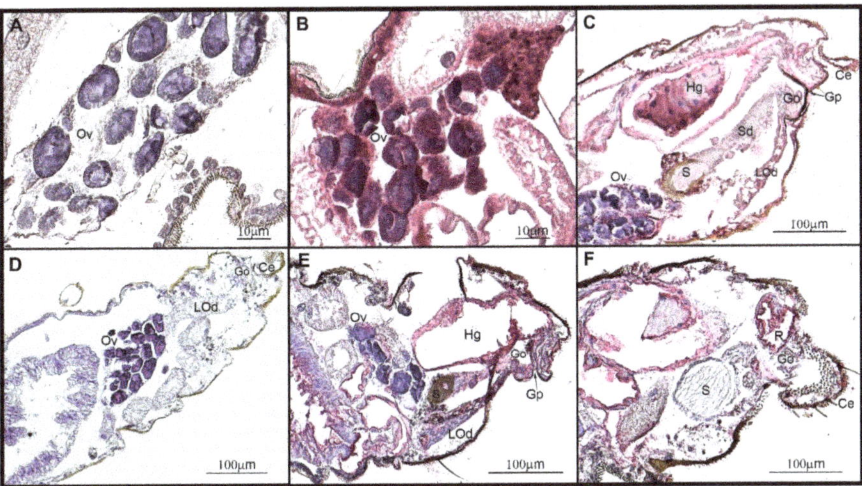

Figure 6. Immunohistochemical VSV-staining of orally infected *C. sonorensis* females. (**A**) Ovaries (Ov) and (**D**) abdomen of non-infected (negative control) females. (**B**) Ovaries (Ov) and (**C,E,F**) abdominal sections from females 9–13 days post feeding with VSV-positive antigen staining of viral nucleocapsid in red and counterstained with hematoxylin (blue). (**B**) VSV antigen staining in the ovarial sheaths and trachea. (**C**) VSV antigen staining in hindgut (Hg), gonopore (Gp), gonotreme (Go), lateral oviduct (LOd), spermathecal duct (Sd), and ovaries (Ov). (**E**) VSV antigen staining in hindgut (Hg), gonopore (Gp), gonotreme (Go), and lateral oviduct (LOd), with negative ovaries (Ov) and spermatheca (S). (**F**) VSV antigen staining in rectum (R) and gonotreme (Go), with negative spermatheca (S).

Table 5. Organs with VSV-positive staining of sections of orally infected *C. sonorensis* females (8–13 dpf).

Ovaries	Oviduct	Spermathecal Duct	Gonotreme	Gonopore
7/19 (36.8%)	3/19 (16.8%)	5/19 (26.3%)	6/19 (31.6%)	11/19 (57.8%)

2.5. Behavioral Observations of C. sonorensis Copulation

There is little information on the mating behavior, comparative function of the sex organs during mating, and the timing for efficient sperm transmission in *Culicoides* midges. In order to better understand the mating behavior of *C. sonorensis* and provide insight into the mechanism of VSV venereal transfer, we conducted observations of midge-matings under laboratory conditions.

C. sonorensis copulation occurred on the bottom of the cages without a swarming flight. When specimens of both sexes were introduced at 1 to 3 days post-emergence, they often rested for long periods with frequent antennal and wing movements. Mating attempts were initiated by males following the females with rapid walking movements and continual antennal vibrations culminating with the efforts of the male to climb onto the back of the female. If a blood meal was not offered, most females showed resistance behavior when males approached. The resistance behavior consisted of the females running rapidly, kicking males with their hind legs, and anteriorly curving the dorsal segments of the abdomen to avoid contact with the male claspers. However, 10–15 min after a blood meal was offered, fed females were receptive to the multiple attempts of males to establish genital contact. Males approached the female terminalia with curved abdomens and open claspers. After attachment, males would then rotate around the female until they were positioned 180° to the female. *Culicoides nubeculosus* also presents this 180° torsion, but for other species only a gentle torsion has been reported [22]. The genital contact (attachment) time was 420 ± 15 s. Detachment following copulation occurred rapidly by the claspers opening while the female pushed with their hind legs until separation.

From our observations in the laboratory, the presence of males stimulates female blood feeding, and subsequently, blood ingestion incites copulation. Moreover, *C. sonorensis* can repeatedly mate within each gonotrophic cycle, as previously reported for the variipennis complex (which includes *C. sonorensis*) [22]. Together these behaviors may impact VSV epidemiology by increasing viral exposure opportunities by blood-feeding females. Furthermore, the relatively long copulation time, and the possibility of multiple matings in a lifetime, makes the implications of VSV venereal transmission from females to males and males to females more likely to impact viral maintenance in *C. sonorensis* midge populations.

2.6. Anatomical Descriptions of C. sonorensis Male Reproductive Tract

Little information exists in the literature relative to the internal anatomy of males of the family Ceratopogonidae. Thus, to better understand how the venereal transmission of VSV is occurring from *C. sonorensis* males to females, we describe our morphological and anatomical observations of the male reproductive tract.

The outer male abdomen is slender with the 9th segment, the tergum, and sternum fused in the shape of a sclerotized ring to which a prominent terminalia is attached (Figures 5C and 7A). Two distinct sclerotized curved claspers are formed by a basal basistyle (Bs) and a claw-like apical dististyle (Ds) (Figure 5C). The aedeagus (Ae) is Y-shape and is held by a sclerotized structure on the ventral side (Figure 5C). The testes (Ts) are elongated pyriform concerted to a tubular vas deferens (Vd) (Figure 5A,B and Figure 7B). Each vas deferens is connected to the base of the accessory gland (Ag) (Figures 5C and 7B) and directs the sperm through to the distal portion of the Ag which contains two circular seminal vesicles (Sv) on each side (Figures 5C and 7B). Each Sv is surrounded by a layer of large secretory cells. Below each seminal vesicle is a pair of broadly ovoid glutinous glands (Gg) (Figures 5C and 7B), which in other *Culicoides* species are known to contain secretory cells [23,24]. The spermatozoa and ejaculatory secretions (most likely transferred to a female in a spermatophore as reported in *Culicoides nubeculosus* [25] and *Culicoides melleus* [22,23]) are released through a common ejaculatory duct (Ed) at the base of the accessory gland which is connected to the aedeagus (Figure 7B).

The sperm transferred during mating consists of a mix of proteins, lipids, carbohydrates, salts, and steroid hormones produced in the male accessory glands, and possibly in the testes [25]. Despite the potential impact that this ejaculatory complex may have on the reproductive physiology and behavior of females, the sperm ejaculation route (Figure 7B) contributes to VSV maintenance in midge populations by allowing the efficient transmission of virus particles from the male into the female upon copulation.

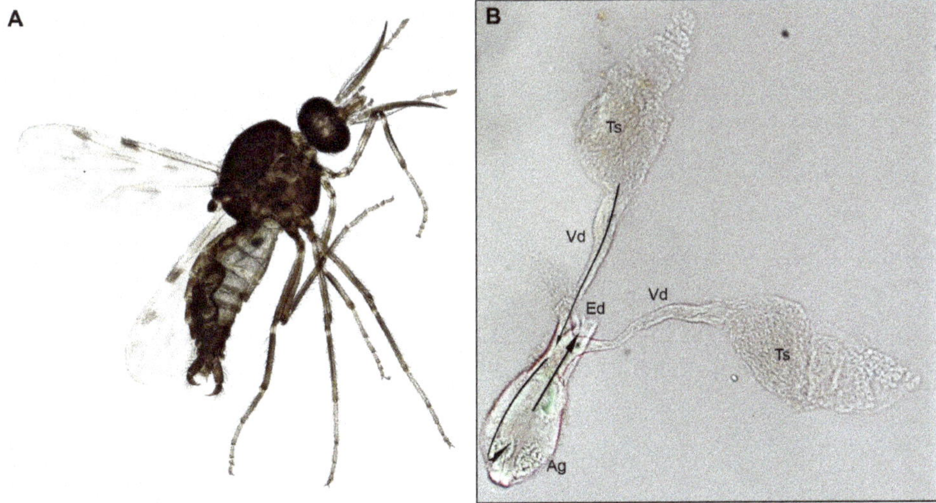

Figure 7. Anatomy of *Culicoides sonorensis* male (**A**) External male morphology. (**B**) Male reproductive tract detached from the terminalia. Arrows indicate sperm ejaculation pathway. Abbreviations: Ag: accessory gland, Ed: ejaculatory duct, Ts: testis, Vd: vas deferens.

2.7. Anatomical Descriptions of C. sonorensis Female Reproductive Tract

Venereal transmission from females to males has rarely been reported in the literature [26], mainly because the male produces all of the secretions that are exchanged during copulation. To better understand how the venereal transmission of VSV is occurring from *C. sonorensis* females to males, we describe the morphology and anatomy of the female reproductive tract.

The outer morphology of the female reproductive system is relatively simple, with a stout abdomen ending in a pair of small rounded cerci (Ce) with long sensory hairs visible below the 9th tergum (IX) (Figures 5H and 8A). Internally, the female reproductive system is complex, presenting two ovaries (Ov) located at the anterior end of the female's abdomen, usually internally located between the 5th and the 6th abdominal segments (Figures 5H and 8B). The ovaries contain oocytes at similar stages of development. Each oocyte is surrounded by follicular cells and contain 3 to 5 nurse cells. The oocytes are held together by an epithelial sheath surrounded by a network of fine, branching tracheae. At the base of each ovary, there is a lateral oviduct (LOd) that fuses into a common oviduct (Od) which is attached to the 8th sternum (Figure 8B). The common oviduct posteriorly enlarges to a gonotreme (Go), which receives the sperm during copulation. The posterior end of the gonotreme is bifurcated into a hyaline duct known as the spermathecal duct (Sd), which contains a minute globular spermathecal gland (Sg) (also known as a female accessory gland) and ending in one sclerotized mushroom-shaped (convex or campanulate) spermatheca (S) (Figures 5H and 8C) for sperm storage. The last section of the posterior end of the gonotreme, located in close proximity to the rectum (R), exits the female reproductive tract at the distal end of the female terminalia, the gonopore (Gp), which also serves to oviposit eggs. The polyandric behavior observed in *C. sonorensis* females, combined by the rich virus load in the anal region and gonopore, contribute to maintenance of VSV in midge populations by favoring virus transmission from females to multiple males in a lifetime of copulation events.

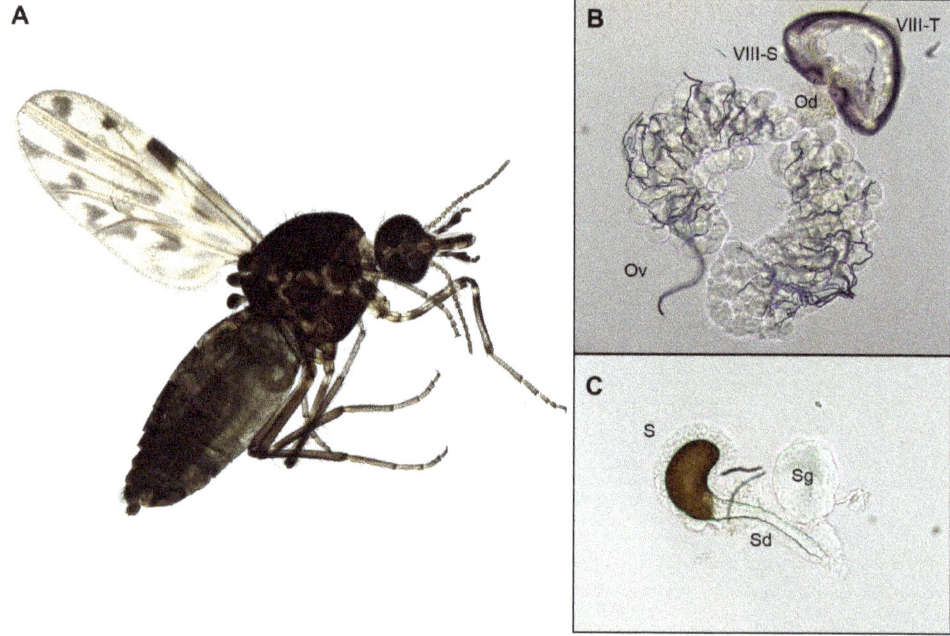

Figure 8. Anatomy of *Culicoides sonorensis* female (**A**) External female morphology. (**B**) Ovaries (Ov) joined by common oviduct (Od) attached to the 8th sternum (VIII-S). (**C**) Detailed portion of the spermatheca (S), spermathecal gland (Sg), and spermatheca duct (Sd).

3. Discussion

Vesicular stomatitis outbreaks in temperate regions peak during summer and fall and typically stop after the first hard freeze, corresponding with the decrease in the number of vectors in affected areas [2]. Outbreak viruses can overwinter with the same viral genotype re-emerging for a second-year outbreak [3,27], as occurred in 2004–2005, 2005–2006, 2014–1015, and in the most recent 2019–2020 outbreaks. Several hypotheses have been proposed to explain the maintenance of VSV during inter-epidemic periods, mainly by suggesting the presence of a yet to be identified, natural mammalian reservoir [28,29]. However, from a vector perspective, the vertical and venereal transmission of arboviruses are possible maintenance mechanisms during inter-epidemic periods in which the virus is maintained in a vector population independent of feeding on viremic animals [30]. Among hematophagous Diptera, the venereal transmission of viruses of human and veterinary importance has been observed with bunyaviruses [31–33], flaviviruses [34–41], rhabdoviruses [42,43], and togaviruses [44,45] in mosquitoes and sand flies. However, no previous studies have reported venereal transmission by *Culicoides* biting midges for any arbovirus, nor for any insect species with VSV.

In this study, we have demonstrated the presence of VSV RNA and infectious virus in previously uninfected midges of both sexes following cohabitation with VSV-infected mates. Our study also revealed the location of viral antigen in the reproductive tracts of both males and females. For the first time, VSV infection of female *C. sonorensis* midges has been shown to occur not only during blood feeding but also during copulation. Venereal transmission of VSV from orally infected females to naïve males and from venereally infected males to naïve females suggests the virus could be maintained in midge populations at a low threshold during inter-epidemic periods and then reinitiate an outbreak when conditions for bite transmission are once again ideal. The testing of heads from positive VNT females was used as an indication of both dissemination and transmission potential. Although virus detected in the heads would include infected neural and optic tissues, previous in situ hybridization staining of infected midges showing significant VSV replication in salivary gland epithelium, and immunohistochemical staining showing VSV in the lumen of salivary glands ready for excretion [10], strongly suggest positive midge heads correlate to bite transmission potential.

Venereal transmission of VSV from females to males was shown to occur at higher rates during the females third gonotrophic cycle (8 to 12 days post-infectious feeding). This particular timing might be explained by the addition of subsequent blood meals which induce midgut expansion, increasing the number of midgut basal lamina micro-perforations and enhancing the likelihood of virus dissemination [46] as well as the stretching of follicular epithelial cells that facilitate higher viral infection rates of reproductive tracts [36,47].

The VSV-positive staining in female reproductive tracts suggests the mechanism of sexual transmission from infected females to males occurs by transmission of virus particles located at the distal end of the female terminalia (gonopore and rectum) upon contact during copulation. The VSV-positive fluorescent puncta in male reproductive tracts suggests the mechanism of sexual transmission of VSV from infected males to females likely follows the sperm ejaculation pathway in which virus particles are released from the base of the testis into the accessory gland where the spermatozoa are mixed with ejaculatory secretions passing through the ejaculatory duct and are released into the female reproductive tract upon copulation.

It has been demonstrated in mosquitoes that males and copulation activities have a synergistic effect on transmission by influencing the female vectorial capacity [20,48]. To date, the role of male midges in the transmission dynamics of VSV or other midge-transmitted viruses has not been reported. Our results of venereal VSV transmission between males and females suggests a potentially important role for males in the natural survival and maintenance of interepidemic VSV. Although male midges are not hematophagous, our study shows they can acquire virus and become infected after copulation with infected females and then transmit VSV to naïve females during subsequent mating. From the ecological and epidemiological perspective, males not only contribute to the overall virus overwintering and transmission, but from our behavioral observations, they also increase the percentage of females that successfully blood feed.

A caveat to mating midges in cages requires the consideration that VSV could have been transmitted between the sexes not only during mating but also through other types of contact with salivary or anal secretions. To undoubtedly determine if infection occurred during the act of mating, we tested if VSV could be transmitted when an induced mating technique was used [49]. This way, only a brief contact of the genitalia would be allowed. Unfortunately, successful copulation was not achieved due to the complex *C. sonorensis* mating physiology and the long duration of attachment required for successful copulation, which is possibly due to spermatophore formation and transfer observed in other *Culicoides* species [23,24].

While there are limitations to confined laboratory experiments in induced mating, our research shows VSV midge-to-midge transmission after cohabitation with orally infected, microinjected, or venereally infected midges of the opposite sex. Our description of *C. sonorensis* mating behavior and the morphological descriptions of the internal reproductive systems of both sexes extends the knowledge of *Culicoides* midges and relates to previous studies on *Culicoides melleus* [23,24,50] and *Culicoides nubeculosus* [51]. This research shows the importance of males in VSV transmission dynamics and in the maintenance of VSV in nature. Additionally, the significant VSV-positive staining of female reproductive tissues suggests vertical transmission may also play a role in VSV maintenance. While further studies are needed to determine the effects of VSV vertical transmission, venereal transmission to oviposition, mating behavior, and mate choices of infected/uninfected midges, these results highlight the need to incorporate alternative routes of transmission in understanding arbovirus outbreaks.

4. Materials and Methods

4.1. Virus and Cells

The New Jersey serotype of VSV (1982 bovine field isolate) was grown in porcine epithelial cells (AG08113; Coriell Institute, Camden, NJ, USA)) in Eagles MEM with Earle's salts (Sigma, St. Louis, MO, USA) and 199E Media (2% FBS, 100U penicillin/streptomycin sulfate) at 37 °C with 5% CO_2. Vero MARU cells (VM; Middle America Research Unit, Panama, Panama) grown in 199E media at 37 °C with 5% CO_2 were used to detect and titer infectious virus in midge samples by standard plaque assay.

4.2. VSV Infection of Culicoides sonorensis Midges

Adult *C. sonorensis* midges used were from the AK colony maintained by USDA, Agricultural Research Service, Arthropod-Borne Animal Diseases Research Unit at the Center for Grain and Animal Health Research in Manhattan, KS, USA. Midges were reared as previously described [52]. Virgin female *C. sonorensis* midges (1–3 days post emergence) were allowed to feed on a glass, 37 °C water-jacketed bell feeder with a parafilm membrane/cage interface for 60 min. The VSV-blood meal consisted of defibrinated sheep blood (Lampire Biological Products, Pipersville, PA, USA) containing 4.25×10^8 PFU VSV-NJ. Fully engorged blood-fed females were sorted from unfed and partially fed females and placed in cardboard maintenance cages. For positive controls, 1–3 day post emergence virgin midges were anesthetized with CO_2, intrathoracically inoculated with VSV-NJ, and placed in maintenance cages. Intrathoracic injections were made dorsally at the prescutellar area using a volume of 46 nl (1.4×10^4 PFU) for males and 60 nl (1.8×10^4 PFU) for females using a Nanoject II injector (Drummond Scientific Company, Broomall, PA, USA). Injected volumes were determined as the maximum capacity for the male and female body size. Adult midges were maintained in environmental chambers at 25 ± 1 °C and 80% relative humidity with a 13:11 light: dark cycle and offered 10% sucrose solution ad libitum.

4.3. Venereal Transmission Assays

Venereal transmission of VSV from infected females to naïve males was tested (Figure 1A) for each of the three-blood meal-induced gonotrophic cycles (1–3GC). VSV-blood-fed virgin females were placed in cages with age-matched naïve males at a ratio of 2:1 females to males. Four days post-mating (dpm), all surviving males were individually collected in 300 μL of TRIzol (Invitrogen; Thermo Fisher Scientific, Inc., Waltham, MA, USA) for RT-qPCR testing, or in 500 μL of antibiotic medium (199E cell culture medium, 200 U/mL penicillin, 200 μg/mL streptomycin, 100 μg/mL gentamycin, and 5 μg/mL amphotericin B) for plaque assays and CPE. Collected midges were stored at −80 °C until further processing. Surviving females were moved to new cages and age-matched naïve males were again added at a ratio of 2:1. Midges were offered non-infectious blood meals for 60 min at the start of each cohabitation to initiate a gonotrophic cycle.

To test venereal transmission of VSV from infected males to naïve females (Figure 1B), four days post intrathoracic injection (dpi), males were transferred to a cage containing age-matched naïve females in a ratio of 2:1 females to males. Midges were offered a non-infectious blood meal for 60 min to initiate a gonotrophic cycle. Following three days of cohabitation, all surviving males were collected as above, and all surviving females were transferred to a new cage and kept for an additional four days. Seven days after cohabitation/mating (7 dpm), all surviving females were collected as above. To determine the VSV titer in males during the cohabitation period, a subsample of the infected males, 4 dpi and 7 dpi, was tested for virus by plaque assay.

To test venereal transmission of VSV from venereally infected males to naïve females (Figure 1C), we used surviving males at the end of the cohabitation period with orally infected females at the third blood meal-induced gonotrophic cycle. Venereally infected males were transferred to a cage containing naïve females in a ratio of 2:1 females to males. Midges were offered a non-infectious blood meal for 60 min to initiate a gonotrophic cycle. Following three days of cohabitation, all surviving males were collected as above, and all surviving females were transferred to a new cage and kept for an additional four days. Seven days after initial cohabitation, all surviving females were individually collected in 300 µL of TRIzol for RT-qPCR testing.

4.4. RNA Extraction and RT-qPCR for VSV Detection

Frozen TRIzol midge samples were thawed on ice and homogenized by high-speed shaking with a Bead Mill Homogenizer (Omni, Kennesaw, GA, USA) for 2 min at 3.1 m/s. Samples were centrifuged at 12,000× g for 6 min to pellet the debris. Total RNA was extracted using Trizol-BCP (1-bromo-3chloropropane; ThermoFisher Life Technologies, Waltham, MA). RNA was precipitated using isopropanol, washed in 75% ethanol, and eluted in 50 µL of nuclease-free water. RNA extracts were analyzed using TaqMan Fast Virus 1-Step MasterMix (Applied Biosystems; Thermo Fisher Scientific, Inc., Waltham, MA, USA) in a RT-qPCR targeting the L segment [53]: forward primer VSVNJ7274: 5'-TGATTCAATATAATTATTTTGGGAC-3; reverse primer VSVNJ7495: 5'-AGG CTCAGAGGCATGTTCAT-3'; probe: FAM-TTGCACACCAGAACATTCAA-3'-BHQ1. For amplification, the following temperature profile was used: Reverse-transcription 1 cycle at 50 °C for 5 min, denaturing and polymerase activation at 95 °C for 20 s, and amplification: 40 cycles of 95 °C for 15 s and 60 °C for 60 s. Samples were initially tested in pools containing RNA of 5 individual midges followed by testing of individual samples from positive pools to determine the exact number of positive individuals. Based on Ct values reported for VNT of dengue, Zika, and chikungunya viruses in mosquitoes [36,37,41,45], RT-qPCR reactions with Ct ≤ 36 were considered positive for VSV RNA. Additionally, to determine if virus acquired by venereal transmission could disseminate into salivary glands of females, pools of bodies and heads (containing the proximal region of the salivary glands) were assayed separately. Subsequent testing of individual bodies and heads from positive pools was conducted to determine the exact number of positive samples.

Standard curves and calculation of Ct values were carried out with the 7500 Fast Dx software (Applied Biosystems; Thermo Fisher Scientific, Inc., Waltham, MA, USA). Ct values were plotted against the log of VSV genome equivalents. The linear regression (y = −3.30578x + 11.02683) was used to determine the amount of viral genomic ssRNA per midge. Genome equivalents were calculated with the published VSV genome molecular weight [54] and the NEBioCalulator (https://nebiocalculator.neb.com/#!/ssrnaamt).

4.5. Cytopathic Effect and Plaque Assays

To isolate infectious virus, frozen midges stored in 500 µL antibiotic media were thawed on ice and individually homogenized as above. Samples were centrifuged at 12,000× g for 6 min to pellet debris. A subsample of cleared supernatant (200 µL) was added to a monolayer of VM cells in a 24-well plate and incubated at 37 °C for seven days at 5% CO_2. Observation of cytopathic effects (CPE) after one or two passages were used as an indicator of infectious virus within that sample. All CPE+ wells were confirmed as VSV+ by testing for viral RNA by RT-qPCR, as described above. All homogenates with positive CPE at the first passage were further analyzed to determine infectious virus titer by standard plaque assay inoculating 200 µL of the original cleared supernatant sample on VM cells in 6-well plates and incubating at 37 °C for three days at 5% CO_2.

4.6. Statistical Analysis

Data were pooled from independent replicates of each experiment. Statistical methods were not used to predetermine the sample size. The proportion of infected midges was calculated by dividing the number of infected whole midges by the total number of midges tested. A female with virus found in the body but not in the head was considered as a non-disseminated infection. When the virus was found in both the body and the head, the midge was determined to have a disseminated infection. The proportion of females with disseminated infection was calculated as the number of midges with positive heads divided by the total number of infected midges. Non-parametric tests were used to compare Ct distributions between gonotrophic cycles. GraphPad Prism version 8 (GraphPad Software Inc., La Jolla, CA, USA) was used for statistical analysis and creation of graphs.

4.7. Fluorescent Immune Assay and Immunohistochemistry

To determine VSV localization in the reproductive tissues of infected *C. sonorensis* midges that might allow VNT, intrathoracically injected males and females (4 dpi) were CO_2 anesthetized and reproductive tracts were dissected in PBST (PBS + 0.5% of Triton X 100; PH 7.4). Tissues were fixed in 4% paraformaldehyde for 4 h and washed in PBST. All reproductive tracts were blocked in 1% Normal Goat Serum for one hour and sequentially incubated with rabbit anti-VSV-NJ nucleocapsid protein antibody (dilution 1:500 or 1:1000) at room temperature (RT). After a 1-day incubation, the reproductive tracts were washed in PBST for 4 h. Binding of primary antibodies was detected by incubating tissues with 1:300 or 1:500 dilution of Alexa Fluor 488 IgG Alpaca anti-Rabbit (Jackson Immuno Research, West Grove, PA, USA). Following a 4 h incubation in the dark at RT, samples were washed in PBST for 4 h. Cell nuclei were stained with 10 ug/mL of DAPI (40,6-diamidino-2-phenylindole) (Invitrogen; Thermo Fisher Scientific, Inc., Waltham, MA, USA). Reproductive tracts were mounted on slides using 100% glycerol and examined in confocal microscope LSM700 (Zeiss International, Oberkochen, Germany). Images were captured using ZEN software (Zeiss International, Oberkochen, Germany). Reproductive tracts of non-injected midges were treated similarly and served as negative controls.

To better determine the virus localization in the reproductive organs of orally infected *C. sonorensis* females, sagittal sections from sequentially sampled VSV-fed females (9–13 dpf) from a previous study [10] were examined and captured using an All-in-One Fluorescence Microscope (BZ-X810; Keyence Corporation, Itasca, IL, USA).

4.8. Behavioral Observations of C. sonorensis Copulation

All assays were conducted in round carton cages of 9 cm diameter and 5 cm height. Observations of mating were made using a Nikon SMZ-1500 binocular stereo zoom microscope (Nikon Instruments Inc., Melville, NY, USA). All observations were carried out under laboratory conditions (25 ± 2 °C, 65% RH).

Author Contributions: Conceptualization, P.R.-L. and B.S.D.; data curation, P.R.-L.; formal analysis, P.R.-L., B.L.-R. and B.S.D.; funding acquisition, B.L.-R. and B.S.D.; investigation, P.R.-L.; methodology, P.R.-L. and B.S.D.; project administration, B.S.D.; resources, B.S.D.; supervision, B.L.-R. and B.S.D.; validation, P.R.-L. and B.S.D.; visualization, P.R.-L. and B.S.D.; writing—original draft, P.R.-L.; writing—review and editing, P.R.-L., B.L.-R. and B.S.D. All authors have read and agreed to the published version of the manuscript.

Funding: Funding was provided by USDA, ARS-Kansas State University Cooperative Agreement #58-3020-7-025 funded by the USDA, ARS, NP103 Animal Health National Program, Project Numbers 3020-32000-010 and 3020-32000-013.

Acknowledgments: We thank Yoonseong Park and the course of Insect Anatomy and Histology taught at the Department of Entomology at Kansas State University for providing theoretical and practical knowledge about whole organs fluorescent immune assay. We also thank L. Reister-Hendricks, D. Jasperson, D. Swanson, and W. Yarnell, at the Arthropod-Borne Animal Diseases Research Unit, USDA, ARS, Manhattan, KS, for their technical assistance.

Conflicts of Interest: The authors declare no conflict of interest with respect to the authorship or the publication of this article.

References

1. Letchworth, G.J.; Rodriguez, L.L.; Del cbarrera, J. Vesicular stomatitis. *Vet. J.* **1999**, *157*, 239–260. [CrossRef]
2. Hanson, R.P.; Estupinan, J.; Castaneda, J. Vesicular stomatitis in the Americas. *Bull. Off. Int. Epizoot.* **1968**, *70*, 37–47.
3. Rodriguez, L.L. Emergence and re-emergence of vesicular stomatitis in the United States. *Virus Res.* **2002**, *85*, 211–219. [CrossRef]

4. USDA-APHIS. Vesicular Stomatitis 2014–2015. FINAL Situation Report—March 13. 2015. Available online: https://www.aphis.usda.gov/animal_health/downloads/animal_diseases/vsv/Sitrep_031315.pdf (accessed on 21 October 2018).

5. Patterson, W.C.; Jenney, E.W.; Holbrook, A.A. Experimental Infections with vesicular stomatitis in swine I. Transmission by direct contact and feeding infected meat scraps. *U. S. Livest. Sanit. Assoc. Proc.* **1955**, *59*, 368–378.

6. Rozo-Lopez, P.; Drolet, B.S.; Londono-Renteria, B. Vesicular Stomatitis Virus Transmission: A Comparison of Incriminated Vectors. *Insects* **2018**, *9*, 190. [CrossRef] [PubMed]

7. Brandly, C.A.; Hanson, R.P. Epizootiology of vesicular stomatitis. *Am. J. Public Health Nations Health* **1957**, *47*, 205–209. [CrossRef] [PubMed]

8. Holbrook, F.R.; Tabachnick, W.J.; Schmidtmann, E.T.; McKinnon, C.N.; Bobian, R.J.; Grogan, W.L. Sympatry in the *Culicoides variipennis* complex (Diptera: Ceratopogonidae): A taxonomic reassessment. *J. Med. Entomol.* **2000**, *37*, 65–76. [CrossRef] [PubMed]

9. McGregor, B.L.; Stenn, T.; Sayler, K.A.; Blosser, E.M.; Blackburn, J.K.; Wisely, S.M.; Burkett-Cadena, N.D. Host use patterns of *Culicoides* spp. biting midges at a big game preserve in Florida, U.S.A., and implications for the transmission of orbiviruses. *Med. Vet. Entomol.* **2018**. [CrossRef]

10. Drolet, B.S.; Campbell, C.L.; Stuart, M.A.; Wilson, W.C. Vector competence of *Culicoides sonorensis* (Diptera: Ceratopogonidae) for vesicular stomatitis virus. *J. Med. Entomol.* **2005**, *42*, 409–418. [CrossRef]

11. Nunamaker, R.A.; Perez de Leon, A.A.; Campbell, C.C.; Lonning, S.M. Oral infection of *Culicoides sonorensis* (Diptera: Ceratopogonidae) by vesicular stomatitis virus. *J. Med. Entomol.* **2000**, *37*, 784–786. [CrossRef]

12. Perez De Leon, A.A.; O'Toole, D.; Tabachnick, W.J. Infection of guinea pigs with vesicular stomatitis New Jersey virus Transmitted by *Culicoides sonorensis* (Diptera: Ceratopogonidae). *J. Med. Entomol.* **2006**, *43*, 568–573. [CrossRef]

13. Perez de Leon, A.A.; Tabachnick, W.J. Transmission of vesicular stomatitis New Jersey virus to cattle by the biting midge *Culicoides sonorensis* (Diptera: Ceratopogonidae). *J. Med. Entomol.* **2006**, *43*, 323–329. [CrossRef]

14. Walton, T.E.; Webb, P.A.; Kramer, W.L.; Smith, G.C.; Davis, T.; Holbrook, F.R.; Moore, C.G.; Schiefer, T.J.; Jones, R.H.; Janney, G.C. Epizootic vesicular stomatitis in Colorado, 1982: Epidemiologic and entomologic studies. *Am. J. Trop. Med. Hyg.* **1987**, *36*, 166–176. [CrossRef] [PubMed]

15. Kramer, W.L.; Jones, R.H.; Holbrook, F.R.; Walton, T.E.; Calisher, C.H. Isolation of arboviruses from *Culicoides* midges (Diptera: Ceratopogonidae) in Colorado during an epizootic of vesicular stomatitis New Jersey. *J. Med. Entomol.* **1990**, *27*, 487–493. [CrossRef]

16. Thurmond, M.C.; Ardans, A.A.; Picanso, J.P.; McDowell, T.; Reynolds, B.; Saito, J. Vesicular stomatitis virus (New Jersey strain) infection in two California dairy herds: An epidemiologic study. *J. Am. Vet. Med. Assoc.* **1987**, *191*, 965–970.

17. Reis, J.L., Jr.; Rodriguez, L.L.; Mead, D.G.; Smoliga, G.; Brown, C.C. Lesion development and replication kinetics during early infection in cattle inoculated with Vesicular stomatitis New Jersey virus via scarification and black fly (*Simulium vittatum*) bite. *Vet. Pathol.* **2011**, *48*, 547–557. [CrossRef]

18. Redelman, D.; Nichol, S.; Klieforth, R.; Van Der Maaten, M.; Whetstone, C. Experimental vesicular stomatitis virus infection of swine: Extent of infection and immunological response. *Vet. Immunol. Immunopathol.* **1989**, *20*, 345–361. [CrossRef]

19. Perez de Leon, A.A.; Lloyd, J.E.; Tabachnick, W.J. Sexual dimorphism and developmental change of the salivary glands in adult *Culicoides variipennis* (Diptera: Ceratopogonidae). *J. Med. Entomol.* **1994**, *31*, 898–902. [CrossRef]

20. Dahalan, F.A.; Churcher, T.S.; Windbichler, N.; Lawniczak, M.K.N. The male mosquito contribution towards malaria transmission: Mating influences the *Anopheles* female midgut transcriptome and increases female susceptibility to human malaria parasites. *PLoS Pathog* **2019**, *15*, e1008063. [CrossRef]

21. Mitchell, S.N.; Kakani, E.G.; South, A.; Howell, P.I.; Waterhouse, R.M.; Catteruccia, F. Mosquito biology. Evolution of sexual traits influencing vectorial capacity in anopheline mosquitoes. *Science* **2015**, *347*, 985–988. [CrossRef]

22. Wirth, W.W.; Blanton, F.S. *The West Indian Sandflies of the Gunus Culicoides (Diptera: Ceratopogonidae)*; U.S. Government Printing Office: Washington, DC, USA, 1974.

23. Linley, J.R. Ejaculation and spermatophore formation in *Culicoides melleus* (Coq.) (Diptera: Ceratopogonidae). *Can. J. Zool.* **1981**, *59*, 332–346. [CrossRef]

24. Linley, J.R. Emptying of the spermatophore and spermathecal filling in *Culicoides melleus* (Coq.) (Diptera: Ceratopogonidae). *Can. J. Zool.* **1981**, *59*, 347–356. [CrossRef]

25. Meuti, M.E.; Short, S.M. Physiological and Environmental Factors Affecting the Composition of the Ejaculate in Mosquitoes and Other Insects. *Insects* **2019**, *10*, 74. [CrossRef]

26. Knell, R.J.; Webberley, K.M. Sexually transmitted diseases of insects: Distribution, evolution, ecology and host behaviour. *Biol. Rev. Camb. Philos. Soc.* **2004**, *79*, 557–581. [CrossRef]

27. Perez, A.M.; Pauszek, S.J.; Jimenez, D.; Kelley, W.N.; Whedbee, Z.; Rodriguez, L.L. Spatial and phylogenetic analysis of vesicular stomatitis virus over-wintering in the United States. *Prev. Vet. Med.* **2010**, *93*, 258–264. [CrossRef] [PubMed]

28. Mesquita, L.P.; Diaz, M.H.; Howerth, E.W.; Stallknecht, D.E.; Noblet, R.; Gray, E.W.; Mead, D.G. Pathogenesis of Vesicular Stomatitis New Jersey Virus Infection in Deer Mice (*Peromyscus maniculatus*) Transmitted by Black Flies (*Simulium vittatum*). *Vet. Pathol.* **2017**, *54*, 74–81. [CrossRef] [PubMed]

29. Cornish, T.E.; Stallknecht, D.E.; Brown, C.C.; Seal, B.S.; Howerth, E.W. Pathogenesis of experimental vesicular stomatitis virus (New Jersey serotype) infection in the deer mouse (Peromyscus maniculatus). *Vet. Pathol.* **2001**, *38*, 396–406. [CrossRef] [PubMed]

30. Agarwal, A.; Parida, M.; Dash, P.K. Impact of transmission cycles and vector competence on global expansion and emergence of arboviruses. *Rev. Med. Virol.* **2017**. [CrossRef]

31. Thompson, W.H.; Beaty, B.J. Venereal transmission of La Crosse (California encephalitis) arbovirus in *Aedes triseriatus* mosquitoes. *Science* **1977**, *196*, 530–531. [CrossRef]

32. Thompson, W.H.; Beaty, B.J. Venereal transmission of La Crosse virus from male to female *Aedes triseriatus*. *Am. J. Trop. Med. Hyg.* **1978**, *27*, 187–196. [CrossRef]

33. Thompson, W.H. Higher venereal infection and transmission rates with La Crosse virus in *Aedes triseriatus* engorged before mating. *Am. J. Trop. Med. Hyg.* **1979**, *28*, 890–896. [CrossRef] [PubMed]

34. Mourya, D.T.; Soman, R.S. Venereal transmission of Japanese encephalitis virus in *Culex bitaeniorhynchus* mosquitoes. *Indian J. Med. Res.* **1999**, *109*, 202–203. [PubMed]

35. Rosen, L.; Shroyer, D.A.; Tesh, R.B.; Freier, J.E.; Lien, J.C. Transovarial transmission of dengue viruses by mosquitoes: *Aedes albopictus* and *Aedes aegypti*. *Am. J. Trop. Med. Hyg.* **1983**, *32*, 1108–1119. [CrossRef] [PubMed]

36. Sanchez-Vargas, I.; Harrington, L.C.; Doty, J.B.; Black, W.C.t.; Olson, K.E. Demonstration of efficient vertical and venereal transmission of dengue virus type-2 in a genetically diverse laboratory strain of *Aedes aegypti*. *PLoS Negl. Trop. Dis.* **2018**, *12*, e0006754. [CrossRef]

37. Rosen, L. Sexual transmission of dengue viruses by Aedes albopictus. *Am. J. Trop. Med. Hyg.* **1987**, *37*, 398–402.

38. Shroyer, D.A. Venereal transmission of St. Louis encephalitis virus by *Culex quinquefasciatus* males (Diptera: Culicidae). *J. Med. Entomol.* **1990**, *27*, 334–337. [CrossRef]

39. Pereira-Silva, J.W.; Nascimento, V.A.D.; Belchior, H.C.M.; Almeida, J.F.; Pessoa, F.A.C.; Naveca, F.G.; Rios-Velasquez, C.M. First evidence of Zika virus venereal transmission in *Aedes aegypti* mosquitoes. *Mem. Inst. Oswaldo Cruz.* **2018**, *113*, 56–61. [CrossRef]

40. Kerr, J.A.; Hayne, T.B. On the Transfer of Yellow Fever Virus from Female to Male *Aedes aegypti* 1. *Am. J. Trop. Med. Hyg.* **1932**, *s1-12*, 255–261. [CrossRef]

41. Campos, S.S.; Fernandes, R.S.; Dos Santos, A.A.C.; de Miranda, R.M.; Telleria, E.L.; Ferreira-de-Brito, A.; de Castro, M.G.; Failloux, A.B.; Bonaldo, M.C.; Lourenco-de-Oliveira, R. Zika virus can be venereally transmitted between *Aedes aegypti* mosquitoes. *Parasit. Vectors* **2017**, *10*, 605. [CrossRef]

42. Mavale, M.S.; Fulmali, P.V.; Geevarghese, G.; Arankalle, V.A.; Ghodke, Y.S.; Kanojia, P.C.; Mishra, A.C. Venereal transmission of Chandipura virus by Phlebotomus papatasi (Scopoli). *Am. J. Trop. Med. Hyg.* **2006**, *75*, 1151–1152. [CrossRef]

43. Mavale, M.S.; Geevarghese, G.; Ghodke, Y.S.; Fulmali, P.V.; Singh, A.; Mishra, A.C. Vertical and venereal transmission of Chandipura virus (Rhabdoviridae) by *Aedes aegypti* (Diptera: Culicidae). *J. Med. Entomol.* **2005**, *42*, 909–911. [CrossRef] [PubMed]

44. Ovenden, J.R.; Mahon, R.J. Venereal transmission of Sindbis virus between individuals of *Aedes australis* (Diptera: Culicidae). *J. Med. Entomol.* **1984**, *21*, 292–295. [CrossRef] [PubMed]

45. Mavale, M.; Parashar, D.; Sudeep, A.; Gokhale, M.; Ghodke, Y.; Geevarghese, G.; Arankalle, V.; Mishra, A.C. Venereal transmission of chikungunya virus by *Aedes aegypti* mosquitoes (Diptera: Culicidae). *Am. J. Trop. Med. Hyg.* **2010**, *83*, 1242–1244. [CrossRef] [PubMed]

46. Armstrong, P.M.; Ehrlich, H.Y.; Magalhaes, T.; Miller, M.R.; Conway, P.J.; Bransfield, A.; Misencik, M.J.; Gloria-Soria, A.; Warren, J.L.; Andreadis, T.G.; et al. Successive blood meals enhance virus dissemination within mosquitoes and increase transmission potential. *Nat. Microbiol.* **2020**, *5*, 239–247. [CrossRef] [PubMed]

47. Agarwal, A.; Dash, P.K.; Singh, A.K.; Sharma, S.; Gopalan, N.; Rao, P.V.; Parida, M.M.; Reiter, P. Evidence of experimental vertical transmission of emerging novel ECSA genotype of Chikungunya Virus in *Aedes aegypti*. *PLoS Negl. Trop. Dis.* **2014**, *8*, e2990. [CrossRef] [PubMed]

48. Goodger, W.J.; Thurmond, M.; Nehay, J.; Mitchell, J.; Smith, P. Economic impact of an epizootic of bovine vesicular stomatitis in California. *J. Am. Vet. Med. Assoc.* **1985**, *186*, 370–373.

49. Baker, R.H.; French, W.L.; Kitzmiller, J.B. Induced copulation in Anopheles mosquitoes. *Mosq. News* **1962**, *22*, 16–17.

50. Linley, J.R.; Adams, G.M. A study of the mating behaviour of *Culicoides melleus* (Coquillett) (Diptera: Ceratopogonidae). *Trans. R. Entomol. Soc. Lond.* **1972**, *124*, 81–121. [CrossRef]

51. Mair, J.; Blackwell, A. Mating behavior of *Culicoides nubeculosus* (Diptera:Ceratopogonidae). *J. Med. Entomol.* **1996**, *33*, 856–858. [CrossRef] [PubMed]

52. Jones, R.H.; Foster, N.M. Oral infection of *Culicoides variipennis* with bluetongue virus: Development of susceptible and resistant lines from a colony population. *J. Med. Entomol.* **1974**, *11*, 316–323. [CrossRef] [PubMed]

53. Hole, K.; Velazquez-Salinas, L.; Clavijo, A. Improvement and optimization of a multiplex real-time reverse transcription polymerase chain reaction assay for the detection and typing of Vesicular stomatitis virus. *J. Vet. Diagn. Investig.* **2010**, *22*, 428–433. [CrossRef] [PubMed]

54. Ball, L.A.; White, C.N. Order of transcription of genes of vesicular stomatitis virus. *Proc. Natl. Acad. Sci. USA* **1976**, *73*, 442–446. [CrossRef] [PubMed]

Article

Entomological Surveillance for Zika and Dengue Virus in *Aedes* Mosquitoes: Implications for Vector Control in Thailand

Nathamon Kosoltanapiwat [1], Jarinee Tongshoob [1], Preeraya Singkhaimuk [2], Chanyapat Nitatsukprasert [2], Silas A. Davidson [2,3] and Alongkot Ponlawat [2,*]

[1] Department of Microbiology and Immunology, Faculty of Tropical Medicine, Mahidol University, Bangkok 10400, Thailand; nathamon.kos@mahidol.ac.th (N.K.); jarinee.pan@mahidol.ac.th (J.T.)
[2] Department of Entomology, Armed Forces Research Institute of Medical Sciences (AFRIMS), Bangkok 10400, Thailand; preerayas.ca@afrims.org (P.S.); chanyapat.nit@gmail.com (C.N.); silas.davidson@westpoint.edu (S.A.D.)
[3] Department of Chemistry and Life Science, United States Military Academy West Point, West Point, NY 10996, USA
*   Correspondence: alongkotp.fsn@afrims.org; Tel.: +66-2-696-2816

Received: 15 May 2020; Accepted: 3 June 2020; Published: 4 June 2020

Abstract: Entomological surveillance for arthropod-borne viruses is vital for monitoring vector-borne diseases and informing vector control programs. In this study, we conducted entomological surveillance in Zika virus endemic areas. In Thailand, it is standard protocol to perform mosquito control within 24 h of a reported dengue case. *Aedes* females were collected within 72 h of case reports from villages with recent Zika–human cases in Kamphaeng Phet Province, Thailand in 2017 and 2018. Mosquitoes were bisected into head-thorax and abdomen and then screened for Zika (ZIKV) and dengue (DENV) viruses using real-time RT-PCR. ZIKV RNA was detected in three samples from two female *Ae. aegypti* (1.4%). A partial envelope sequence analysis revealed that the ZIKV sequences were the Asian lineage identical to sequences from ZIKV-infected cases reported in Thailand during 2016 and 2017. Dengue virus-1 (DENV-1) and dengue virus-4 (DENV-4) were found in four *Ae. aegypti* females (2.8%), and partial capsid sequences were nearly identical with DENV-1 and DENV-4 from Thai human cases reported in 2017. Findings in the current study demonstrate the importance of entomological surveillance programs to public health mosquito-borne disease prevention measures and control.

Keywords: zika virus; dengue virus; *Aedes aegypti*; mosquito surveillance; Thailand

1. Introduction

Mosquitoes are vectors of various flaviviruses, which are enveloped, positive-sense single stranded RNA viruses (Family *Flaviviridae*, Genus *Flavivirus*). The most important mosquito transmitted flaviviruses are the dengue (DENV), Zika (ZIKV), yellow fever (YFV), West Nile (WNV) and Japanese encephalitis (JEV) viruses. Among them, DENV, ZIKV and JEV circulate widely in Southeast Asia, including Thailand [1–3]. DENV and ZIKV are primarily transmitted to humans by *Aedes* mosquitoes, with *Aedes (Stegomyia) aegypti* serving as the major vector and *Aedes (Stegomyia) albopictus* as a secondary vector; whereas JEV and WNV are predominantly transmitted by *Culex* mosquitoes [1,4,5]. Both dengue and Zika viral infections in humans range from asymptomatic to symptoms of fever, headache, myalgia, arthralgia and maculopapular rash at the initial stage of illnesses [2,6]. DENV infections can be classified as dengue fever (DF), dengue hemorrhagic fever (DHF) or dengue shock syndrome (DSS) according to the severity of symptoms. DHF and DSS can be fatal if appropriate treatments are not provided, making the disease a significant public health concern, especially in endemic areas. Due to the self-limiting nature of the illness and low mortality rates, infection with ZIKV has generally caused less of a significant public health impact than dengue, since its first identification. Although typically milder than infection with DENV, there are two severe ZIKV disease complications of note: Guillain–Barre syndrome and meningoencephalitis. After the Brazilian ZIKV outbreak in 2015, public health scientist noted Guillain–Barre syndrome and a drastic increase in the number of infants with microcephaly born to mothers with the ZIKV infection meningoencephalitis, causing widespread alarm, especially among pregnant women [7–9].

DENV has been postulated to have emerged about 1000 years ago in an infectious cycle involving non-human primates and mosquitoes [10]. The first human case of DENV (DENV-1) was isolated in Japan in 1943 [11]. Afterwards, outbreaks of four DENV serotypes (DENV-1, DENV-2, DENV-3 and DENV-4) have been reported globally [12]. As reported by the World Health Organization (WHO), the global incidence of dengue has increased 30-fold during the past five decades, with 50–100 million new infections estimated to occur annually in more than 100 endemic countries. These infections generally occur in tropical and sub-tropical regions [13]. Although more recently, DENV has been reported from previously unaffected, more temperate areas such as France, Croatia, Portugal and other European countries [1]. DENV incidence in Thailand has been reported since the 1950s [12] and several major outbreaks with high morbidity have been documented since then. According to the WHO, during 2004–2010, Thailand was ranked sixth among the 30 most highly dengue-endemic countries in the world [13]. One of the largest DENV epidemics in Thailand was in 1987 with 174,285 cases and 1008 deaths. Recently, more than 72,000 dengue cases with a fatality rate of 0.13% were reported in 2018 to the Ministry of Public Health, Thailand. All four DENV serotypes (DENV-1, DENV-2, DENV-3 and DENV-4) are reported to circulate in Thailand with a recurrent circulation of each serotype [12,14,15]. The Department of Disease Control—part of the Ministry of Public Health (MOPH), Thailand—reported that the major DENV serotypes causing illness in 2017 were DENV-1 and DENV-4. In 2018, DENV-1 remained the dominant serotype and the incidence of DENV-4 decreased.

ZIKV was first isolated from a sentinel rhesus monkey in the Zika forest, Uganda in 1947 and later from a pool of *Ae. africanus* mosquitoes collected in the same forest [16]. Afterwards, serological evidence of ZIKV infections in humans was reported for various African and Asian countries [6,17]. In 1975, Moore et al. reported the first ZIKV isolation from a patient with febrile illness from Nigeria, as well as the isolation of the virus from specimens collected in 1968 [18]. In Southeast Asia, ZIKV circulation was suspected as early as 1954, due to reactivity in a seroprevalence study [5]. Starting in 2009, active ZIKV circulation has been identified in Cambodia [19], Indonesia [20], Malaysia [21], the Philippines [22], Singapore [23], Thailand [2], and Vietnam [24]. The first report of ZIKV in Thailand was in 2013 from reports of infections in Canadian, German and Japanese travelers returning from the Southern part of Thailand [25–27]. Later, a retrospective investigation by the Thai MOPH confirmed autochthonous ZIKV cases in several different provinces across Thailand during 2012–2014, indicating wide-spread distribution of ZIKV in the country [2]. ZIKV isolated from this study was from the Asian lineage and is closely related to an isolate from the 2013 French Polynesia outbreak [2]. According to the Bureau of Emerging Infectious Diseases, Department of Disease Control, Thai MOPH, in 2016 and 2017, more than 1600 confirmed ZIKV-infected cases were reported from different regions across the central, north, east and west of Thailand [28].

Although many human cases of ZIKV and DENV infections have been reported in Thailand, there are few reports of ZIKV circulation in wild-caught mosquitoes [29]. In this study, entomological surveillance for arthropod-borne viruses was conducted in Kampheang Phet (KPP) Province, Thailand, a dengue-endemic area where human Zika cases had been recently reported. *Aedes* mosquitoes were screened for ZIKV and DENV and the virus sequences were analyzed. Findings from this study can be used as the critical first step toward developing a routine entomological surveillance and vector control program in Thailand.

2. Results

2.1. Mosquito Collection

Over 86 trap-days, 488 mosquitoes belonging to four genera, *Aedes*, *Culex*, *Armigeres* and *Anopheles*, were captured (Table 1). The majority (40.8%) of mosquitoes collected inside houses belonged to the genus *Aedes* with 187 *Ae. aegypti* (130 females and 57 males) and 12 *Ae. albopictus* (11 females and one male).

Table 1. Total number of collected mosquitoes in Kampheang Phet between July 2017 and February 2018.

Species	Female	Male	Total
Ae. aegypti	130	57	187
Ae. albopictus	11	1	12
An. peditaeniatus	1	-	1
An. tessellatus	1	-	1
An. sp.	1	-	1
Ar. subalbatus	4	-	4
Cx. brevipalpis	3	-	3
Cx. quinquefasciatus	62	123	185
Cx. vishnui	88	5	93
Cx. sp	1	-	1

2.2. Virus Detection

A total of 282 samples from 141 *Aedes* females (130 *Ae. aegypti* and 11 *Ae. albopictus*) were screened for ZIKV and DENV. Each female mosquito was separated in two parts (head-thorax and abdomen) to determine virus dissemination. Both samples were tested from each mosquito. ZIKV and DENV RNA was detected only in *Ae. aegypti* collected from houses that had recent human cases of ZIKV. No ZIKV or DENV RNA was detected in *Ae. aegypti* from surveyed households within a 50 m radius from detected human cases. A TaqMan® real-time RT-PCR and ZDC multiplex real-time RT-PCR assay detected a ZIKV genome in three samples (R012, R013 and R015) from two *Ae. aegypti* females (1.4%, 2/141). A partial ZIKV envelope gene was amplified from two samples from the same mosquito (R012, head-thorax and R013, abdomen) by a conventional RT-PCR. DNA sequencing was performed and analyzed by a BLAST search; sequences were identified as ZIKV and shared 100% sequence identity. Four *Aedes* females (2.8%, 4/141) were DENV positive by real-time RT-PCR. Analysis of partial capsid/prM sequences identified three DENV-1 (R248, R251 and R290) and one DENV-4 (R097). The DNA sequences obtained from this study were submitted to the NCBI database and the accession numbers were assigned as MN527290 for R012 (ZIKV), and MN527286, MN527287, MN527288 and MN527289 for R097, R248, R251 and R290 (DENV), respectively.

2.3. Phylogenetic Relationship of ZIKV

A partial envelope sequence of R012 (MN527290) was compared with other ZIKV strains using multiple sequence alignment. The genetic relationship was analyzed by construction of a phylogenetic tree (Figure 1). The result showed that the R012 sequence belonged to the Asian lineage ZIKV clade (nucleotide sequence identity ≥ 96%) and the alignment is shown in Supplementary Figure S1. In addition, the E gene sequence was 100% identical to the sequence from a ZIKV detected in a citizen from Thailand, reported in 2016 (MG548660.1), from patient serum collected in 2017 from Thailand (MK237998.1), and the sequence from a ZIKV-infected Japanese tourist, who visited Thailand in 2017 (LC369584.1) [30].

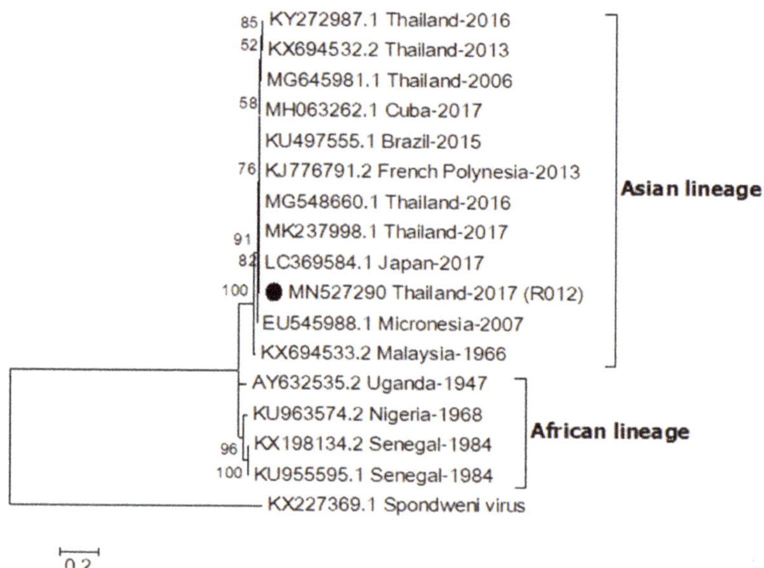

Figure 1. Maximum likelihood phylogenetic tree of Zika virus (ZIKV) partial envelope gene (331 bp). The E gene sequences of ZIKV (R012) found in this study are indicated with black circles. Accession numbers of all reference sequences are provided in the phylogenetic tree. Bootstrap values > 50% are indicated at nodes. A scale bar represents nucleotide substitutions per site. A sequence of Spondweni virus was used as an outlier.

2.4. Phylogenetic Relationship of DENV

One sequence from an *Ae. aegypti* collected in 2017 (MN527286) was identified as DENV-4 based on sequence analysis of a partial DENV capsid/prM (nucleotide sequence identity > 98%), whereas the other three sequences from *Ae. aegypti* females caught in 2018 (MN527287-MN527289) were DENV-1 (nucleotide sequence identity > 96%, Figure 2). They were closely related to DENV-1 (LC410183.1) and DENV-4 (LC410203.1) detected in patients from Thailand reported in 2017 [14]. The sequence alignment is shown in Supplementary Figure S2.

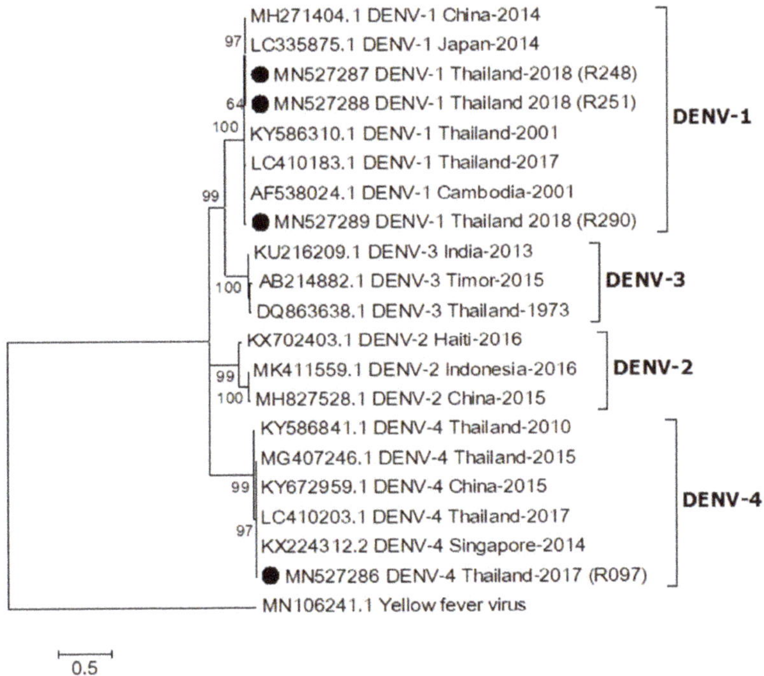

Figure 2. Maximum likelihood phylogenetic tree of dengue virus (DENV) partial capsid/prM gene (395 bp). All DENV capsid/prM sequences found in this study were indicated with black circles. All accession numbers of reference sequences are provided in the phylogenetic tree. Bootstrap values > 50 are indicated at nodes. A scale bar represents nucleotide substitutions per site. Yellow fever virus was used as an outlier.

3. Discussion

Mosquito-based ZIKV and DENV surveillance in KPP was conducted in areas where Zika–human cases were reported recently. Among the four endophilic mosquito genera caught, the majority were *Aedes* mosquitoes (40.8%). Only female *Ae. aegypti* and *Ae. albopictus* were subsequently tested for ZIKV and DENV, as they are considered the major vectors to transmit these viruses. A total of 488 mosquitoes were captured from 86 trap-days (5.7 mosquitoes/trap). This is considerably high since mosquito collections were performed within 72 h after receiving information about human cases of ZIKV from the local hospitals. Local regulations required all houses with confirmed DENV or ZIKV cases to be sprayed with an adulticide within this time period. Surprisingly, both ZIKV and DENV infected mosquitoes were found in adulticide sprayed houses (index houses).

ZIKV was detected in two *Ae. aegypti* samples from the same mosquito specimen collected from a Zika index house. Based on a sequence analysis of the E gene, which has been used to classify ZIKV into African and Asian lineages, the ZIKV detected in this study was of the Asian lineage [31]. Analysis showed the ZIKV sequence was 100% identical to previous human cases of ZIKV reported in Thailand during 2016–2017 [30]. This confirms that the same ZIKV lineage was still circulating in Thailand. The Asian lineage is presently responsible for various global outbreaks, including outbreaks in Thailand [32]. No virus was detected in *Ae. albopictus* specimens but this may be due to the low number of specimens (*n* = 11) collected in this study. Since the mosquito traps were set inside houses, it was not surprising that *Ae. albopictus* was collected less frequently since it is more exophilic than *Ae. aegypti* [5]. In a recent report from Thailand, both *Ae. aegypti* and *Ae. albopictus* were screened for

ZIKV and ZIKV was detected in female, male and larvae of field-caught *Ae. aegypti* collected around active ZIKV patients' houses [29]. That study also did not detect ZIKV in *Ae. albopictus*.

Additionally, we detected DENV in mosquitoes collected from ZIKV patient homes, although no DENV outbreak or dengue cases were reported in the study area during this period. It is possible that the mosquitoes may have acquired DENV from unreported, asymptomatic infected humans, or by vertical transmission [33]. The two DENV serotypes, DENV-1 and DENV-4, found in this study were identical (more than 96% identity) to DENV from human cases reported in 2017 [14], which suggests a connection to an outbreak of those DENV serotypes in Thailand during that time.

This study demonstrates the importance of entomological surveillance for public health programs and to provide guidance to mosquito-borne disease prevention and control. The entomological surveillance is crucial for the planning, implementation, and monitoring of vector control programs. Both adult and larval control measures should be conducted focusing on index houses which can serve as major hubs for transmission of the virus during an outbreak. In this study, disease vectors and infected mosquitoes were captured in index houses although vector control, including adulticide and larvicide application, was performed in homes within 24 h after cases of human disease were confirmed. Therefore, a surveillance system for efficacy of vector control operations (e.g., mosquito sentinel, larval bioassay, etc.) is warranted. This system will provide a warning to local public health officers regarding the effectiveness of the current vector control program.

4. Conclusions

The presence of ZIKV and DENV in *Ae. aegypti* mosquitoes collected from houses recently sprayed with adulticide has implications for public health and vector control measures in areas at risk for flavivirus transmission. Since there are no effective vaccines or specific therapeutic treatments for DENV and ZIKV infections, vector control is the only effective tool to mitigate disease transmission. Vector surveillance before and after vector control is essential to validate the efficacy of the control measures employed. Our study suggests that a single application of non-residual adulticide alone may not be sufficient to reduce the risk of further spread of ZIKV by controlling the *Aedes* vector. Routine entomological surveillance is essential for the planning and implementation of an effective vector control program.

5. Materials and Methods

5.1. Study Site

From July 2017 to February 2018, mosquito surveillance was performed in 5 districts in KPP, which is located approximately 360 km northeast of Bangkok. The climate in KPP is tropical with three seasons: rainy (June to October), winter (November to February), and summer (March to May). These five districts include Khanu Woralaksaburi, Khlong Lan, Kosamphi Nakhon, Mueang, and Sai Ngam (Figure 3). These are DENV endemic areas but have recently reported human cases of ZIKV. During the study period, there were nine human cases of ZIKV in these study villages reported to the Provincial Public Health Office. Almost all Zika cases (8/9 cases) were reported in the rainy season.

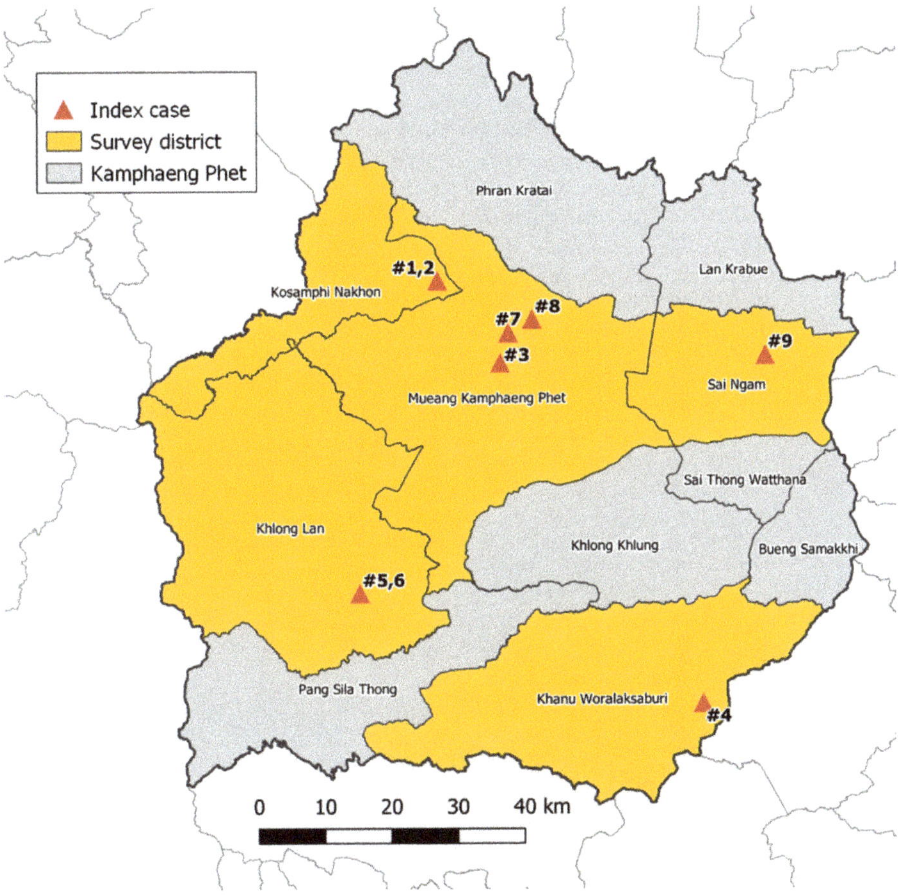

Figure 3. Mosquito collections from 9 Zika index case houses and 25 neighboring houses (within a 50 m radius from the Zika index cases) located in 5 districts, Kamphaeng Phet Province, Thailand during July 2017–February 2018.

5.2. Mosquito Collection

BG Sentinel traps (Biogents AG, Regensburg, Germany) baited with 1 kg of dry ice and BG-lure containing ammonia, lactic acid and caproic acid, were used to collect mosquitoes. Traps were placed inside 34 houses, including houses with ZIKV reported cases ($n = 9$) and neighboring houses within a 50 m radius of Zika index houses ($n = 25$). Based on the feeding behavior of *Aedes* mosquitoes, traps were continuously operated to collect daytime biting mosquitoes for 8 h (0800–1600). Mosquito collection was conducted no longer than 72 h after receiving information about human cases of ZIKV from the local hospitals. Mosquito samples in each trap were stored in a cooler containing dry ice for transportation to the Armed Forces Research Institute of Medical Sciences (AFRIMS) Entomology laboratory in Bangkok, where the samples were further processed. All collected mosquitoes were morphologically identified to species under the stereomicroscope [34]. They were sorted by sex and location prior to keeping in a −80 °C freezer until use.

5.3. Detection of ZIKV and DENV by Real-Time RT-PCR

Both *Ae. aegypti* and *Ae. albopictus* females were screened for ZIKV and DENV by real-time RT-PCR. *Aedes* head-thorax parts were dissected from the abdominal parts and separately transferred into microfuge tubes containing RPMI medium. Both mosquito parts (head-thorax and abdomen) were homogenized in RPMI medium using a Bullet Blender® Storm (Next Advance, Inc., Troy, NY, USA). Supernatant was collected for a total nucleic acid extraction using a PureLinkTM Viral RNA/DNA Mini Kit (Thermo Fisher Scientific, Waltham, MA, USA) according to the manufacturer's instruction.

A TaqMan® Fast Virus 1-Step Master Mix (Applied Biosystems, Foster City, CA, USA) and specific primers and probes were used in the real-time RT-PCR for detection of ZIKV [35] as shown in Table 2. The real-time RT-PCR reaction contained 1× TaqMan® Fast Virus 1-Step Master Mix, 0.5 µM each of forward (ZIKV835 or ZIKV1086) and reverse (ZIKV911c or ZIKV1162c) primers, 0.125 µM probe (ZIKV860-FAM or ZIKV1107-FAM), and 5 µL of the total RNA in a final volume of 20 µL. Thermal cycling conditions were: 50 °C for 15 min, 95 °C for 30 s, followed by 40 cycles of 95 °C for 10 s and 60 °C for 30 s in a CFX96TM Real-Time System (Bio-Rad, Hercules, CA, USA).

Table 2. Primer and probe sequences used in this study.

Primer	Sequence (5′→3′)	Position (nt) [a]	Virus	Region
Real-Time RT-PCR				
ZIKV835	TTGGTCATGATACTGCTGATTGC	835–857	ZIKV	Matrix
ZIKV911c	CCTTCCACAAAGTCCCTATTGC	911–890		Envelope
ZIKV860-FAM [b]	CGGCATACAGCATCAGGTGCATAGGAG	860–886		Envelope
ZIKV1086	CCGCTGCCCAACACAAG	1086–1102	ZIKV	Envelope
ZIKV1162c	CCACTAACGTTCTTTTGCAGACAT	1162–1139		Envelope
ZIKV1107-FAM [b]	AGCCTACCTTGACAAGCAGTCAGACACTCAA	1107–1137		Envelope
RT-PCR for Sequencing				
ZIKENVF	GCTGGDGCRGACACHGGRACT	1643–1663	ZIKV	Envelope
ZIKENVR	RTCYACYGCCATYTGGRCTG	1989–2008		Envelope
D1	TCAATATGCTGAAACGCGCGAGAAACCG	132–159	DENV	Capsid/prM
D2	TTGCACCAACAGTCAATGTCTTCAGGTTC	614–642		Capsid/prM

[a] The nucleotide numbering corresponds to that of the published ZIKV sequence [GenBank: AY632535.2] or DENV sequence [GenBank: DQ863638.1]. [b] Probes are labeled with 5′ 6-FAM and 3′ TAMRA-N.

The real-time RT-PCR detection of ZIKV and DENV was done using a ZDC Multiplex RT-PCR Assay (Bio-Rad) according to the manufacturer's instruction. The real-time RT-PCR reaction contained 1× iTaq™ Universal Probes One-Step Reaction Mix, 0.6 µL of iScript™ Reverse Transcriptase, 1× ZDC Multiplex PCR Assay Mix and 5 µL of the total nucleic acid in a final volume of 25 µL. The thermal cycler (CFX96TM Real-Time System) with conditions of 50 °C for 15 min, 94 °C for 2 min, followed by 45 cycles of 94 °C for 15 s, 55 °C for 40 s and 68 °C for 30 s was set.

5.4. DNA Sequencing and Phylogenetic Analysis

For nucleotide sequencing, the total RNA extracted from the ZIKV and DENV real-time RT-PCR positive samples were subjected to the RT-PCR with sequencing primers (Table 2) using a SuperScript® III One-Step RT-PCR System with a Platinum® Taq DNA Polymerase kit (Invitrogen, Carlsbad, CA, Bio-Rad). The reaction mixture contained 1× reaction mix, 0.4 µM of each forward (ZIKENVF or D1) and reverse primer (ZIKENVR or D2), 1 µL of SuperScript® III RT/Platinum® Taq Mix, and 5 µL of the total RNA in a final volume of 25 µL. The thermal cycling conditions were set as a reverse transcription at 50 °C for 30 min, followed by an initial denaturation of cDNA at 94 °C for 2 min, 40 cycles of denaturation at 94 °C for 15 s, annealing at 54 °C for 30 s, and an extension at 68 °C for 45 s, followed by a final elongation step at 68 °C for 10 min. The PCR products of 365 bp and 511 bp, for ZIKV and DENV, respectively, were resolved on a 1.5% agarose gel, stained with ethidium bromide, and visualized under a gel documentation system (Gel Doc™ XR+ imaging system, Molecular Imager®, Bio-Rad). The PCR products were extracted from gel using the QIAquick gel extraction kit (Qiagen,

Hilden, Germany) following the manufacturer's instructions. Purified DNA was sequenced using the forward and reverse primers. Sequencing chromatograms were inspected and processed using BioEdit 7.0.4.1. [36]. Nucleotide sequences were queried against the NCBI database using BLAST and aligned with published reference sequences using ClustalW in BioEdit. Phylogenetic trees were constructed using the Molecular Evolutionary Genetics Analysis software (MEGA, version 7.0.21) [37]. The maximum likelihood method was applied based on the phylogenetic model analysis. Bootstrap resampling analysis of 1000 replicates was used.

Supplementary Materials: The following are available online at http://www.mdpi.com/2076-0817/9/6/442/s1, Figure S1: Alignment of 331 bp ZIKV partial envelop sequences, Figure S2: Alignment of 395 bp DENV partial capsid/prM sequences.

Author Contributions: Conceptualization, A.P. and N.K.; methodology, N.K.; J.T.; C.N.; P.S.; validation, S.A.D. and A.P.; formal analysis, N.K. and P.S.; writing—original draft preparation, N.K. and A.P.; writing—review and editing, S.A.D. and A.P.; visualization, N.K.; supervision, A.P.; funding acquisition, A.P. All authors have read and agree to the published version of the manuscript.

Funding: This work was financially supported by the Armed Forces Health Surveillance Branch, Global Emerging Infections Surveillance and Response System (AFHSB-GEIS), Silver Spring, Maryland, USA (ProMIS ID: P0081_18_AF_03.03, Funding Year: 2018).

Acknowledgments: We thank Dr. Darunee Buddhari, staff at Kamphaeng Phet-AFRIMS Virology Research Unit (KAVRU), and villagers in Kamphaeng Phet, Thailand for their support. We also thank Dr. Thanyalak Fansiri, Arissara Pongsiri, Udom Kijchalao, Somsak Tiang-trong, and all staff in Vector Biology and Control section, Department of Entomology, AFRIMS for mosquito collection. Material has been reviewed by the Walter Reed Army Institute of Research. There is no objection to its publication. The opinions or assertions contained herein are the private views of the author, and are not to be construed as official, or as reflecting true views of the Department of the Army or the Department of Defense.

Conflicts of Interest: The authors declare no conflict of interest.

References

1. World Health Organization. Dengue and severe dengue. Available online: https://www.who.int/news-room/fact-sheets/detail/dengue-and-severe-dengue (accessed on 3 April 2020).

2. Buathong, R.; Hermann, L.; Thaisomboonsuk, B.; Rutvisuttinunt, W.; Klungthong, C.; Chinnawirotpisan, P.; Manasatienkij, W.; Nisalak, A.; Fernandez, S.; Yoon, I.K.; et al. Detection of Zika virus infection in Thailand, 2012–2014. *Am. J. Trop. Med. Hyg.* **2015**, *93*, 380–383. [CrossRef] [PubMed]

3. Caldwell, J.P.; Chen, L.H.; Hamer, D.H. Evolving epidemiology of Japanese Encephalitis: Implications for vaccination. *Curr. Infect. Dis. Rep.* **2018**, *20*, 30. [CrossRef] [PubMed]

4. Lee, H.; Halverson, S.; Ezinwa, N. Mosquito-borne diseases. *Prim. Care.* **2018**, *45*, 393–407. [CrossRef]

5. Boyer, S.; Calvez, E.; Chouin-Carneiro, T.; Diallo, D.; Failloux, A.B. An overview of mosquito vectors of Zika virus. *Microbes. Infect.* **2018**, *20*, 646–660. [CrossRef] [PubMed]

6. Wikan, N.; Smith, D.R. Zika virus: History of a newly emerging arbovirus. *Lancet. Infect. Dis.* **2016**, *16*, e119–e126. [CrossRef]

7. Cao-Lormeau, V.M.; Blake, A.; Mons, S.; Lastere, S.; Roche, C.; Vanhomwegen, J.; Dub, T.; Baudouin, L.; Teissier, A.; Larre, P.; et al. Guillain-Barré syndrome outbreak associated with Zika virus infection in French Polynesia: A case-control study. *Lancet.* **2016**, *387*, 1531–1539. [CrossRef]

8. Carteaux, G.; Maquart, M.; Bedet, A.; Contou, D.; Brugières, P.; Fourati, S.; Cleret de Langavant, L.; de Broucker, T.; Brun-Buisson, C.; Leparc-Goffart, I.; et al. Zika virus associated with meningoencephalitis. *N. Engl. J. Med.* **2016**, *374*, 1595–1596. [CrossRef]

9. Kleber de Oliveira, W.; Cortez-Escalante, J.; De Oliveira, W.T.; do Carmo, G.M.; Henriques, C.M.; Coelho, G.E.; Araújo de França, G.V. Increase in reported prevalence of microcephaly in infants born to women living in areas with confirmed Zika virus transmission during the first trimester of pregnancy—Brazil, 2015. *MMWR. Morb. Mortal. Wkly. Rep.* **2016**, *65*, 242–247. [CrossRef]

10. Holmes, E.C.; Twiddy, S.S. The origin, emergence and evolutionary genetics of dengue virus. *Infect. Genet. Evol.* **2003**, *3*, 19–28. [CrossRef]

11. Hotta, S. Experimental studies on dengue. I. Isolation, identification and modification of the virus. *J. Infect. Dis.* **1952**, *90*, 1–9. [CrossRef]

12. Messina, J.P.; Brady, O.J.; Scott, T.W.; Zou, C.; Pigott, D.M.; Duda, K.A.; Bhatt, S.; Katzelnick, L.; Howes, R.E.; Battle, K.E.; et al. Global spread of dengue virus types: Mapping the 70 years history. *Trends. Microbiol.* **2014**, *22*, 138–146. [CrossRef]

13. World Health Organization. Global strategy for dengue prevention and control 2012–2020. Available online: http://apps.who.int/iris/bitstream/handle/10665/75303/9789241504034_eng.pdf;jsessionid=8BD7C5C13E1CC23A760ADD649728C11C?sequence=1 (accessed on 3 April 2020).

14. Phadungsombat, J.; Lin, M.Y.; Srimark, N.; Yamanaka, A.; Nakayama, E.E.; Moolasart, V.; Suttha, P.; Shioda, T.; Uttayamakul, S. Emergence of genotype cosmopolitan of dengue virus type 2 and genotype III of dengue virus type 3 in Thailand. *PLoS. ONE.* **2018**, *13*, e0207220. [CrossRef]

15. Sittivicharpinyo, T.; Wonnapinij, P.; Surat, W. Phylogenetic analyses of DENV-3 isolated from field-caught mosquitoes in Thailand. *Virus. Res.* **2018**, *244*, 27–35. [CrossRef] [PubMed]

16. Dick, G.W.; Kitchen, S.F.; Haddow, A.J. Zika virus. I. Isolations and serological specificity. *Trans. R. Soc. Trop. Med. Hyg.* **1952**, *46*, 509–520. [CrossRef]

17. Hayes, E.B. Zika virus outside Africa. *Emerg. Infect. Dis.* **2009**, *15*, 1347–1350. [CrossRef]

18. Moore, D.L.; Causey, O.R.; Carey, D.E.; Reddy, S.; Cooke, A.R.; Akinkugbe, F.M.; David-West, T.S.; Kemp, G.E. Arthropod-borne viral infections of man in Nigeria, 1964–1970. *Ann. Trop. Med. Parasitol.* **1975**, *69*, 49–64. [CrossRef] [PubMed]

19. Duong, V.; Ong, S.; Leang, R.; Huy, R.; Ly, S.; Mounier, U.; Ou, T.; In, S.; Peng, B.; Ken, S.; et al. Low circulation of Zika virus, Cambodia, 2007–2016. *Emerg. Infect. Dis.* **2017**, *23*, 296–299. [CrossRef]

20. Perkasa, A.; Yudhaputri, F.; Haryanto, S.; Hayati, R.F.; Ma'roef, C.N.; Antonjaya, U.; Yohan, B.; Myint, K.S.; Ledermann, J.P.; Rosenberg, R.; et al. Isolation of Zika virus from febrile patient, Indonesia. *Emerg. Infect. Dis.* **2016**, *22*, 924–925. [CrossRef]

21. Tappe, D.; Nachtigall, S.; Kapaun, A.; Schnitzler, P.; Günther, S.; Schmidt-Chanasit, J. Acute Zika virus infection after travel to Malaysian Borneo, September 2014. *Emerg. Infect. Dis.* **2015**, *21*, 911–913. [CrossRef]

22. Alera, M.T.; Hermann, L.; Tac-An, I.A.; Klungthong, C.; Rutvisuttinunt, W.; Manasatienkij, W.; Villa, D.; Thaisomboonsuk, B.; Velasco, J.M.; Chinnawirotpisan, P.; et al. Zika virus infection, Philippines, 2012. *Emerg. Infect. Dis.* **2015**, *21*, 722–724. [CrossRef]

23. Singapore Zika Study Group. Outbreak of Zika virus infection in Singapore: An epidemiological, entomological, virological, and clinical analysis. *Lancet. Infect. Dis.* **2017**, *17*, 813–821. [CrossRef]

24. Moi, M.L.; Nguyen, T.T.T.; Nguyen, C.T.; Vu, T.B.H.; Tun, M.M.N.; Pham, T.D.; Pham, N.T.; Tran, T.; Morita, K.; Le, T.Q.M.; et al. Zika virus infection and microcephaly in Vietnam. *Lancet. Infect. Dis.* **2017**, *17*, 805–806. [CrossRef]

25. Fonseca, K.; Meatherall, B.; Zarra, D.; Drebot, M.; MacDonald, J.; Pabbaraju, K.; Wong, S.; Webster, P.; Lindsay, R.; Tellier, R. First case of Zika virus infection in a returning Canadian traveler. *Am. J. Trop. Med. Hyg.* **2014**, *91*, 1035–1038. [CrossRef] [PubMed]

26. Tappe, D.; Rissland, J.; Gabriel, M.; Emmerich, P.; Gunther, S.; Held, G.; Smola, S.; Schmidt-Chanasit, J. First case of laboratory-confirmed Zika virus infection imported into Europe, November 2013. *Eur. Surveill.* **2014**, *19*, 20685. [CrossRef] [PubMed]

27. Shinohara, K.; Kutsuna, S.; Takasaki, T.; Moi, M.L.; Ikeda, M.; Kotaki, A.; Yamamoto, K.; Fujiya, Y.; Mawatari, M.; Takeshita, N.; et al. Zika fever imported from Thailand to Japan, and diagnosed by PCR in the urines. *J. Travel. Med.* **2016**, *23*, tav011. [CrossRef]

28. Khongwichit, S.; Wikan, N.; Auewarakul, P.; Smith, D.R. Zika virus in Thailand. *Microbes. Infect.* **2018**, *20*, 670–675. [CrossRef]

29. Phumee, A.; Buathong, R.; Boonserm, R.; Intayot, P.; Aungsananta, N.; Jittmittraphap, A.; Joyjinda, Y.; Wacharapluesadee, S.; Siriyasatien, P. Molecular epidemiology and genetic diversity of Zika virus from field-caught mosquitoes in various regions of Thailand. *Pathogens* **2019**, *8*, 30. [CrossRef]

30. Wongsurawat, T.; Athipanyasilp, N.; Jenjaroenpun, P.; Jun, S.R.; Kaewnapan, B.; Wassenaar, T.M.; Leelahakorn, N.; Angkasekwinai, N.; Kantakamalakul, W.; Ussery, D.W.; et al. Case of microcephaly after congenital infection with Asian lineage Zika virus, Thailand. *Emerg. Infect. Dis.* **2018**, *24*, 1758. [CrossRef]

31. Faye, O.; Freire, C.C.; Iamarino, A.; Faye, O.; de Oliveira, J.V.; Diallo, M.; Zanotto, P.M.; Sall, A.A. Molecular evolution of Zika virus during its emergence in the 20(th) century. *PLoS. Negl. Trop. Dis.* **2014**, *8*, e2636. [CrossRef]

32. Haddow, A.D.; Schuh, A.J.; Yasuda, C.Y.; Kasper, M.R.; Heang, V.; Huy, R.; Guzman, H.; Tesh, R.B.; Weaver, S.C. Genetic characterization of Zika virus strains: Geographic expansion of the Asian lineage. *PLoS. Negl. Trop. Dis.* **2012**, *6*, e1477. [CrossRef]

33. Ferreira-de-Lima, V.H.; Lima-Camara, T.N. Natural vertical transmission of dengue virus in *Aedes aegypti* and *Aedes albopictus*: A systematic review. *Parasit. Vectors.* **2018**, *11*, 77. [CrossRef]

34. Rattanarithikul, R.; Harbach, R.E.; Harrison, B.A.; Panthusiri, P.; Coleman, R.E.; Richardson, J.H. Illustrated keys to the mosquitoes of Thailand. *VI. Tribe Aedini. Southeast. Asian. J. Trop. Med. Public. Health.* **2010**, *41*, 1–225.

35. Lanciotti, R.S.; Kosoy, O.L.; Laven, J.J.; Velez, J.O.; Lambert, A.J.; Johnson, A.J.; Stanfield, S.M.; Duffy, M.R. Genetic and serologic properties of Zika virus associated with an epidemic, Yap State, Micronesia, 2007. *Emerg. Infect. Dis.* **2008**, *14*, 1232–1239. [CrossRef]

36. Hall, T.A. BioEdit: A user-friendly biological sequence alignment editor and analysis program for Window 95/98/NT. *Nucl. Acids. Symp. Ser.* **1999**, *41*, 95–98.

37. Kumar, S.; Stecher, G.; Tamura, K. MEGA7: Molecular Evolutionary Genetics Analysis version 7.0 for bigger datasets. *Mol. Biol. Evol.* **2016**, *33*, 1870–1874. [CrossRef]

Article

Vector Competence of *Aedes aegypti, Aedes albopictus* and *Culex quinquefasciatus* from Brazil and New Caledonia for Three Zika Virus Lineages

Rosilainy S. Fernandes [1],[†], Olivia O'Connor [2],[†], Maria Ignez L. Bersot [1], Dominique Girault [2], Marguerite R. Dokunengo [2], Nicolas Pocquet [3], Myrielle Dupont-Rouzeyrol [2],[*] and Ricardo Lourenço-de-Oliveira [1],[*]

[1] Laboratório de Mosquitos Transmissores de Hematozoários, Instituto Oswaldo Cruz-FIOCRUZ, Rio de Janeiro 21040-900, Brazil; rosilainysf@gmail.com (R.S.F.); ignez@ioc.fiocruz.br (M.I.L.B.)
[2] Institut Pasteur de Nouvelle-Calédonie, URE Dengue et Arboviroses, Réseau International des Instituts Pasteur, 98800 Noumea, New Caledonia; ooconnor@pasteur.nc (O.O.); dgirault@pasteur.nc (D.G.); rowynadokunengo@outlook.fr (M.R.D.)
[3] Institut Pasteur de Nouvelle-Calédonie, URE Entomologie Médicale, Réseau International des Instituts Pasteur, 98800 Noumea, New Caledonia; npocquet@pasteur.nc
*   Correspondence: mdupont@pasteur.nc (M.D.-R.); lourenco@ioc.fiocruz.br (R.L.-d.-O.)
†   These authors contributed equally.

Received: 18 June 2020; Accepted: 11 July 2020; Published: 16 July 2020

Abstract: Zika virus (ZIKV) has caused severe epidemics in South America beginning in 2015, following its spread through the Pacific. We comparatively assessed the vector competence of ten populations of *Aedes aegypti* and *Ae. albopictus* from Brazil and two of *Ae. aegypti* and one of *Culex quinquefasciatus* from New Caledonia to transmit three ZIKV isolates belonging to African, Asian and American lineages. Recently colonized mosquitoes from eight distinct sites from both countries were orally challenged with the same viral load (10^7 TCID$_{50}$/mL) and examined after 7, 14 and 21 days. *Cx. quinquefasciatus* was refractory to infection with all virus strains. In contrast, although competence varied with geographical origin, Brazilian and New Caledonian *Ae. aegypti* could transmit the three ZIKV lineages, with a strong advantage for the African lineage (the only one reaching saliva one-week after challenge). Brazilian *Ae. albopictus* populations were less competent than *Ae. aegypti* populations. *Ae. albopictus* generally exhibited almost no transmission for Asian and American lineages, but was efficient in transmitting the African ZIKV. Viral surveillance and mosquito control measures must be strengthened to avoid the spread of new ZIKV lineages and minimize the transmission of viruses currently circulating.

Keywords: transmission efficiency; vector capacity; susceptibility

1. Introduction

Zika virus (ZIKV) is an arbovirus (Flaviviridae, Flavivirus) that originated in Africa, where it is transmitted either in the wild or modified environments by *Aedes* mosquitoes. The virus was little known for many decades after its discovery in the 1940s. However, beginning in 2007, when it caused the first outbreak detected outside Africa (Yap Island, Federated States of Micronesia, Pacific region) the virus gained notoriety [1,2]. In 2013, ZIKV emerged in French Polynesia [3], leading to more than 8700 suspected cases and 30,000 medical consultations reported by the sentinel surveillance network [4]. A recent seroprevalence study estimated that more than half of the population was infected by ZIKV in French Polynesia [5]. From French Polynesia, ZIKV spread to New Caledonia in 2013 [6,7], affecting the whole territory. The New Caledonia Health Authorities estimated the number of cases at about 11,000 [8]. From 2014 to 2017, ZIKV was detected in the Cook Islands, Easter Islands, Vanuatu, Fiji, Samoa, Solomon Islands, Tonga and American Samoa [8,9]. During the same period, ZIKV spread to the American continent, being first detected in northeastern Brazil in 2015 [10]. Subsequently, countrywide epidemics were reported in Brazil as well as several South and Central American localities. In 2015, Brazil reported 37,011 probable Zika symptomatic infections and the first cases of microcephaly associated with ZIKV [10,11]. The number of cases in Brazil peaked in 2016 with nearly 215,320 probable Zika cases [12], followed by 17,593 and 10,768 reported cases annually between 2017 and 2019 [13,14]. The occurrence of microcephaly and other congenital neurological malformations associated with ZIKV infections has been described worldwide [15–17], and the virus continues to be considered an important threat [18].

Although ZIKV can be directly transmitted between humans, vector borne transmission is believed to play a major role in virus spread [19,20]. *Aedes aegypti* is generally considered the primary vector of ZIKV in all surveyed areas [20–22]. Substantial variation in vector competence, however, has been described in populations of *Ae. aegypti* of different geographical origins when challenged with distinct ZIKV strains controlling for titer in infectious blood meal [20,23]. In ZIKV epidemic and endemic areas, *Ae. aegypti* cooccurs with other domestic and peridomestic mosquitoes such as *Culex quinquefasciatus* and *Aedes albopictus* that frequently bite humans. Natural ZIKV infections have been reported in these species, and thus, they came under suspicion as alternative vectors [24–27]. Concerning *Ae. albopictus*, results of vector competence evaluations to date have shown that this mosquito is susceptible to ZIKV but with a transmission efficiency that is significantly inferior to that of *Ae. aegypti* [20,25]. In contrast, domestic *Culex* species tested worldwide to date have consistently been shown to be essentially refractory to ZIKV, and thus, their potential role in ZIKV transmission has become controversial [28–31].

Phylogenetic studies have shown that Brazilian ZIKV isolates clustered with other American isolates in the so-called American lineage, sharing common ancestry with the Asian genotype that includes strains that circulated in French Polynesia and other Pacific areas such as New Caledonia in 2013–2014 [32–34]. Interestingly, *Ae. aegypti* and *Ae. albopictus* from Brazil exhibited lower vector competence for a ZIKV from New Caledonia than for isolates from Brazil [28,35]. And curiously, *Ae. aegypti* from Singapore was also less competent in experimentally transmitting a French Polynesian isolate than a Brazilian one [36].

It is clear that vector competence is dependent on the specific combination of mosquito and virus genotypes from different geographic regions [37,38]. Determining vector competence of human biting mosquito populations from endemic ZIKV territories is required to inform control measures. Here, we comparatively assessed the vector competence of 13 mosquito populations of *Ae. aegypti*, *Ae. albopictus* and *Cx. quinquefasciatus* from New Caledonia and Brazil to African, Asian and American ZIKV strains.

2. Results

2.1. Aedes aegypti from All Brazilian Regions Can Experimentally Transmit ZIKV Isolated from Africa, Asia and America, but with Different Competence

Five Brazilian populations of *Ae. aegypti* were orally challenged with three strains of ZIKV; the population represented the five geographic regions of the country. While infection was detected in mosquitoes of all populations, regardless of the incubation time and the virus tested (Figure 1A, Table S1), quantitative infection rates (IR) varied depending on the virus strain and the incubation time. In general, IR did not increase with incubation time, regardless of the tested ZIKV isolates; the results from the African isolate were a very representative example of this trend (Figure 1). Except for the population of Natal, the IRs with the African genotype were always above 90%, being 100% for most populations, regardless of the incubation time. There was greater heterogeneity in the IR with the different viruses at 7 d.p.i. than at 14 and 21 d.p.i., except for the population of Rio de Janeiro, where significant differences were detected between essentially all viruses in all incubation times

(Figure 1A). At 7 d.p.i., the IRs were significantly lower for the ZIKV isolate of the Asian lineage in all challenged Brazilian *Ae. aegypti* except for Natal. In general, IRs tended to be similar among viruses at 14 d.p.i., except for the Rio de Janeiro *Ae. aegypti* population.

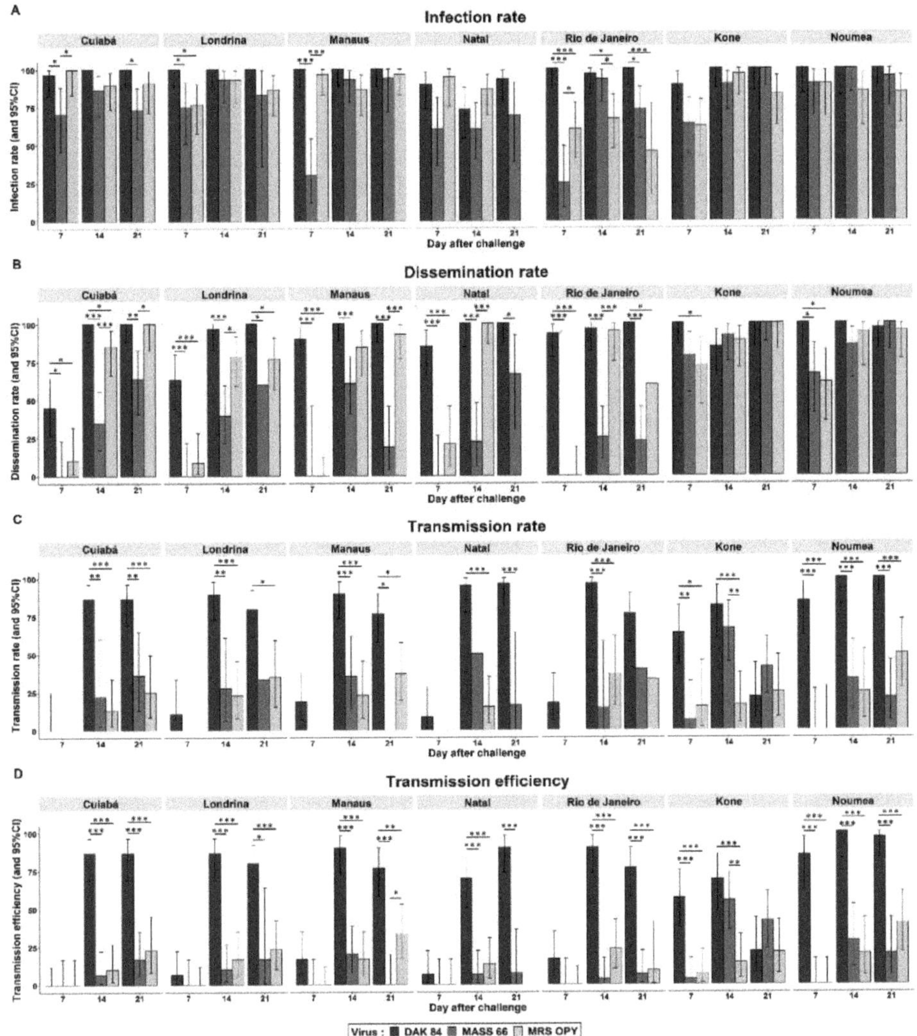

Figure 1. Vector competence results for *Aedes aegypti* from different cities of Brazil (Cuiabá,Londrina, Manaus, Natal and Rio de Janeiro) and New Caledonia (Kone and Noumea) orally challenged with three ZIKV isolates: DAK 84 (African lineage), MASS 66 (Asian lineage), MRS OPY (American lineage). (**A**) Infection rate, (**B**) dissemination rate, (**C**) transmission rate and (**D**) transmission efficiency at 7, 14 and 21 days after challenge. Error bars represent 95% confidence intervals. Significant differences are indicated by asterisks (Fisher's Exact test: * $p < 0.05$; ** $p < 0.01$; *** $p < 0.001$).

All three viruses tested were able to disseminate to secondary tissues (head) in all challenged Brazilian *Ae. aegypti* populations, albeit to varying extents (Figure 1B, Table S1). However, in contrast to IR, the dissemination rates (DR) were quite heterogeneous among the tested viruses and Brazilian populations. At 7 d.p.i., DR were generally null or nonsignificant for all mosquito populations challenged with the two ZIKV isolates of the Asian genotype. In contrast, at this same incubation time, the African genotype isolate had already surprisingly disseminated to the heads of more than 80% of *Ae. aegypti* from Rio de Janeiro, Manaus and Natal. Albeit with lower rates, the African genotype also disseminated in significantly higher rates and more rapidly than the other viruses in the mosquitoes of Londrina and Cuiaba. At 14 d.p.i., all three tested viruses disseminated to the heads of mosquitoes of all five Brazilian populations, but with heterogeneous DR. Again, regardless of the mosquito population, the DRs of the Asian lineage were lower than those detected for African and American lineages at 14 and 21 d.p.i. (Figure 1B). In general, from two weeks after the infectious meal, these two viral isolates disseminated in more than 75% of infected mosquitoes. The DR values increased over the incubation time ($p < 0.05$) in the Cuiabá and Londrina *Ae. aegypti* infected with any of the ZIKV isolates and in the Manaus and Natal populations infected with the Asian and American lineages (Tables S3–S5).

Interestingly, from 14 d.p.i. on, transmission was achieved in all combinations of Brazilian *Ae. aegypti* and virus lineages, except for the isolate belonging to the African lineage which was transmitted as early as 7 d.p.i. by four out of the five challenged populations (Figures 1C and 2D, Table S1). It is noteworthy that from 14 d.p.i. on, infectious particles of the ZIKV of the African genotype were detected in over 75% of individuals from all five mosquito populations in which the virus had disseminated, and the transmission rates (i.e., the proportion of mosquitoes with infectious viral particles in saliva among those with disseminated infection, or TR) increased with incubation time ($p < 0.05$). In general, transmission efficiency values (the proportion of mosquitoes with virus in saliva among all orally challenged mosquitoes, or TE) were higher in the African lineage compared to the other two lineages. With few exceptions (TE for Manaus, Londrina and Rio de Janeiro; $p < 0.05$), there was no significant difference between the TRs or TEs of the two ZIKV isolates of the Asian and American lineages in any population of *Ae. aegypti* from Brazil.

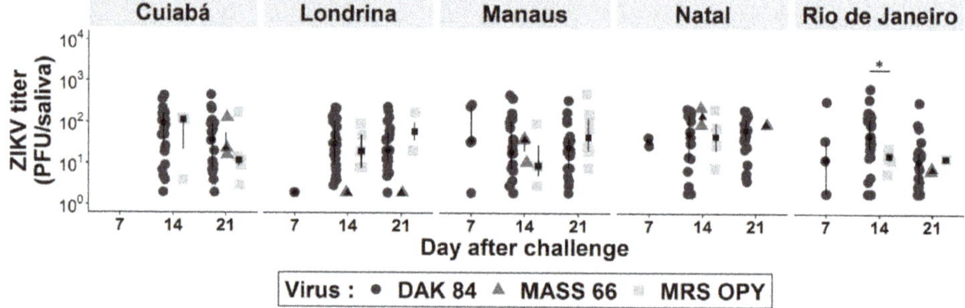

Figure 2. ZIKV load in saliva of *Ae. aegypti* from five Brazilian cities at 7, 14 and 21 days after oral challenge with three isolates: DAK 84 (African lineage), MASS 66 (Asian lineage), MRS OPY (American lineage). Virus was detected by plaque forming unit (PFU) assays on Vero cells. Significant difference is indicated by asterisk (Wilcoxon test: * $p < 0.05$).

The median viral load in Brazilian *Ae. aegypti* saliva ranged from 2 to 36, 2 to 160.5 and 2 to 97 PFU at 7, 14 and 21 d.p.i., respectively (Table S2). The viral load per saliva sample reached values as high as 685, 502 and 230 PFU when infected with the African, American and Asian lineage, respectively. No difference was found when comparing viral load in positive saliva expectorated by *Ae. aegypti* individuals belonging to the same population infected with the three different viruses, regardless of incubation time, except for those from Rio de Janeiro infected with the African genotype compared to the American lineage at 14 d.p.i. ($p = 0.02$) (Figure 2).

Comparing *Ae. aegypti* populations, only the viral titers in the saliva of mosquitoes from Natal were higher than those from Rio de Janeiro ($p = 0.001$) infected with the African ZIKV at 21 d.p.i, whereas the pairwise comparisons with the American and Asian lineages and incubation times did not differ (Tables S3–S5).

2.2. Two Populations of Aedes aegypti from New Caledonia Can Experimentally Transmit ZIKV Isolated from Africa, Asia and America, but with Different Competence

Two New Caledonian populations of *Ae. aegypti* were orally challenged with three strains of ZIKV. Infection was detected in mosquitoes of both populations, regardless of the incubation time and the virus tested (Figure 1A, Table S1). In general, IR was already high at 7 d.p.i, and did not increase with incubation time with any of the three ZIKV isolates. The IRs with the African genotype were always above 85%, regardless of the incubation time. Some heterogeneity between the three ZIKV strains could be observed with *Ae. aegypti* from Kone, with the IR ranging from 62% to 89% at 7 d.p.i.

All three viruses tested were able to disseminate to secondary tissues in both New Caledonian *Ae. aegypti* populations (Figure 1B, Table S1). The DRs increased over time and ranged from 61 to 100% across the three virus isolates and two New Caledonian mosquito populations. Only at 7 d.p.i., a significantly higher DR for the African ZIKV genotype was observed compared to the Asian genotype for both *Ae. aegypti* populations (p = 0.01 for Kone, and p = 0.01 and =0.01 for Noumea, compared to ZIKV American and Asian lineages respectively).

Transmission was achieved in all combinations of New Caledonian *Ae. aegypti* and virus lineages assayed from 14 d.p.i. on, except for the isolate belonging to the African genotype, which was transmitted as early as 7 d.p.i. (Figure 1C,D, Table S1). For the African genotype, TR increased over time for both *Ae. aegypti* populations (57 to 82%) except for *Ae. aegypti* from Kone at 21 d.p.i, with a lower level of infectious saliva detected in individuals (17%). In general, the African genotype was also by far the most efficiently transmitted for both *Ae. aegypti* populations compared to the Asian one (p < 0.001). There was no significant difference between the TR or TE of the two ZIKV isolates of the Asian genotype in any population of *Ae. aegypti* from New Caledonia, except at 14 d.p.i. for Kone (p < 0.01).

Comparing the two New Caledonian *Ae. aegypti* populations, no significant differences were detected for IR and DR, irrespective of the d.p.i. and the ZIKV strains. Regarding the TR, *Ae. aegypti* from Noumea was higher at 21 d.p.i. for the African genotype compared to *Ae. aegypti* from Kone (p < 0.001). Likewise, the TE of African genotype was higher for *Ae. aegypti* from Noumea compared to those from Kone at 21 d.p.i. (p < 0.001). Unfortunately, due to technical issues, it was not possible to estimate the viral load in the saliva of New Caledonia *Ae. aegypti*.

2.3. Aedes aegypti from Brazil and New Caledonia Have Similar Transmission Efficiency for ZIKV of the American Lineage, but Differ Regarding the African and Asian Lineages at Some Incubation Periods

In comparisons of *Ae. aegypti* populations, no significant differences (p > 0.05) were found between the TR and TE values exhibited by the seven Brazilian and two New Caledonian *Ae. aegypti* populations challenged with ZKV of the American lineage, irrespective of incubation time (Tables S3–S5). On the other hand, some significant differences were found in TR and/or TE with the African and Asian ZIKV lineages. Interestingly, with the African ZIKV isolate, TR and TE values at 7 d.p.i were significantly much higher for both New Caledonian populations (Noumea: p < 0.001 and Kone: p < 0.05) when compared with any Brazilian *Ae. aegypti*. However, at 14 d.p.i., TR and TE values were statistically similar (p > 0.05) for all comparisons with the African isolate. At 21 d.p.i, TR and TE values were also essentially similar for all comparisons, with one exception: the New Caledonian *Ae. aegypti* from Kone were significantly much less efficient (p < 0.01) in transmitting the African isolate than all other tested populations. When considering transmission of the Asian lineage, TR values did not differ in any pairwise comparisons at 7, 14 and 21 d.p.i. In contrast, when considering TE at 14 d.p.i., values for the Asian lineage were much higher (p < 0.01) for the New Caledonian Kone population compared to all Brazilian ones except Manaus. At 21 d.p.i., TE values for Kone continued to be lower than Manaus, Rio (p < 0.05) when challenged with ZIKV of the Asian lineage.

2.4. Brazilian Ae. albopictus Poorly Transmits ZIKV Isolates of the Asian Genotype Compared to the African One

We assessed vector competence for the same three strains of ZIKV in five Brazilian populations of *Ae. albopictus* sympatric to the tested *Ae. aegypti* populations. All orally challenged *Ae. albopictus* populations became infected with the three virus isolates, but with significantly higher IR for the African lineage compared to the Asian and American ones (Figure 3A, Table S1). This difference is quite evident for all tested *Ae. albopictus* populations, except for that of Natal, where no significant differences were found between IRs with the American and African isolates at 14 d.p.i. As for the viruses of the Asian ZIKV genotype, the population of Natal could not be challenged with the Asian lineage due to an insufficient number of mosquitoes.

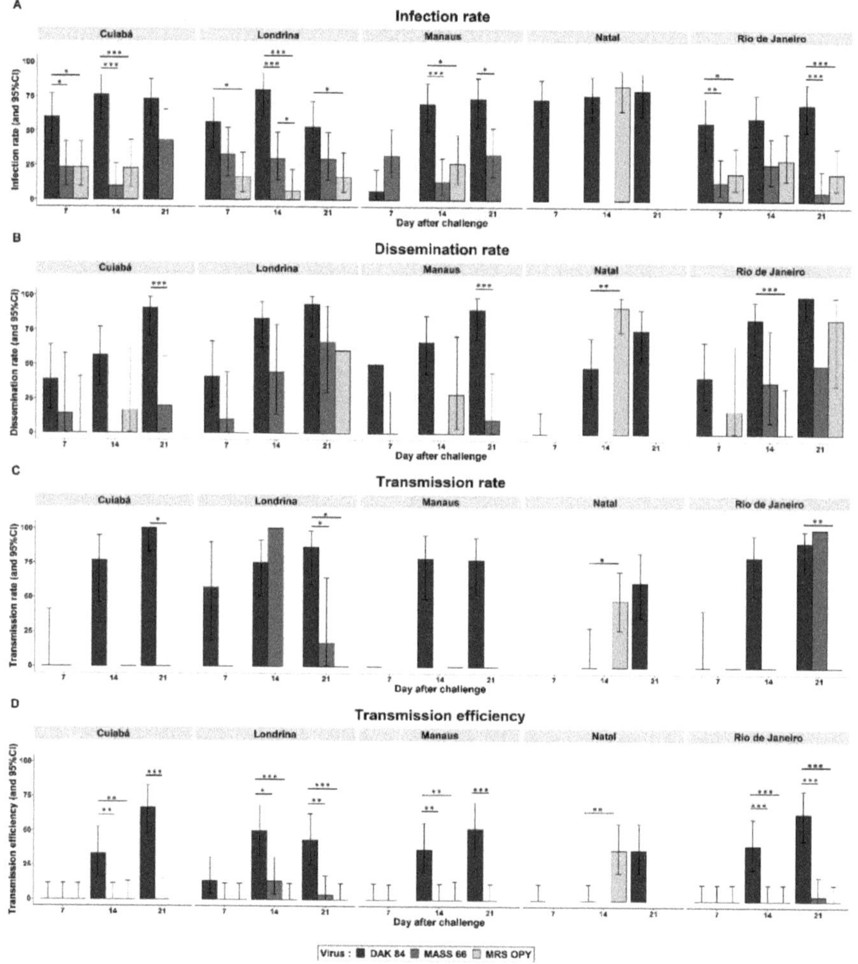

Figure 3. Vector competence results for *Aedes albopictus* from different Brazilian cities orally challenged with three ZIKV three isolates: DAK 84 (African lineage), MASS 66 (Asian lineage), MRS OPY (American lineage). (**A**) Infection rate, (**B**) dissemination rate, (**C**) transmission rate and (**D**) transmission efficiency at 7, 14 and 21 days after oral challenge. Error bars represent 95% confidence intervals. Significant differences are indicated by asterisks (Fisher's Exact test: * $p < 0.05$; ** $p < 0.01$; *** $p < 0.001$).

All three ZIKVs disseminated in all *Ae. albopictus* populations at some time point (Figure 3B, Table S1). Values of DRs significatively increased over time ($p < 0.05$) in all *Ae. albopictus* populations infected with the African ZIKV, except in the mosquitoes from Manaus. No difference in DR values was detected at 7 d.p.i. At 14 or 21 d.p.i., the African virus disseminated at a significantly higher rate than the isolates of the Asian genotype, except for the Londrina and Natal populations. The ZIKV isolate of the American lineage took longer than the other lineages to disseminate to secondary tissues in *Ae. albopictus* from Cuiabá, Manaus (14 d.p.i.) and Londrina (21 d.p.i.).

Transmission was detected from 14 d.p.i. in all tested *Ae. albopictus* populations. With the population of Londrina infected with the African virus, infective particles were detected in saliva even earlier, i.e., at 7 d.p.i. (Figure 3C,D, Table S1). Only the African ZIKV was transmitted by all *Ae. albopictus* populations. Transmission of the Asian and American lineages was heterogeneous among *Ae. albopictus*, and very low to null TR and TE were usually recorded. The populations of Cuiabá and Manaus were unable to transmit any virus of the Asian genotype, while those from Londrina and Rio de Janeiro were able to transmit only the Asian isolate but not the American one; the latter isolate was transmitted by Natal *Ae. albopictus*. When considering the initial number of challenged *Ae. albopictus*, TE values were zero to less than 20% for almost all combinations of mosquito population, virus isolate and incubation time, with a few exceptions with the African genotype. For instance, transmission was detected in ~70% of *Ae. albopictus* from Rio de Janeiro and Cuiabá challenged with ZIKV of the African genotype.

No difference was detected between viral load in saliva expectorated by Brazilian *Ae. albopictus* of the same population infected with the three different isolates (Figure 4). The maximum viral load per saliva of *Ae. albopictus* were 638, 337 and 127 PFU when infected with the African, American and Asian lineages, respectively. Among all *Ae. albopictus* saliva samples at 7 d.p.i, the only positive ones were from four individuals from Londrina with a median of only 2 PFU of ZIKV of the African genotype. The median viral loads varied from 6.5 to 80 and 3 to 60 PFU at 14 and 21 d.p.i., respectively (Table S6).

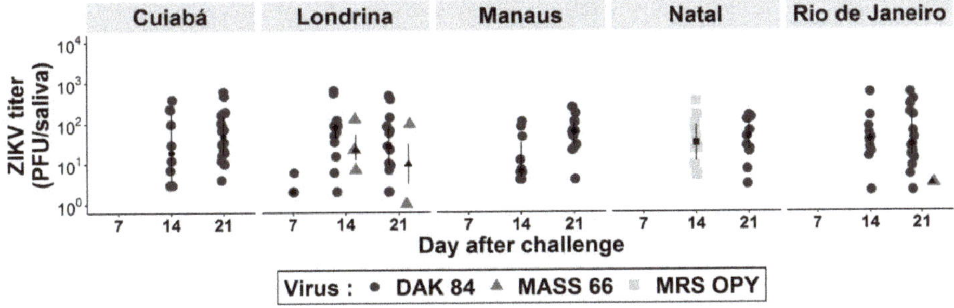

Figure 4. ZIKV load in saliva of *Ae. albopictus* from five Brazilian cities at 7, 14 and 21 days after oral challenge with three isolates: DAK 84 (African lineage), MASS 66 (Asian lineage), MRS OPY (American lineage). Virus was detected by plaque forming unit (PFU) assays on Vero cells.

When comparing viral load per positive saliva between populations of Brazilian *Ae. albopictus* challenged with the same isolate, no significant difference was detected (Tables S3–S5).

2.5. Brazilian Ae. aegypti Are Superior in Vector Competence to All Tested ZIKV Compared to Sympatric Ae. albopictus

The comparison of vector competence of sympatric *Aedes* species from the five Brazilian localities revealed that *Ae. aegypti* is generally much more permissive than *Ae. albopictus* for infection with all three ZIKV lineages. The IR values for *Ae. aegypti* were significantly higher than for sympatric *Ae. albopictus* challenged with the same ZIKV isolate and incubation period in 62% of the 37 possible pairwise comparisons; differences were expressively high ($p < 0.001$ or <0.01) in 15 out of the 37 comparisons (see Table S1). In general, the IRs for Londrina Cuiabá, Manaus and Rio de Janeiro *Ae. aegypti* were significantly higher than for sympatric *Ae. albopictus* despite the ZIKV isolates and incubation times (Tables S3–S5). Curiously, no difference in IR ($p > 0.05$) was found between mosquito species from Natal, although the number of possible comparisons in this case is relatively small. There was a clear tendency for a superiority of *Ae. aegypti* compared to sympatric Brazilian *Ae. albopictus* in the dissemination and transmission of the tested virus, although we were unable to demonstrate this trend statistically in most cases due to the sample size. In fact, the number of individuals with infectious heads among infected mosquitoes, and of individuals with infectious saliva among those with disseminated infection, in *Ae. albopictus* populations was extremely low or nonexistent, greatly decreasing the usefulness of making interspecies comparisons. Despite this limitation, significant differences ($p < 0.05$) in DR and TE between sympatric *Ae. aegypti* and *Ae. albopictus* could be demonstrated, specially concerning the challenges with the African ZIKV isolate, and to a lesser extent, with the American lineage (Table S1). Despite being heterogeneous, all populations of *Ae. aegypti* were able to transmit the two ZIKV isolates of the Asian genotype at 14 and/or 21 d.p.i, while the sympatric *Ae. albopictus* were often unable to transmit or transmitted with very low efficiency. No difference was detected when comparing viral load in positive salivas of sympatric *Ae. albopictus* and *Ae. aegypti* infected with the same ZIKV isolate (Tables S3–S5).

2.6. Culex quinquefasciatus from New Caledonia Is Refractory to ZIKV

No infection was observed for *Cx. quinquefasciatus* from New Caledonia, irrespective of the d.p.i. and the ZIKV strain tested (Table S1).

3. Discussion

We evaluated vector competence to three ZIKV lineages among 13 mosquito populations from Brazil and New Caledonia including *Ae. aegypti*, *Ae. albopictus* and *Cx. quinquefasciatus*. This study is unique in including a number of mosquito populations of three mosquito species from two continents and testing multiple ZIKV strains from three continents all under identical experimental conditions. The two territories focused on, i.e., Brazil and New Caledonia, were at the origin of the recent ZIVK pandemic.

With the exception of *Cx. quinquefasciatus,* all challenged New Caledonian and Brazilian *Aedes* were susceptible to infection with all tested ZIKV strains. However, viral dissemination to secondary tissues and transmission was highly variable among the 12 *Aedes* populations; ZIVK lineage and incubation time strongly influenced differences. The African ZIKV disseminated faster and at higher rates, regardless of incubation time in all 12 *Aedes* populations, compared to the Asian and American ZIKV lineages with a single exception, i.e., Kone *Ae. aegypti* at 21 d.p.i. The decrease in susceptibility to infection in these mosquitoes suggests that *Ae. aegypti* from Kone could be more resistant to infection by African ZIKV than those from Noumea. The African strain was the only one to have been transmitted as early as 7 d.p.i. in both *Ae. aegypti* and *Ae. albopictus,* with the two New Caledonian *Ae. aegypti* populations being significantly more efficient in transmitting this strain than the Brazilian populations. These advantages in transmission rates of the African ZIKV genotype relative to the Asian genotype were similarly described for Guadalupian *Ae. aegypti*, challenged with the same isolates and viral load [39]. Previous studies with other African ZIKV strains have shown greater transmission rates in *Ae. aegypti* from Mexico, USA, Brazil and New Caledonia, as well as higher fitness in vitro compared to ZIKV of the Asian genotype [8,31,40,41]. Curiously, it has been shown that New Caledonian *Ae. aegypti* are able to transmit African ZIKV isolates earlier than both the American and Asian ZIKV isolates [42]. It is worth noting that all seven New Caledonian and Brazilian *Ae. aegypti* tested here transmitted all virus lineages from two weeks or more after oral challenge, with the viral load in positive Brazilian *Ae. aegypti* salivas not differing among the tested viruses, except for those from Rio de Janeiro, which presented higher viral loads when infected with the African isolate compared to the American one at 14 d.p.i. When considering the performance of the two ZIKV strains belonging to the Asian genotype in *Ae. aegypti*, we found that the dissemination rates of the Asian lineage were lower than those detected for the American lineage, independent of the mosquito population origin, while rates of transmission did not differ. Interestingly, the seven Brazilian and New Caledonian *Ae. aegypti* did not differ in their efficiency of transmitting the American ZKV lineage at 14 and 21 days' extrinsic incubation. The American ZIKV lineage, the only one circulating in Brazil, has caused disease outbreaks with distinct incidences in the five Brazilian regions that were the origin of *Ae. aegypti* populations assessed herein. During the 2016 outbreak, the Central-western region had the highest incidence of probable Zika cases (222.0/100 thousand inhabitants), followed by the Northeast (134.4/100 thousand inhab.), Southeast (106.2/100 thousand inhab.) and South (3.4/100 thousand inhab.) [12]. Annual incidences in all regions have decreased considerably in recent years, but the Northeast continues to report the highest rates (9.5/100 thousand inhab. in 2019) [14]. In this context, it is important to point out that vector competence is only one component determining the vectorial capacity of *Ae. aegypti* populations, the accurate assessment of which requires analyses of epidemiological, environmental and climate factors to help better understand the vector transmission dynamics and extent and intensity of urban arbovirus transmission [20,43,44].

The vector competence of Brazilian *Ae. albopictus* has been poorly evaluated [35,45]. Here, we assessed the vector competence of sympatric *Ae. aegypti* and *Ae. albopictus* from five Brazilian cities. In agreement with previous works [25,35,46,47], *Ae. aegypti* displayed greater susceptibility to infection and likelihood of transmitting the three tested ZIKVs compared to *Ae. albopictus* from Brazil. In the present study, the dissemination rates and transmission efficiency in *Ae. aegypti* were significantly higher than in *Ae. albopictus*, especially for the African ZIKV, although this isolate could be detected in the saliva of all five tested *Ae. albopictus* populations. Ciota et al. (2017) [48] found that USA *Ae. albopictus* were more susceptible to infection than *Ae. aegypti*, but that transmission efficiency was lower than *Ae. aegypti*, a result that suggested the existence of important transmission barriers in *Ae. albopictus*. In contrast to the African ZIKV, Brazilian *Ae. albopictus* were incapable of transmitting ZIKV of the Asian and American lineages in essentially all cases. The inefficiency in transmitting American strains of ZIKV was previously reported [45]. Importantly, in spite of its potential to transmit ZIKV, including reports of natural ZIKV infections elsewhere, there is no solid evidence that *Ae. albopictus* has caused human ZIKV transmission in Brazil [20,21,23,25].

The epidemiological histories of ZIKV differ in Brazil and New Caledonia, but they are quite related by the fact that the rapid spread of the virus in Brazil was preceded by the epidemic passage of the same genotype in the Pacific, particular in New Caledonia [49,50]. On the other hand, the histories of colonization by the two *Aedes* species tested herein differed considerably. While *Ae. aegypti* is very widespread in Brazil as well as in most of the Pacific islands, the invasive Asian mosquito *Ae. albopictus* has not yet colonized New Caledonia [51]. The invasion and colonization of the Americas by the African mosquito *Ae. aegypti* dates back more than four centuries; the species was eradicated in Brazil in the 1950s and later reintroduced from populations from neighboring countries [52–54], while the Pacific area and New Caledonia were colonized more recently, mostly at the end of the XIXth and beginning of the XXth century by founders of Asian, Australian and American origins [51,55]. In addition, the establishment of populations in distinct ecosystems, climates and environments, as well as exposure to control measures, differently influenced natural history, microbiota and population genetics, which, in turn, modulated phenotypes like vector competence [38,56].

The Brazilian mosquito populations tested here originated from five cities with differing demographic and infrastructural conditions, as well as being environmentally distinct, varying from semiarid to equatorial, with primary vegetation coverage from the savanna-like *cerrado* to the Atlantic and Amazon rain forest. The relatively low human population densities of the inland cities of Cuiabá, Manaus and Londrina (163.88, 158.06 and 306.49 inhabitants/Km^2, respectively) contrast with high densities in the coastal Natal and Rio de Janeiro (4808.20 and 5265.81 inhab./Km^2) [57]. Insecticide-induced changes in resistant mosquitoes may be associated with phenotypic variations in vector competence, either through sublethal exposure to insecticides or indirectly through changes in the environment [56]. Genes controlling resistance to insecticides can have pleiotropic effects and result in changes in the insect's vector capacity, such as longevity, behavior, and vector competence [58]. The Brazilian *Aedes* mosquito populations tested here differed considerably in terms of their resistance to insecticides such as Temephos and Deltramethrin which were employed in the National Dengue Control Program [59–66]. The Natal and Rio de Janeiro *Ae. aegypti* populations exhibit greater resistance rates to the larvicide Temephos (an organophosphate), while Londrina and Manaus exhibit higher resistance to the adulticide Deltramethrin (a pyrethroid) [65]. The high frequency of *kdr* alleles that contribute to pyrethroid resistance is widespread in Brazilian *Ae. aegypti*, including the populations challenged here [65], and the Phe1534Cys *kdr* mutation has been detected in *Ae. albopictus* populations, including in the nearby of Londrina [64]. In New Caledonia, important resistant *kdr* mutations were not detected in Noumea *Ae. aegypti*, which display low resistance levels to Deltramethrin [67]. Large genetic differentiation has been reported among Brazilian *Ae. aegypti* [68–72]. In contrast, lower genetic differentiation has been recorded between New Caledonian *Ae. aegypti* from the main island (Noumea and Poindimie) than from a smaller island (from Ouvea) [51]. Additionally, except for differences in demographic densities, Noumea (2186.6 inhabitants/Km^2) and Kone (19.6 inhabitants/Km^2) are quite similar with regard to climate (http://www.meteo.nc/nouvelle-caledonie/climat/fiches-climatologiques). Together, these observations help to explain similarities in the infection and dissemination rates between the two New Caledonian *Ae. aegypti* populations, regardless of the tested viruses. However, transmission rates and efficiency differed, which may suggest different frequencies in transmission barriers. In sum, heterogeneity in vector competence to different ZIKV lineages in *Ae. aegypti* and *Ae. albopictus* is not surprising, considering their differing origins and histories [20,23,45].

As expected, the New Caledonian *Cx. quinquefasciatus* was refractory to all three ZIKV strains. This result agrees with the repeated vector competence evaluations performed with ZIKV isolates and *Cx. quinquefasciatus*, as well as *Cx. pipiens* from several origins [28,30,40,48,73–81]. In view of these data, no role in the human transmission of ZIKV can be attributed to *Cx. quinquefasciatus* from New Caledonian or elsewhere [30,74,81].

There are likely genetic, technical and environmental factors that contribute to the variations in vector competence in *Aedes* mosquitoes observed in the laboratory that may not always be relevant to phenotypic variations in nature [38,40,45,82,83]. For example, mosquitoes fed directly on viremic ZIKV vertebrates tend to become more infected and transmit better. Thus, it is possible that Brazilian and New Caledonian *Aedes* are more competent than what we observed in the laboratory, which indirectly suggests a potentially greater risk of transmission. While population origin influences transmission efficiency, Brazilian and New Caledonian *Ae. aegypti* are competent vectors for ZIKV of African, Asian and American lineages, with advantages for the African lineage, which can also be transmitted by Brazilian *Ae. albopictus*. Reinforcing viral surveillance and improving and strengthening control measures to reduce infestation by domestic and peridomestic *Aedes* are imperative if endemic Brazilian and New Caledonian territories are to avoid the spread of new ZIKV lineages and mitigate the transmission of locally circulating ZIKV in Brazil.

4. Materials and Methods

4.1. Ethic Statements

This study was done in compliance with New Caledonia Ethic regulations and the Institutional Ethics Committee on Animal Use (CEUA-IOC license LW-34/14) at the Instituto Oswaldo Cruz. This study did not involve endangered or protected species.

4.2. Viral Strains

Three ZIKV strains were used, each representing a viral lineage: DAK 84 (Senegal 1984, African genotype), MRS_OPY_Martinique_PaRi_2015 (Martinique 2015, Asian genotype, American lineage) and MASS 66 (Malaysia 1966, Asian genotype, Asian lineage). Viruses were provided by the Emergence

Virus Unit (Marseilles, France) via the European EVAg project. Lyophilizates were resuspended in sterile distilled water and inoculated onto Vero cells (ATCC, ref. CCL-81) for viral production using a multiplicity of infection of 0.1 and DMEM medium supplemented with 2% fetal bovine serum (FBS). Supernatants were collected after three days of growth and stored at −80 °C prior to vector competence experiments. The viral titer of each virus strain was determined by serial 10-fold dilutions of viral stock on Vero cells, and was expressed as 50% tissue culture infective dose per milliliter ($TCID_{50}$/mL). In the case of MRS_OPY_Martinique_PaRi_2015 ZIKV strain, virus stock was centrifuged using Vivaspin 6 centrifugal concentrator (Sartorius, Stonehouse, UK) to achieve the final concentration at 10^7 $TCID_{50}$/mL used in the oral challenge experiments.

4.3. Mosquito Populations

Three mosquito species of eight populations were orally challenged: two of *Ae. aegypti* (F1 generation) and one of *Cx. quinquefasciatus* (F0 generation) from New Caledonia, and five of *Ae. aegypti* (F2–F4 generation) and *Ae. albopictus* (F2–F4 generation) from Brazil (Figure 5). In New Caledonia, *Ae. aegypti* (larvae) and *Cx. quinquefasciatus* (eggs rafts) were collected in Noumea (Southwestern), Kone (Northwestern) and Dumbea (Southwestern) during the hot season (April and March 2018) (Figure 1). The Brazilian *Ae. aegypti* and *Ae. albopictus* were derived from eggs collected with ovitraps set in Manaus (Northern, Amazon), Natal (Northeastern), Rio de Janeiro (Southeastern), Cuiabá (West-Central), Londrina (Southern) at different periods (Figure 5). Larvae were reared in 2 L plastic pans, with a density of 150–200 larvae per pan containing dechlorinated tap water, supplemented with brewer's yeast which was renewed every 2–3 days. Pupae were then transferred to rearing cages where adults were maintained at 28 ± 1 °C, 70–80% relative humidity and a 12:12 h light:dark cycle, with access to 10% sucrose solution ad libitum.

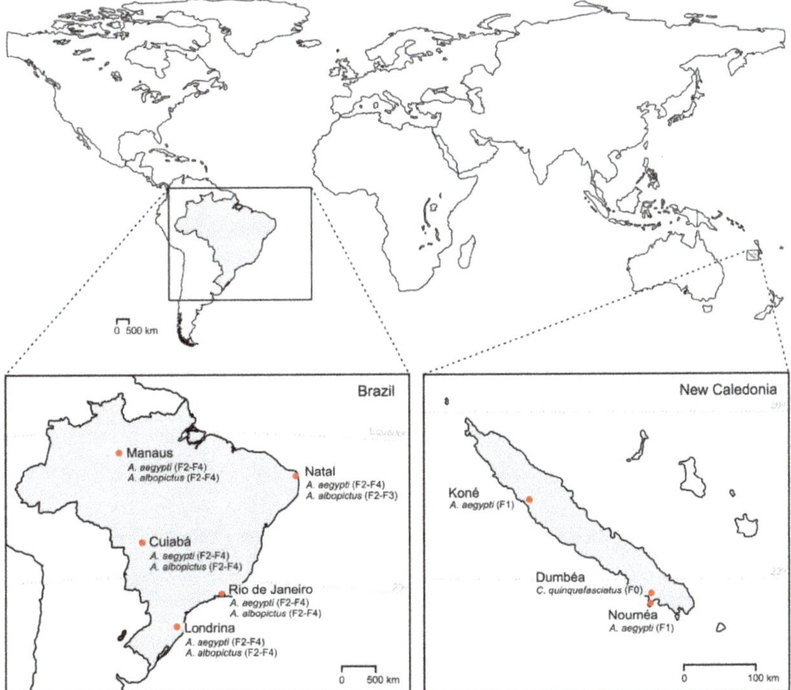

Figure 5. Species and localization of mosquitoes from Brazil and New Caledonia challenged with Zika virus. The generations of tested mosquitoes are in parenthesis.

4.4. Oral Challenge

Seven- to ten-day-old nulliparous females were used for oral infection, with a preliminary starvation of 24 h and 48 h for *Aedes* species and *Cx. quinquefasciatus*, respectively. The infectious blood meal consisted of a mix (2:1) of washed rabbit erythrocytes and viral suspension at 3×10^7 $TCID_{50}$/mL (final concentration at 10^7 $TCID_{50}$/mL) supplemented with a phagostimulant (5 mM ATP). Mosquito feeding was performed for 20 min with a Hemotek system (Hemotek Limited, Great Harwood, UK). Fully engorged females were transferred into cardboard containers and maintained with 10% sucrose solution at 28 ± 1 °C, 70–80% relative humidity and a 12:12 h light:dark cycle for further analysis.

4.5. Infection, Dissemination and Transmission Analyses

For each combination of viral strain and mosquito population, 20–30 females were randomly analyzed at 7, 14 and 21 days after oral challenge (hereinafter abbreviated as d.p.i.). Mosquitoes were individually processed using disposable and disinfected supplies to avoid contamination between individuals and between tissues of the same mosquito, as previously described [84]. After brief anesthesia by exposure to cold (ice bath) and removal of the wings and legs, mosquito saliva was collected in 5 μL of FBS for 30 min at room temperature. Each saliva sample was added to 45 μL of DMEM medium and stored at −80 °C until use. For the determination of viral infection and dissemination, each mosquito body (abdomen and thorax) and head was separately ground in 350 μL of DMEM medium supplemented with 2% FBS and antibiotics/antifungals (100 units/mL of penicillin, 0.1 mg/mL of streptomycin and 0.25 μg/mL amphotericin B). The obtained homogenates were then centrifuged at 10,000 g for 5 min at 4 °C before inoculation in cell culture. The infectious status was determinate by the presence of cytopathogenic effect (CPE). Briefly, 100 μL of diluted samples were inoculated onto confluent monolayer Vero cells in 96-well plates, and incubated for 7 days at 37 °C with 5% CO_2 under 2.4% CMC (carboxymethyl cellulose) or agarose. Plates were then stained with a 0.2% crystal violet solution (in 10% formaldehyde and 20% ethanol). The presence of ZIKV particles in saliva and viral titer were determined by plaque assay. Saliva samples were inoculated in six-well plates seeded with Vero cells, and incubated and stained as described above. Saliva titers were expressed as plaque forming units (PFU) per saliva sample. The infection rate corresponded to the proportion of mosquitoes with infected bodies among all the tested mosquitoes. The dissemination rate represented the proportion of mosquitoes with infectious heads among infected mosquitoes. The transmission rate corresponded to the proportion of mosquitoes with infectious viral particles in saliva among mosquitoes with infected heads. The transmission efficiency corresponded to the proportion of mosquitoes with virus in saliva among all the orally challenged mosquitoes.

4.6. Statistical Analyses

The different rates and efficiencies calculated either by mosquito populations, viral strains or days postchallenge were compared by a Fisher's exact test. Quantitative data corresponding to viral titer per saliva were compared using Wilcoxon test. All tests were corrected for multiple comparisons using the Holm method. All statistical analyses were performed with the R v3.6.1 software [85] considering p-values < 0.05 as significant.

Supplementary Materials: The following are available online at http://www.mdpi.com/2076-0817/9/7/575/s1, Table S1: Infection rates, dissemination rates, transmission rates and transmission efficiencies obtained for all the mosquito populations tested with the three Zika virus isolates at 7, 14 and 21 days after oral challenge. Table S2: Medians and interquartile rages of viral load in saliva of *Aedes aegypti* Brazilian populations challenged with the three Zika virus isolates at 7, 14 and 21 days after challenge. Table S3: Summarize of the p-values obtained with the Fisher'Exact test and the Wilcoxon test corrected with the Holm method of infection rates, dissemination rates, transmission rates, transmission efficiencies and viral loads in saliva obtained for all the mosquito populations tested with the Zika virus of the African lineage at 7, 14 and 21 days after oral challenge. Table S4: Summarize of the p-values obtained with the Fisher'Exact test and the Wilcoxon test corrected with the Holm method of infection rates, dissemination rates, transmission rates, transmission efficiencies and viral loads in saliva obtained for all the mosquito populations tested with the Zika virus of the Asian lineage at 7, 14 and 21 days after oral challenge. Table S5: Summarize of the p-values obtained with the Fisher'Exact test and the Wilcoxon test corrected with the

Holm method of infection rates, dissemination rates, transmission rates, transmission efficiencies and viral loads in saliva obtained for all the mosquito populations tested with the Zika virus of the American lineage at 7, 14 and 21 days after oral challenge. Table S6: Medians and interquartile rages of viral load in saliva of *Aedes albopictus* Brazilian populations challenged with the three Zika virus isolates at 7, 14 and 21 days after challenge.

Author Contributions: Conceptualization: R.L.-d.-O., M.D.-R.; Data Curation: R.S.F., O.O., M.R.D., D.G.; Formal Analysis: O.O., N.P., R.S.F.; Funding Acquisition: R.L.-d.-O., M.D.-R.; Investigation: R.L.-d.-O., M.D.-R.; Methodology: R.S.F., M.I.L.B., D.G., M.R.D., N.P.; Supervision: M.D.-R., R.L.-d.-O.; Visualization: O.O.; Writing—Original Draft: R.S.F., R.L.-d.-O., O.O., M.D.-R.; Writing—Review & Editing: R.S.F., R.L.-d.-O., O.O., M.D.-R., N.P., M.R.D., D.G., M.I.L.B. All authors have read and agreed to the published version of the manuscript.

Funding: National Institut of Health (Grant no. 1UO1 AI115595-01), the European Union's Horizon 2020 Research and Innovation Programme under ZIKAlliance Grant Agreement no. 734548; Conselho Nacional de Desenvolvimento Científico e Tecnológico (Grant no. 312446/2018), Preventing and Combating the Zika Virus MCTIC/FNDCT-CNPq/MEC-CAPES/MS-Decit. (Grant no. 440929/2016-4), Fundação Carlos Chagas Filho de Amparo à Pesquisa do Estado do Rio de Janeiro (Grants nos. E-26/203.064/2016, E-26/202.431/2019 and E-26/201.335/2016). MRD was supported by "Cadre Avenir" and "Vivaldi, South Province", New Caledonia.

Acknowledgments: We thank Jeffrey Powell for the critical review and English editing of the manuscript, Sosiasi Kilama for field sampling, Marine Minier and Jordan Tutagata for mosquito rearing, Iule de Souza Bonelly and Marcelo Quintela Gomes for technical support and Heloisa Diniz for preparing the map.

Conflicts of Interest: The authors declare no conflict of interest.

References

1. Duffy, M.R.; Guillaumot, L.; Biggerstaff, B.J.; Hancock, W.T.; Lanciotti, R.S.; Holzbauer, S.; Lambert, A.J.; Chen, T.-H.; Bel, M.; Hayes, E.B.; et al. Zika virus outbreak on Yap Island, Federated States of Micronesia. *N. Engl. J. Med.* **2009**, *360*, 2536–2543. [CrossRef]
2. Musso, D.; Gubler, D.J. Zika virus. *Clin. Microbiol. Rev.* **2016**, *29*, 487–524. [CrossRef]
3. Cao-Lormeau, V.; Roche, C.; Teissier, A.; Robin, E.; Berry, A.; Mallet, H.; Sall, A.A.; Musso, D. Zika virus, French Polynesia, South Pacific, 2013. *Emerg. Infect. Dis.* **2014**, *20*, 1085–1086. [CrossRef]
4. Cao-Lormeau, V.M.; Blake, A.; Mons, S.; Lastère, S.; Roche, C.; Vanhomwegen, J.; Dub, T.; Baudouin, L.; Teissier, A.; Larre, P.; et al. Guillain-Barré Syndrome outbreak associated with Zika virus infection in French Polynesia: A case-control study. *Lancet* **2016**, *387*, 1531–1539. [CrossRef]
5. Aubry, M.; Teisser, A.; Huart, M.; Merceron, S.; Vanhomwegen, J.; Roche, C.; Vial, A.L.; Teururai, S.; Sicard, S.; Paulous, S.; et al. ZIKV seroprevalence French Polynesia 2014–2015. *Emerg. Infect. Dis.* **2017**, *23*, 669–672. [CrossRef]
6. Dupont-Rouzeyrol, M.; O'Connor, O.; Calvez, E.; Daures, M.; John, M.; Grangeon, J.P.; Gourinat, A.C. Co-infection with Zika and dengue viruses in 2 patients, New Caledonia, 2014. *Emerg. Infect. Dis.* **2015**, *21*, 381–382. [CrossRef] [PubMed]
7. Gourinat, A.C.; O'Connor, O.; Calvez, E.; Goarant, C.; Dupont-Rouzeyrol, M. Detection of Zika virus in urine. *Emerg. Infect. Dis.* **2015**, *21*, 84–86. [CrossRef] [PubMed]
8. Calvez, E.; Mousson, L.; Vazeille, M.; O'Connor, O.; Cao-Lormeau, V.M.; Mathieu-Daudé, F.; Pocquet, N.; Failloux, A.B.; Dupont-Rouzeyrol, M. Zika virus outbreak in the Pacific: Vector competence of regional vectors. *PLoS Negl. Trop. Dis.* **2018**, *12*, 1–12. [CrossRef] [PubMed]
9. Roth, A.; Mercier, A.; Lepers, C.; Hoy, D.; Duituturaga, S.; Benyon, E.; Guillaumot, L.; Souarès, Y. Concurrent outbreaks of dengue, chikungunya and Zika virus infections—An unprecedented epidemic wave of mosquito-borne viruses in the Pacific 2012–2014. *Eurosurveillance* **2014**, *19*, 1–8. [CrossRef]
10. Possas, C.; Brasil, P.; Marzochi, M.C.; Tanuri, A.; Martins, R.M.; Marques, E.T.; Bonaldo, M.C.; Ferreira, A.G.; Lourenço-de-Oliveira, R.; Nogueira, R.M.R.; et al. Zika puzzle in Brazil: Peculiar conditions of viral introduction and dissemination—A review. *Mem. Inst. Oswaldo Cruz* **2017**, *112*, 319–327. [CrossRef] [PubMed]
11. MS—Ministério da Saúde/Secretaria de Vigilância em Saúde. Boletim Epidemiológico 47: Situação Epidemiológica da Infecção Pelo Vírus Zika no Brasil, de 2015 a 2017. Available online: https://www.saude.gov.br/images/pdf/2018/novembro/12/2018-034.pdf (accessed on 11 June 2020).

12. MS—Ministério da Saúde/Secretaria de Vigilância em Saúde. Boletim Epidemiológico—Até a Semana Epidemiológica 52. 2016. Available online: http://portalarquivos2.saude.gov.br/images/pdf/2017/abril/06/2017-002-Monitoramento-dos-casos-de-dengue--febre-de-chikungunya-e-febre-pelo-v--rus-Zika-ate-a-Semana-Epidemiologica-52--2016.pdf (accessed on 11 June 2020).

13. MS—Ministério da Saúde/Secretaria de Vigilância em Saúde. Monitoramento dos Casos de Arboviroses Urbanas Transmitidas pelo Aedes (Dengue, Chikungunya e Zika), Semanas Epidemiológicas 1 a 34. 2019. Available online: http://www.saude.gov.br/images/pdf/2019/setembro/11/BE-arbovirose-22.pdf (accessed on 3 May 2020).

14. MS—Ministério da Saúde/Secretaria de Vigilância em Saúde. Monitoramento dos Casos de Arboviroses Urbanas Transmitidas Pelo Aedes (Dengue, Chikungunya e Zika), Semanas Epidemiológicas 01 a 52. 2019. Available online: https://portalarquivos2.saude.gov.br/images/pdf/2020/janeiro/20/Boletim-epidemiologico-SVS-02-1-.pdf (accessed on 3 May 2020).

15. Coelho, A.V.; Crovella, S. Microcephaly prevalence in infants born to Zika virus-infectedwomen: A systematic review and meta-analysis. *Int. J. Mol. Sci.* **2017**, *18*, 1714. [CrossRef] [PubMed]

16. Kuadkitkan, A.; Wikan, N.; Sornjai, W.; Smith, D.R. Zika virus and microcephaly in Southeast Asia: A cause for concern? *J. Infect. Public Health* **2020**, *13*, 11–15. [CrossRef] [PubMed]

17. Brady, O.J.; Osgood-Zimmerman, A.; Kassebaum, N.J.; Ray, S.E.; De Araùjo, V.E.M.; Da Nóbrega, A.A.; Frutuoso, L.C.V.; Lecca, R.C.R.; Stevens, A.; De Oliveira, B.Z.; et al. The association between Zika virus infectionand microcephaly in Brazil 2015–2017: An observational analysis of over 4 million births. *PLoS Med.* **2019**, *16*, 1–21. [CrossRef] [PubMed]

18. Keane, Í.; Dias, R.; Luciana, C.; Maria, R.; Martins, G.; Fernanda, K.; Santana, S.; Vieira, S.; Joventino, E.S.; Corina, M.; et al. Review Article Zika virus:—A review of the main aspects of this type of arbovirosis. *Rev. Soc. Bras. Med. Trop.* **2018**, *51*, 261–269. [CrossRef]

19. Petersen, L.R.; Jamieson, D.J.; Powers, A.M.; Honein, M.A. Zika virus. *N. En gl. J. Med.* **2016**, *374*, 1552–1563. [CrossRef]

20. Gutiérrez-Bugallo, G.; Piedra, L.A.; Rodriguez, M.; Bisset, J.A.; Lourenço-de-Oliveira, R.; Weaver, S.C.; Vasilakis, N.; Vega-Rúa, A. Vector-borne transmission and evolution of Zika virus. *Nat. Ecol. Evol.* **2019**, *3*, 561–569. [CrossRef]

21. Ferreira-de-Brito, A.; Ribeiro, I.P.; De Miranda, R.M.; Fernandes, R.S.; Campos, S.S.; Antunes, K.; Da Silva, B.; De Castro, M.G.; Bonaldo, M.C.; Brasil, P.; et al. First detection of natural infection of *Aedes aegypti* with Zika virus in Brazil and throughout South America. *Mem. Inst. Oswaldo Cruz* **2016**, *11*, 655–658. [CrossRef]

22. Weaver, S.C.; Costa, F.; Garcia-Blanco, M.A.; Ko, A.I.; Ribeiro, G.S.; Saade, G.; Shi, P.Y.; Vasilakis, N. Zika virus: History, emergence, biology, and prospects for control. *Antivir. Res.* **2016**, *130*, 69–80. [CrossRef]

23. Epelboin, Y.; Talaga, S.; Epelboin, L.; Dusfour, I. Zika virus: An updated review of competent or naturally infected mosquitoes. *PLoS Negl. Trop. Dis.* **2017**, *11*, 1–22. [CrossRef]

24. Grard, G.; Caron, M.; Mombo, M.; Nkoghe, D.; Ondo, S.M.; Jiolle, D.; Fontenille, D.; Paupy, C.; Leroy, E.M. Zika virus in Gabon (Central Africa)–2007: A new threat from *Aedes albopictus*? *PLoS Negl. Trop. Dis.* **2014**, *8*, 1–6. [CrossRef]

25. Pereira-dos-Santos, T.; Roiz, D.; Lourenço-de-Oliveira, R.; Paupy, C. A systematic review: Is *Aedes albopictus* an efficient bridge vector for zoonotic arboviruses? *Pathogens* **2020**, *9*, 266. [CrossRef] [PubMed]

26. Ayres, C.F.J.; Guedes, D.R.D.; Paiva, M.H.S.; Morais-Sobral, M.C.; Krokovsky, L.; Machado, L.C.; Melo-Santos, M.A.V.; Crespo, M.; Oliveira, C.M.F.; Ribeiro, R.S.; et al. Zika virus detection, isolation and genome sequencing through Culicidae sampling during the epidemic in Vitória, Espírito Santo, Brazil. *Parasites Vectors* **2019**, *12*, 1–9. [CrossRef] [PubMed]

27. Moutailler, S.; Yousfi, L.; Mousson, L.; Devillers, E.; Vazeille, M.; Vega-Rúa, A.; Perrin, Y.; Jourdain, F.; Chandre, F.; Cannet, A.; et al. A new high-throughput tool to screen mosquito-borne viruses in Zika virus endemic/epidemic areas. *Viruses* **2019**, *11*, 904. [CrossRef]

28. Fernandes, S.R.; Campos, S.S.; Ferreira-de-Brito, A.; Miranda, M.R.; Silva, A.B.K.; Gonçalves, C.M.; Raphael, L.M.; Brasil, P.; Failloux, A.; Bonaldo, M.C.; et al. Culex quinquefasciatus from Rio de Janeiro is not competent to transmit the local Zika Virus. *PLoS Negl. Trop. Dis.* **2016**, *10*, 1–13. [CrossRef] [PubMed]

29. Leal, W.S. Zika mosquito vectors: The jury is still out. *F1000Research* **2016**, *5*, 1–10. [CrossRef]

30. Lourenço-de-Oliveira, R.; Failloux, A.B. Lessons learned on Zika virus vectors. *PLoS Negl. Trop. Dis.* **2017**, *11*, 15–17. [CrossRef]

31. Roundy, C.M.; Azar, S.R.; Rossi, S.L.; Huang, J.H.; Leal, G.; Yun, R.; Fernandez-Salas, I.; Vitek, C.J.; Paploski, I.A.D.; Kitron, U.; et al. Variation in *Aedes aegypti* mosquito competence for Zika virus transmission. *Emerg. Infect. Dis.* **2017**, *23*, 625–632. [CrossRef]

32. Musso, D.; Nilles, E.J.; Cao-Lormeau, V.M. Rapid spread of emerging Zika virus in the Pacific area. *Clin. Microbiol. Infect.* **2014**, *20*, 595–596. [CrossRef]

33. Zanluca, C.; De Melo, V.C.A.; Mosimann, A.L.P.; Dos Santos, G.I.V.; dos Santos, C.N.D.; Luz, K. First report of autochthonous transmission of Zika virus in Brazil. *Mem. Inst. Oswaldo Cruz* **2015**, *110*, 569–572. [CrossRef]

34. Bonaldo, M.C.; Ribeiro, I.P.; Lima, N.S.; Santos, A.A.C. Isolation of infective Zika virus from urine and saliva of patients in Brazil. *PLoS Negl. Trop. Dis.* **2016**, *10*, 1–17. [CrossRef]

35. Chouin-carneiro, T.; Vega-rua, A.; Vazeille, M.; Yebakima, A. Differential susceptibilities of *Aedes aegypti* and *Aedes albopictus* from the Americas to Zika virus. *PLoS Negl. Trop. Dis.* **2016**, *10*, 1–11. [CrossRef] [PubMed]

36. Pompon, J.; Morales-Vargas, R.; Manuel, M.; Tan, C.H.; Vial, T.; Tan, J.H.; Sessions, O.M.; Vasconcelos, P.D.C.; Ng, L.C.; Missé, D. A Zika virus from America is more efficiently transmitted than an Asian virus by *Aedes aegypti* mosquitoes from Asia. *Sci. Rep.* **2017**, *7*, 1–8. [CrossRef] [PubMed]

37. Lambrechts, L. Quantitative genetics of *Aedes aegypti* vector competence for dengue viruses: Towards a new paradigm? *Trends Parasitol.* **2011**, *27*, 111–114. [CrossRef] [PubMed]

38. Tabachnick, W.J. Nature, nurture and evolution of intra-species variation in mosquito arbovirus transmission competence. *Int. J. Environ. Res. Public Health* **2013**, *10*, 249–277. [CrossRef]

39. Hery, L.; Boullis, A.; Delannay, C.; Vega-Rúa, A. Transmission potential of African, Asian and American Zika virus strains by *Aedes aegypti* and *Culex quinquefasciatus* from Guadeloupe (French West Indies). *Emerg. Microbes Infect.* **2019**, *8*, 699–706. [CrossRef] [PubMed]

40. Weger-Lucarelli, J.; Ruckert, C.; Chotiwan, N.; Nguyen, C.; Weger-lucarelli, J.; Ru, C.; Luna, M.G.; Fauver, J.R.; Foy, B.D.; Perera, R.; et al. Vector competence of American Mosquitoes for three strains of Zika virus. *PLoS Negl. Trop. Dis.* **2016**, *10*, 1–16. [CrossRef]

41. Smith, D.R.; Sprague, T.R.; Hollidge, B.S.; Valdez, S.M.; Padilla, S.L.; Bellanca, S.A.; Golden, J.W.; Coyne, S.R.; Kulesh, D.A.; Miller, L.J.; et al. African and Asian Zika virus isolates display phenotypic differences both in vitro and in vivo. *Am. J. Trop. Med. Hyg.* **2018**, *98*, 432–444. [CrossRef]

42. Calvez, E.; O'Connor, O.; Pol, M.; Rousset, D.; Faye, O.; Richard, V.; Tarantola, A.; Dupont-Rouzeyrol, M. Differential transmission of Asian and African Zika virus lineages by *Aedes aegypti* from New Caledonia. *Emerg. Microbes Infect.* **2018**, *7*, 1–10. [CrossRef]

43. Wilcox, B.A.; Gubler, D.J. Disease ecology and the global emergence of zoonotic pathogens. *Environ. Health Prev. Med.* **2005**, *10*, 263–272. [CrossRef]

44. Tabachnick, W.J. Challenges in predicting climate and environmental effects on vector-borne disease episystems in a changing world. *J. Exp. Biol.* **2010**, *213*, 946–954. [CrossRef]

45. Azar, S.R.; Roundy, C.M.; Rossi, S.L.; Huang, J.H.; Leal, G.; Yun, R.; Fernandez-Salas, I.; Vitek, C.J.; Paploski, I.A.D.; Stark, P.M.; et al. Differential vector competency of *Aedes albopictus* populations from the Americas for Zika virus. *Am. J. Trop. Med. Hyg.* **2017**, *97*, 330–339. [CrossRef] [PubMed]

46. Di Luca, M.; Severini, F.; Toma, L.; Boccolini, D.; Romi, R.; Remoli, M.E.; Sabbatucci, M.; Rizzo, C.; Venturi, G.; Rezza, G.; et al. Experimental studies of susceptibility of Italian *Aedes albopictus* to Zika virus. *Eurosurveillance* **2016**, *21*. [CrossRef] [PubMed]

47. Lozano-Fuentes, S.; Kenney, J.L.; Varnado, W.; Byrd, B.D.; Kristen, L.; Savage, H.M. Susceptibility and vectorial capacity of American *Aedes albopictus* and *Aedes aegypti* (Diptera: Culicidae) to American Zika virus strains. *J. Med. Entomol.* **2020**, *56*, 233–240. [CrossRef]

48. Ciota, A.T.; Bialosuknia, S.M.; Zink, S.D.; Brecher, M.; Ehrbar, D.J.; Morrissette, M.N.; Kramer, L.D. Effects of Zika virus strain and *Aedes* mosquito species on vector competence. *Emerg. Infect. Dis.* **2017**, *23*, 1110–1117. [CrossRef] [PubMed]

49. Musso, D. Zika virus transmission from French Polynesia to Brazil. *Emerg. Infect. Dis.* **2015**, *21*, 1887–1889. [CrossRef] [PubMed]

50. Musso, D.; Cao-Lormeau, V.M.; Gubler, D.J. Zika virus: Following the path of dengue and chikungunya? *Lancet* **2015**, *386*, 243–244. [CrossRef]

51. Calvez, E.; Guillaumot, L.; Millet, L.; Marie, J.; Bossin, H.; Rama, V.; Faamoe, A.; Kilama, S.; Teurlai, M.; Mathieu-Daudé, F.; et al. Genetic diversity and phylogeny of *Aedes aegypti*, the main arbovirus vector in the Pacific. *PLoS Negl. Trop. Dis.* **2016**, *10*, 1–17. [CrossRef] [PubMed]

52. Powell, J.R.; Tabachnick, W.J. History of domestication and spread *of Aedes aegypti*—A review. *Mem. Inst. Oswaldo Cruz* **2013**, *108*, 11–17. [CrossRef] [PubMed]

53. Kotsakiozi, P.; Gloria-Soria, A.; Caccone, A.; Evans, B.; Schama, R.; Martins, A.J.; Powell, J.R. Tracking the return of *Aedes aegypti* to Brazil, the major vector of the dengue, chikungunya and Zika viruses. *PLoS Negl. Trop. Dis.* **2017**, *11*, 1–20. [CrossRef]

54. Powell, J.R.; Gloria-Soria, A.; Kotsakiozi, P. Recent history of *Aedes aegypti*: Vector genomics and epidemiology records. *Bioscience* **2018**, *68*, 854–860. [CrossRef]

55. Guillaumot, L. Arboviruses and their vectors in the Pacific—Status report. *Pac. Health Dialog* **2005**, *12*, 45–52. [PubMed]

56. Alto, B.W. Lounibos Vector competence for arboviruses in relation to the larval environment of mosquitoes. *Ecol. Parasite–Vector Interact.* **2013**, *3*, 81–101. [CrossRef]

57. Instituto Brasileiro de Geografia e Estatística/IBGE. Censo Demográfico 2010. Available online: https://www.ibge.gov.br/estatisticas/sociais/populacao/2098-np-censo-demografico/9662-censo-demografico-2010.html?=&t=sobre (accessed on 31 May 2020).

58. Rivero, A.; Vézilier, J.; Weill, M.; Read, A.F.; Gandon, S. Insecticide control of vector-borne diseases: When is insecticide resistance a problem? *PLoS Pathog.* **2010**, *6*, 5–6. [CrossRef] [PubMed]

59. Lima, J.B.P.; Da-Cunha, M.P.; Da Silva, R.C.; Galardo, A.K.R.; Da Silva Soares, S.D.; Braga, I.A.; Ramos, R.P.; Valle, D. Resistance of *Aedes aegypti* to organophosphates in several municipalities in the state of Rio de Janeiro and Espírito Santo, Brazil. *Am. J. Trop. Med. Hyg.* **2003**, *68*, 329–333. [CrossRef] [PubMed]

60. Da-Cunha, M.P.; Lima, J.B.P.; Brogdon, W.G.; Moya, G.E.; Valle, D. Monitoring of resistance to the pyrethroid cypermethrin in Brazilian *Aedes aegypti* (Diptera Culicid) populations collected between 2001 and 2003. *Mem. Inst. Oswaldo Cruz* **2005**, *100*, 441–444. [CrossRef]

61. Montella, I.R.; Martins, A.J.; Viana-Medeiros, P.F.; Lima, J.B.P.; Braga, I.A.; Valle, D. Insecticide resistance mechanisms of Brazilian *Aedes aegypti* populations from 2001 to 2004. *Am. J. Trop. Med. Hyg.* **2007**, *77*, 467–477. [CrossRef]

62. Martins, A.J.; Lima, J.B.P.; Peixoto, A.A.; Valle, D. Frequency of Val1016Ile mutation in the voltage-gated sodium channel gene of *Aedes aegypti* Brazilian populations. *Trop. Med. Int. Health* **2009**, *14*, 1351–1355. [CrossRef]

63. Bellinato, D.F.; Viana-Medeiros, P.F.; Araújo, S.C.; Martins, A.J.; Lima, J.B.P.; Valle, D. Resistance status to the insecticides temephos, deltamethrin, and diflubenzuron in Brazilian *Aedes aegypti* populations. *Biomed Res. Int.* **2016**, *2016*, 1–12. [CrossRef]

64. Aguirre-Obando, O.A.; Martins, A.J.; Navarro-Silva, M.A. First report of the Phe1534Cys kdr mutation in natural populations of *Aedes albopictus* from Brazil. *Parasites Vectors* **2017**, *10*, 1–10. [CrossRef]

65. Valle, D.; Bellinato, D.F.; Viana-Medeiros, P.F.; Lima, J.B.P.; Martins Junior, A.D.J. Resistance to temephos and deltamethrin in *Aedes aegypti* from Brazil between 1985 and 2017. *Mem. Inst. Oswaldo Cruz* **2019**, *114*, 1–17. [CrossRef]

66. Salgueiro, P.; Restrepo-Zabaleta, J.; Costa, M.; Galardo, A.K.R.; Pinto, J.; Gaborit, P.; Guidez, A.; Martins, A.J.; Dusfour, I. Liaisons dangereuses: Cross-border gene flow and dispersal of insecticide resistance-associated genes in the mosquito *Aedes aegypti* from Brazil and French Guiana. *Mem. Inst. Oswaldo Cruz* **2019**, *114*, 1–9. [CrossRef] [PubMed]

67. Dusfour, I.; Zorrilla, P.; Guidez, A.; Issaly, J.; Girod, R.; Guillaumot, L.; Robello, C.; Strode, C. Deltamethrin resistance mechanisms in *Aedes aegypti* populations from three French Overseas Territories Worldwide. *PLoS Negl. Trop. Dis.* **2015**, *9*, 1–17. [CrossRef] [PubMed]

68. Ayres, C.F.J.; Romao, T.P.A.; Melo-Santos, M.A.V.; Furtado, A.F. Genetic diversity in Brazilian populations of *Aedes albopictus*. *Mem. Inst. Oswaldo Cruz* **2002**, *97*, 871–875. [CrossRef] [PubMed]

69. Ayres, C.F.J.; Melo-santos, M.A.V.; Solé-Cava, A.M.; Furtado, A.F. Genetic differentiation of *Aedes aegypti* (Diptera: Culicidae), the major dengue vector in Brazil. *J. Med. Entomol.* **2003**, *40*, 430–435. [CrossRef]

70. Lourenço-de-Oliveira, R.; Vazeille, M.; De Filippis, A.M.B.; Failloux, A.B. *Aedes aegypti* in Brazil: Genetically differentiated populations with high susceptibility to dengue and yellow fever viruses. *Trans.* **2004**, *98*, 43–45.

71. Da Costa-Ribeiro, M.C.V.; Lourenço-de-Oliveira, R.; Failloux, A.B. Geographic and temporal genetic patterns of *Aedes aegypti* populations in Rio de Janeiro, Brazil. *Trop. Med. Int. Health* **2006**, *11*, 1276–1285. [CrossRef]

72. Monteiro, F.A.; Shama, R.; Martins, A.J.; Gloria-Soria, A.; Brown, J.E.; Powell, J.R. Genetic diversity of Brazilian *Aedes aegypti*: Patterns following an Eradication Program. *PLoS Negl. Trop. Dis.* **2014**, *8*, 1–10. [CrossRef]

73. Kenney, J.L.; Romo, H.; Duggal, N.K.; Tzeng, W.; Kristen, L.; Brault, A.C.; Savage, H.M. Transmission Incompetence of *Culex quinquefasciatus* and *Culex pipiens* pipiens from North America for Zika Virus. *Am. J. Trop. Med. Hyg.* **2017**, *96*, 1235–1240. [CrossRef]

74. Lourenço-de-Oliveira, R.; Marques, J.T.; Sreenu, V.B.; Nten, C.A.; Aguiar, E.R.G.R.; Varjak, M.; Kohl, A.; Failloux, A.B. *Culex quinquefasciatus* mosquitoes do not support replication of Zika virus. *J. Gen. Virol.* **2018**, *99*, 258–264. [CrossRef]

75. Boccolini, D.; Toma, L.; Di Luca, M.; Severini, F.; Romi, R.; Remoli, M.E.; Sabbatucci, M.; Venturi, G.; Rezza, G.; Fortuna, C. Experimental investigation of the susceptibility of Italian *Culex pipiens* mosquitoes to Zika virus infection. *Eurosurveillance* **2016**, *21*. [CrossRef]

76. Hall-Mendelin, S.; Pyke, A.T.; Moore, P.R.; Mackay, I.M.; Mcmahon, L.; Ritchie, S.A.; Taylor, C.T.; Moore, F.A.J.; Van, A.F. Assessment of local mosquito species incriminates *Aedes aegypti* as the potential vector of Zika virus in Australia. *PLoS Negl. Trop. Dis.* **2016**, *10*, 1–14. [CrossRef] [PubMed]

77. Aliota, M.T.; Peinado, S.A.; Osorio, J.E.; Bartholomay, L.C. *Culex pipiens* and *Aedes triseriatus* mosquito susceptibility to Zika virus. *Emerg. Infect. Dis.* **2016**, *22*, 1857–1859. [CrossRef] [PubMed]

78. Fernandes, R.S.; Campos, S.S.; Ribeiro, P.S.; Raphael, L.M.S.; Bonaldo, M.C.; Lourenço-de-Oliveira, R. *Culex quinquefasciatus* from areas with the highest incidence of microcephaly associated with Zika virus infections in the Northeast region of Brazil are refractory to the virus. *Mem. Inst. Oswaldo Cruz* **2017**, *112*, 577–579. [CrossRef] [PubMed]

79. Amraoui, F.; Atyame-Nten, C.; Vega-Rúa, A.; Lourenço-de-Oliveira, R.; Vazeille, M.; Failloux, A.B. *Culex* mosquitoes are experimentally unable to transmit Zika virus. *Eurosurveillance* **2016**, *21*. [CrossRef]

80. Heitmann, A.; Jansen, S.; Lühken, R.; Leggewie, M.; Badusche, M.; Pluskota, B.; Becker, N.; Vapalahti, O.; Chanasit, J.S.; Tannich, E. Experimental transmission of Zika virus By mosquitoes from Central Europe. *Eurosurveillance* **2017**, *22*. [CrossRef]

81. Roundy, C.M.; Azar, S.R.; Brault, A.C.; Ebel, G.D.; Failloux, A.B.; Fernandez-Salas, I.; Kitron, U.; Kramer, L.D.; Lourenço-de-Oliveira, R.; Osorio, J.E.; et al. Lack of evidence for Zika virus transmission by *Culex* mosquitoes. *Emerg. Microbes Infect.* **2017**, *6*, 1–2. [CrossRef]

82. Azar, S.R.; Weaver, S.C. Vector competence: What has Zika virus taught us? *Viruses* **2019**, *11*, 867. [CrossRef]

83. Dickson, L.B.; Merkling, S.H.; Gautier, M.; Ghozlane, A.; Jiolle, D.; Paupy, C.; Ayala, D.; Moltini-Conclois, I.; Fontaine, A.; Lambrechts, L. Exome-wide association study reveals largely distinct gene sets underlying specific resistance to dengue virus types 1 and 3 in *Aedes aegypti*. *PLoS Genet.* **2020**, *16*. [CrossRef]

84. Vazeille, M.; Mousson, L.; Martin, E.; Failloux, A.B. Orally co-infected *Aedes albopictus* from La Reunion Island, Indian Ocean, can deliver both dengue and chikungunya infectious viral particles in their saliva. *PLoS Negl. Trop. Dis.* **2010**, *4*. [CrossRef]

85. R Core T. R: A Language and Environment for Statistical Computing. Available online: http://www.r-project. org/ (accessed on 2 April 2020).

Article

Vector Competence for Dengue-2 Viruses Isolated from Patients with Different Disease Severity

Ronald Enrique Morales-Vargas [1,*], Dorothée Missé [2], Irwin F. Chavez [3] and Pattamaporn Kittayapong [4,5]

[1] Department of Medical Entomology, Faculty of Tropical Medicine, Mahidol University, Bangkok 10400, Thailand
[2] MIVEGEC, University Montpellier, IRD, CNRS, 34394 Montpellier, France; dorothee.misse@ird.fr
[3] Department of Tropical Hygiene, Faculty of Tropical Medicine, Mahidol University, Bangkok 10400, Thailand; irwin.cha@mahidol.edu
[4] Center for Vectors and Vector-Borne Diseases, Faculty of Science, Mahidol University at Salaya, Nakhon Pathom 73170, Thailand; pattamaporn.kit@mahidol.ac.th
[5] Department of Biology, Faculty of Science, Mahidol University, Rama 6 Road, Bangkok 10400, Thailand
* Correspondence: ronald.mor@mahidol.ac.th

Received: 7 August 2020; Accepted: 19 October 2020; Published: 21 October 2020

Abstract: Dynamics of dengue serotype 2 virus isolated from patients with different disease severity, namely flu-like classic dengue fever (DF) and dengue shock syndrome (DSS) were studied in its mosquito vector *Aedes aegypti*. We compared isolate infectivity and vector competence (VC) among thirty two *A. aegypti*-viral isolate pairs. Mosquito populations from high dengue incidence area exhibited overall greater VC than those from low dengue incidence area at 58.1% and 52.5%, respectively. On the other hand, the overall infection rates for the isolates ThNR2/772 (DF, 62.3%) and ThNR2/391 (DSS, 60.9%), were significantly higher than those for isolates ThNR2/406 (DF, 55.2%) and ThNR2/479 (DSS, 54.8%). These results suggest that the efficacy of dengue virus circulation was likely to vary according to the combination between the virus strains and origin of the mosquito strains, and this may have epidemiologic implications toward the incidence of flu-like classic dengue fever (DF) and dengue shock syndrome (DSS).

Keywords: dengue virus; disease severity; vector competence

---

1. Introduction

Flulike classic dengue fever (DF) and dengue hemorrhagic fever (DHF) are increasingly important public health problems issues in the tropical region. Annually, an estimated 390 million dengue infections per year and 96 million of symptomatic dengue infections occur [1]. It has been predicted that population at risk of dengue will rise by 2.25 (1.27–2.80) billion by the year 2080 compared to 2015, which will put over 6.1 (4.7–6.9) billion at risk—nearly 60% of the world's projected 2080 population [2]. DF/DHF and DSS has re-emerged in tropical and subtropical zones, with DHF as the prominent cause of pediatric mortality and hospitalization during the past 20 years in Southeast Asia [3,4]. All dengue viruses (DENV) can result to DHF cases, but the severe form is mostly associated with DENV-2 and DENV-3 [5]. It has been suggested that DHF epidemics occur as a consequence of a very complex mechanism comprising the intersection of three groups of factors (viral, host, and epidemiological) [6]. In addition, some amino acid changes on the premembrane (prM) and envelope (E) proteins of DEN-2 strains have been related to DHF epidemics [7]. Furthermore, Pandy and Igarashi classified Thailand DENV-2, according to the non-synonymous amino acid replacement, in three subtypes (I, II, and III) and proposed that viruses' molecular structures and patients' serological responses influence clinical severity [8].

Aedes aegypti, the primary vector of dengue, is well-adapted to cohabitate with humans by breeding in man-made containers. Several population genetic and vector competence (VC) studies have been done on *A. aegypti* because of its ability to transmit dengue and yellow fever flaviviruses [9–16]. Genetic markers have been used to identify biotypes, biting behavior, and other epidemiologically relevant traits [17]. Vector movement among disease foci or from a site of initial colonization can be investigated through genetic similarity between geographically heterogeneous populations [18]. The genetically intrinsic leniency of a vector to infection, replication, and virus transmission, vector competence [19], present significant variability among *A. aegypti* populations particularly for DENV-2es [10,11,20]. Genes or sets of genes controlling the midgut infection and escape barriers have been found to be associated with the VC of *A. aegypti* for flaviviruses [14–16].

There are several possible mechanisms that can affect the transmission potential of different dengue virus strains. Some strains may infect and replicate target cells more efficiently, sustain higher viremia in the human host, and infect more mosquitoes [21–23]. Studies have shown that dengue virus serotypes and strains within a serotype, between and within a genotype may vary in their ability to infect and disseminate in mosquitoes [10,11,24–26].

Several studies have reported variation in the oral receptivity of *A. aegypti* for DENV-2 at a regional geographic scale [27–29], as well as VC [30]. Furthermore, it has also been reported that DENV-2 vary according to the genotype in their efficiency to infect and disseminate in *A. aegypti* [24–26]. However, these studies were done using only a few mosquito collections or virus strains resulting in the evaluation of patterns of differential infection in a small number of virus–vector pairs.

In the present study, we address the potential variation in VC and patterns of oral infection of dengue virus isolates representing distinct disease severity in sympatric DENV-2-*A. aegypti* pairs on a regional geographic scale in Thailand. In particular, eight geographically heterogeneous mosquito strains collected from areas with high and low dengue incidence were orally challenged with four virus isolates recently isolated from patients exhibiting flu-like classic dengue fever (DF) and dengue shock syndrome (DSS).

2. Results

Disseminated infection rates: Variations in the disseminated infection rates of the virus isolates in *Aedes aegypti* were observed according to the mosquito origin (dengue incidence at the area and location), mosquito strain, disease severity of the patients from whom the virus was isolated, and the virus isolate (Table 1 and Figure 1).

Table 1. Infection and dissemination rates (%) [A] of DENV-2es isolated from patients with different disease severity in orally challenged *Aedes aegypti* strains.

Mosquito Strains	ThNR2/406 (DF) [B]		ThNR2/772 (DF)		ThNR2/391 (DSS)		ThNR2/479 (DSS)	
	Infec	Disse	Infec	Disse	Infec	Disse	Infec	Disse
TAL	66.7	58.3 (60) [C]	57.1	52.4 (63)	56.4	56.4 (55)	55.5	52.4 (63)
KAO	75.0	70.4 (44)	73.1	67.3 (52)	61.3	61.4 (44)	21.6	19.6 (51)
BPA	48.1	42.6 (54)	66.7	63.6 (66)	55.5	53.7 (54)	72.7	72.7 (44)
TAK	40.7	37.0 (54)	73.1	73.1 (52)	80.0	78.0 (50)	85.4	85.4 (48)
KKR	88.6	81.8 (44)	64.0	60.0 (50)	70.1	67.3 (48)	48.1	46.3 (54)
GKE	70.7	69.3 (75)	59.4	56.3 (64)	40.6	40.6 (64)	62.7	59.3 (59)
RBK	18.8	18.8 (69)	40.0	40.0 (60)	61.2	55.1 (49)	51.4	47.2 (72)
RBA	44.8	41.4 (58)	68.3	61.7 (60)	67.2	65.5 (58)	51.7	51.7 (58)

[A] Infection and dissemination: number of mosquitoes for viral RNA in abdomen and head tissue, respectively; [B] DF: dengue fever; DSS: dengue shock syndrome; [C] Total number of mosquitoes tested for each mosquito/virus combination.

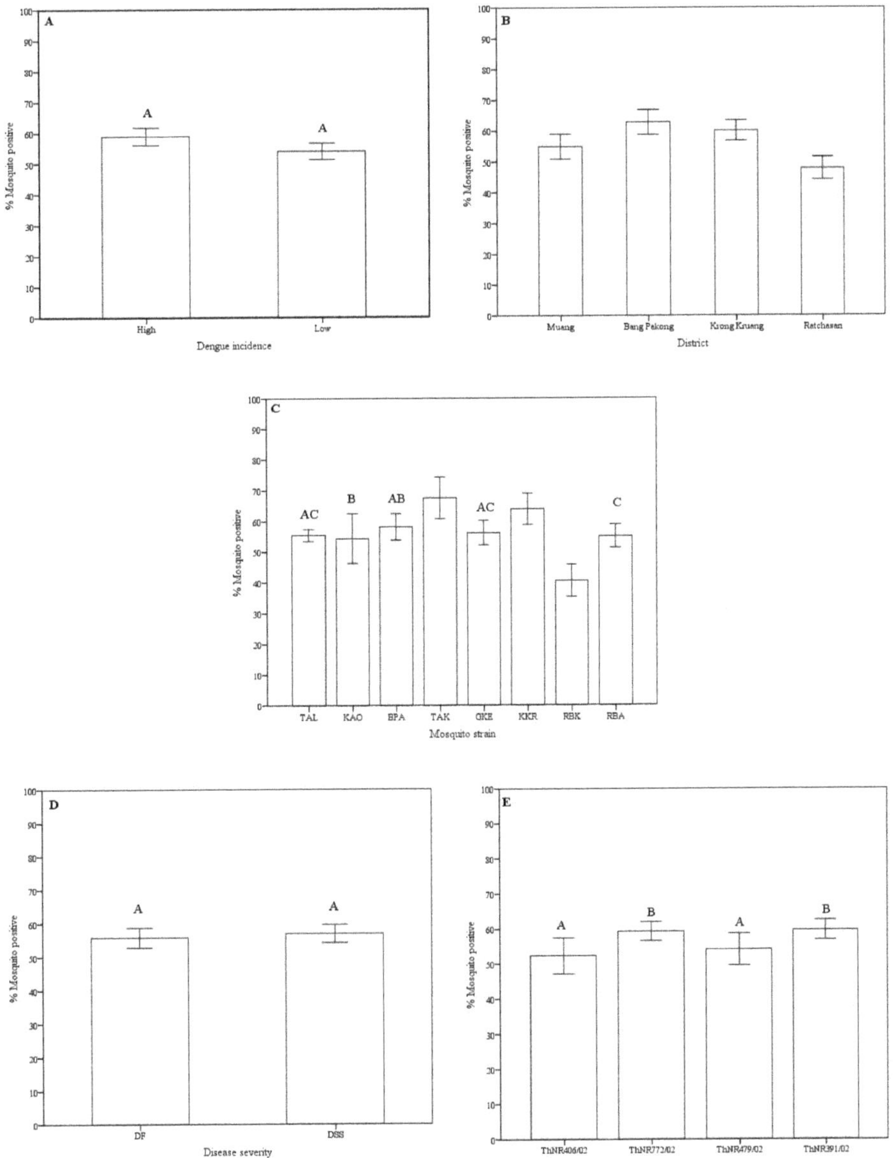

Figure 1. Disseminated infection rate variations of dengue virus serotype 2 isolated from patients with different disease severity in *Aedes aegypti* according to dengue incidence at the mosquito origin area (**A**), mosquito geographic origin and districts (**B**), mosquito strains (**C**), patient disease severity (**D**), and virus isolate (**E**). Mosquitoes were scored positive if RNA dengue virus was detected in head tissue by RT-PCR. Bars and error bars show the mean percentage of positive mosquitoes and the standard error, respectively. Bars with the same letter are not significantly different ($p < 0.05$), while bars without or different letter are significantly different ($p > 0.05$), based on Tukey's test for means comparison.

Overall, the disseminated infection rates of dengue viruses in mosquito strains ranged from 66.7% (TAK, 136 of 204 mosquitoes) to 39.2% (RBK, 98 of 250 mosquitoes). Mosquito strains collected from high dengue incidence areas showed a higher infection rate (58.1% of 854 mosquitoes) compared to mosquitoes from low dengue incidence area (52.5% of 953 mosquitoes). However, no significant statistical difference was observed (Figure 1A). Significant differences in infection rates were observed among geographic origin. Mosquitoes collected from Bang Pakong district exhibited the highest infection rate (62.1% of 422 mosquitoes) while mosquitoes collected from Ratchasan district exhibited the lowest disseminated infection rate (46.9% of 484 mosquitoes). Mosquito strains collected from the Muang district, despite being the most urbanized and with the highest reported dengue incidence, showed lower disseminated infection rate (54.9% of 432 mosquitoes) compared to mosquito strains from the rural Krong Kruang district (59.5% of 459 mosquitoes) that has lower reported dengue incidence (Table 2 and Figure 1B). Among mosquito strains, significantly different disseminated infection rates were observed, except for the BPA-GKE, BPA-TAL, BPA-KAO, GKE-RBA, and GKE-TAL pairs of mosquito strains (Figure 1C). For virus isolates grouped according to the disease severity of the patients from whom they were isolated, no differences in disseminated infection rate was observed between DF and DSS virus isolates (Figure 1D). However, disseminated infection rates of the isolates ThNR2/772 (from DF patient) and ThNR2/391 (from DSS patient), were overall significantly different from isolates ThNR2/406 (DF) and ThNR2/479 (DSS) with $p < 0.005$ (Figure 1E).

Table 2. Demographic information and dengue (DEN) background of the origin of *Aedes aegypti* strains from Chachoengsao province, central eastern Thailand, used in this study.

Strain	At Collection Site			At Collection Area		
	Subdistrict	Demographic [A]	Incidence [B]	District	Demographic [A]	Incidence [B]
TAL KAO	Na Muang	4480.9/1640.4	153.2	Muang	369.2/106.8	141.3
BPA	Bang Paakong	207.4/89.5	284.9	Bang Pakong	306.1/89.6	105.8
TAK	Tha Kam	542.5/345.2	204.2			
KKR	Krong Kruang	114.2/30.9	117.3	Krong Kruang	106.5/25.9	91.7
GKE	Gon Kheo	99.7/21.2	88.45			
RBK RBA	Bang Kla	70.4/18.3	40.3	Ratchasan	92.4/23.4	72.1

[A] Numbers of people/houses per square Km. Based on census of 2002, Public Health Office, Chachoengsao; [B] Dengue (DHF + DSS) annual average incidence, cases/100,000 habitants. Public Health Office, Chachoengsao, 1999–2002.

Vector competence variation: As a proxy of vector competence (VC), the VC was calculated as the number of mosquitoes positive for the viral RNA in head tissues divided by the number of mosquitoes exposed to the respective viruses. To better understand the geographical variation in the *A. aegypti* vector competence for dengue viruses isolated from patients exhibiting different disease severity (DF and DSS), the mosquito strains were grouped by dengue incidence at the area of mosquito origin (sub-district level). High dengue incidence sub-districts comprises TAL, KAO, BPA, and TAK strains, while low incidence sub-districts were KKR, GKE, RBK, and RBA strains. Mosquito strains were further classified location at the district administrative level Muang (TAL and KAO), Bang Pakong (BPA and TAK), Krong Kruang (KKR and GKE) and Ratchasan (RBK and RBA).

For the first grouping, the VC among mosquitoes from high dengue incidence for DSS isolates ranged from 19.6% (KAO mosquito strain for the ThNR2/479 virus isolate) to 85.4% (TAK for the ThNR2/479 virus isolate). DF isolates ranged from 37.0% (TAK mosquito strain for the ThNR2/406 virus isolate) to 73.1% (TAK mosquito strain for the ThNR2/772 virus isolate). Although no statistically significant differences were observed between the overall VC for DF and DSS, mosquito strains collected from high dengue incidence areas showed an overall higher competence for viruses isolated from DSS patients than from DF patients. Whereas mosquito strains collected from low dengue incidence areas exhibited a similar overall VC for either DF or DSS virus isolates, with VC ranging from 40.6% (GKE mosquito strain for the ThNR2/391 virus isolate) to 67.3% (KKR mosquito strain for the ThNR2/391) for DSS isolates, and 18.8% (RBK mosquito strain for the ThNR2/406) to 81.8% (KKR mosquito strain for the ThNR2/406) for DF isolates (Table 1). For the second grouping, mosquito strains from Bang Pakong and Ratchasan districts showed a significantly higher VC for viruses isolated from patients with DSS compared to patients with DF symptoms (Figure 2B). The VC for mosquito strains from Bang Pakong ranged from 53.7% (for ThNR2/391 isolate virus) to 85.4% (ThNR2/479 isolate virus) for DSS isolates, whereas DF isolates ranged from 37.0% (for ThNR2/406 isolate virus) to 73.1% (ThNR2/772). The VC for mosquito strains from Ratchasan ranged from 47.2% (for ThNR2/479) to 65.5% (for ThNR2/391) for DSS isolates, and from 18.8% (for ThNR2/406) to 61.7% (ThNR2/772) for DF isolates (Table 1). In contrast, mosquito strains from Muang and Krong Kruang districts showed a significantly higher VC for viruses isolated from patients with DF than from those with DSS, (Figure 2B). The VC for mosquito strains from Muang ranged from 52.3% (for ThNR2/772) to 70.4% (ThNR2/406) for DF isolates and from 19.6% (for ThNR2/479) to 61.4% (for ThNR2/391) DSS. The VC for mosquito strains from Krong Kruang ranged from 56.2% (for ThNR2/772) to 81.8% (for ThNR2/406 isolate virus) for DF, and from 40.6% (for ThNR2/391) to 67.3% (for ThNR2/391) for DSS isolates (Table 1).

The VC of individual mosquito strains significantly differed between virus isolated from patients with DSS and DF, $p < 0.05$. Females of the TAL, KAO, GKE, and KKR strains exhibited a significantly higher VC for DF virus isolates than for DSS, while BPA, TAK, and RBK showed a significantly higher VC for DSS than for DF virus isolates, ($p < 0.05$) as shown on Figure 2C. As summarized on Table 1, the VC among mosquito strains for the four isolates differed according to the *A. aegypti* strain-virus isolate combinations which ranged from 18.8% (RBK-ThNR2/406) to 85.4% (TAK-ThNR2/479).

Viral isolates infectivity variations: The infectivity of a virus isolate for *A. aegypti* was calculated as the number mosquitoes positive for viral RNA in the abdomen divided by the number exposed to the virus isolate. Overall, ThNR2/772 (DF) and ThNR2/391 (DSS) were the most infectious virus isolates, where 62.3% of 467 and 60.9% of 422 mosquitoes were positive, respectively; while ThNR2/406 (DF) and ThNR2/479 (DSS) were significantly less infectious where 55.2% of 458 and 54.8% of 449 mosquitoes were positive, respectively, ($p < 0.05$; Figure 1E).

Different patterns of infection dynamics among individual virus isolates in mosquitoes were observed. Infection rates for ThNR2/772 (DF) and ThNR2/391 (DSS) isolates were more uniform. ThNR2/772 ranged from 40% (RBK mosquito strain) to 73.1% (KAO mosquito strain), while ThNR2/391 ranged from 40% (GKE mosquito strain) to 80% (TAK mosquito strain). In contrast, virus isolates ThNR2/406 (DF) and ThNR2/479 (DSS), exhibited more phenotypic variation where infection rates ranged from 18.8% (RBK mosquito strain) to 88.6% (KKR mosquito strain), and from 21.6 (KAO mosquito strain) to 85.4% (TAK mosquito strain), respectively, (Table 1). Furthermore, according to the mosquito strains, significant differences in the number of infected mosquitoes were observed among individual viral isolates. The proportion of mosquitoes infected by the ThNR2/406 (DF) virus isolate was not significantly different only between KAO-GKE and TAK-RBA mosquito strain pairs, and by the ThNR2/772 (DF) virus isolate was not significantly different between GKE-KKR-RBA mosquito strains, and by the ThNR2/391 (DSS) virus isolate was not significantly different between TAL-KAO-BPA-RBK mosquito strains, and by the ThNR2/479 (DSS) viral isolate was not significantly different only between KKR-RBA mosquito strains.

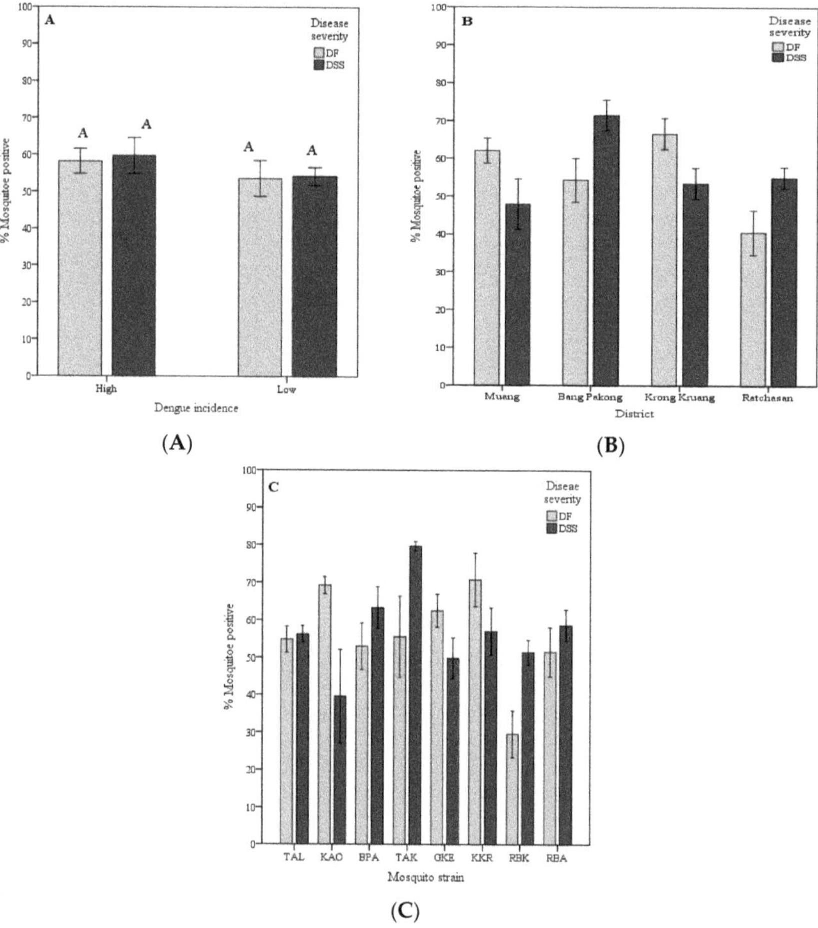

Figure 2. Vector competence of *Aedes aegypti* for dengue serotype 2 virus isolated from patients exhibiting different disease severity according to dengue incidence at the mosquito strain origin area (**A**), mosquito geographic origin, districts (**B**), and mosquito strains (**C**). Vector competence is expressed as the number of mosquitoes positive for viral RNA in head tissues/number tested. Bars and error bars show the mean percentage of positive mosquitoes and the standard error, respectively. Bars with the same letter within the same group are not significantly different ($p < 0.05$) based on Tukey's test for means comparison.

3. Discussion

Vector, viral, and environmental factors determine the capacity of dengue virus to infect and disseminate in its mosquito vector. In the present study, we challenged each mosquito strain with all four virus isolates at the same time and mosquitoes were held together during the same environmental fluctuations of the incubation period (14 days), to minimize the effect of environmental variation. Even though all mosquito strains may not have experienced the same environmental conditions no statistical significant differences were observed when mean temperature and humidity that each mosquito strain experienced during the incubation period were compared among strains (data not shown). Nonetheless, some of the observed mosquito VC and viral isolate infectivity variations may have been influenced by environmental variation among experimental groups.

Variations in infection rates of *A. aegypti* experimentally infected with DEN virus has been reported [10,11,20,27–30]. Differences on the proportion of infected mosquitoes between DENV-2 with SEA and American genotype have also been reported in geographically separated *A. aegypti* populations [24–26]. In contrast to previous studies, the present study used low-passage virus isolates that were recently isolated from patients exhibiting different disease severity (i.e., DF and DSS), which were of sympatric geographic origin with the used *A. aegypti* strains. Although the *A. aegypti* were collected in the same province, considerable variability in overall infection rates was observed according to the dengue incidence in the collection area, geographic origin (districts), mosquito strain, individual virus isolate, and disease severity of the patients from whom the virus was isolated (Figure 1A–E).

From our observations, although neither high or low dengue incidence mosquito strains showed significantly different VC for DF or DSS viral isolates (Figure 2A), the mosquito strains collected from high dengue incidence sites exhibited higher overall VC for virus isolates either isolated from patients exhibiting mild (DF) or severe (DSS) disease severity than mosquito strains from sites with low dengue incidence site (Table 3). However, significant differences in VC for DF and DSS virus isolates were observed among mosquito strains from different geographic origins (districts) (Figure 2B). Furthermore, most individual mosquito strains showed significantly different VC either for DF or DSS virus isolates. Only the RBA mosquito strain did not show significantly different VC between DF and DSS viral isolates (Figure 2C). Therefore, we speculate that factors such as the origins of mosquitoes, the viral strains (DF or DSS) circulating in a given locale, and mosquito factors, per se play an important role in determining *A. aegypti* VC. A detailed genetic analysis is needed to determine if these mosquito strains are actually isolated and constitute independent populations that differ in susceptibility to the infection with dengue virus. In previous studies, it has been shown that mosquitoes from populated and urbanized areas are genetically highly differentiated and exhibit high and heterogeneous infection rates, and this genetic differentiation has been related to the intensity of insecticide control and human population density [16,27–29]. As for mosquito factors, the VC for arboviruses is associated with anatomic barriers to productive vector infection, including a midgut infection barrier (MIB), a midgut escape barrier (MEB), and a salivary gland barrier [30,31]. *A. aegypti* VC for DENV-2 is considered as being universally proportional to the level of MIB and MEB found in mosquito collections, where strong MIBs and MEBs decrease transmission potential [30]. It is also important to remember that there is potential for different loci to affect VC in different populations [14,16]. Thus, the need to look for genetic variation is evident. This study also raised the possibility that VC for a virus isolate is not randomly distributed across the disease severity of the patient from whom the virus was isolated. We found that mosquito strains from the same district exhibited the same VC pattern, namely TAL, KAO, and GKE, KKR mosquito strains collected from Muang and Krong Kruang districts, respectively, were significantly more competent for DF virus isolates than for DSS viral isolates. In addition, the BPA and TAK mosquito strains collected from Bang Pakong districts and RBK and RBA from Rachasan district, though RBA did not show significant different VC, exhibited the same VC pattern, indeed, were more competent for DSS than for DF viral isolate (Figure 2C).

Dengue viruses' efficiency to infect and disseminate in their vector mosquitoes can vary greatly. In the present study, virus infectivity was evaluated by measuring the proportion of mosquitoes infected by individual virus isolates. Overall, we found variation in the proportion of mosquitoes infected with individual virus isolates. Specifically, the ThNR2/772 and ThNR2/391 isolates were more infective than the ThNR2/406 and ThNR2/479 viral isolates (Figure 1E). However, the virus isolate infectivity seems likely to be differentiated according to mosquito strain–viral isolate pairs, Figure 3. Therefore, we speculate that differences in the proportions of infected mosquitoes among individual virus isolate are probably due to certain genetic elements of the isolated virus. Few genetic differences among dengue virus isolates may have a significant effect on their infection, dissemination and transmission by vector mosquitoes. Previous studies on DENV-2 revealed differences in the proportions of mosquitoes infected by individual virus isolates and have been associated with molecular differences among virus isolates [24–26,32].

Table 3. Medical data and in vitro infectivity results of four DEN-2 isolates virus isolated patients of Nakorn Ratchasima provincial hospital from Nakorn Ratchasima province, northeast Thailand.

Isolate Name	Sex	Age	Clinical Diagnosis	Antibody Response	ELISA Assay (Unit)		Infectivity LLC-MK2 (PFU/mL)
					Den-IgG	Den-IgM	
ThNR02/406	M	16	DF	NA [A]	7	0	1.2×10^5
ThNR02/772	M	14	DF	NA	14	9	8.5×10^3
ThNR02/391	M	12	DSS	NA	29	2	1.38×10^5
ThNR02/479	M	12	DSS	Secondary	3	0	1.0×10^2
					149 [B]	56	

[A] Due to the lack of a second serum sample; [B] Second serum sample was obtained 2 weeks after the first sample.

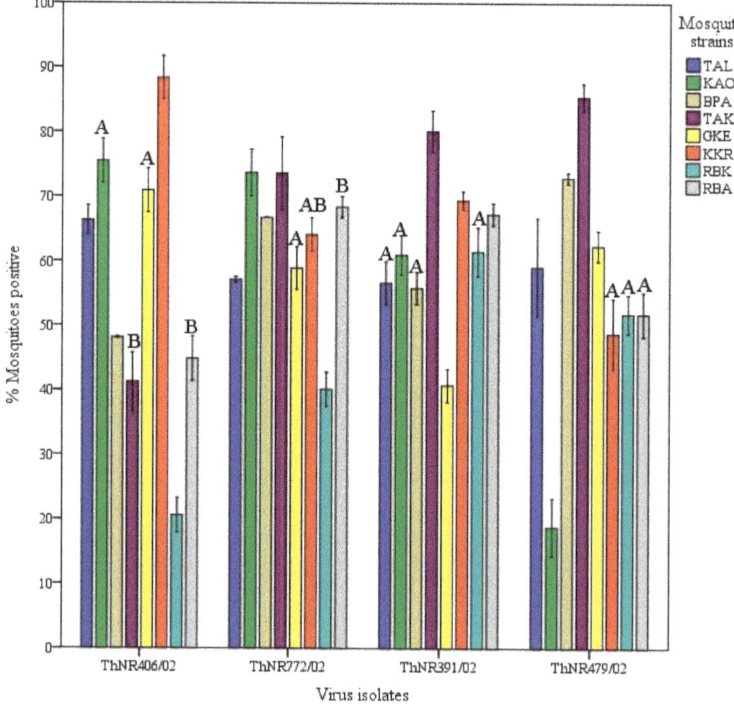

Figure 3. Infectivity of four dengue 2 viruses isolated from patients exhibiting different disease severity for geographically different *Aedes aegypti* strains. Virus infectivity is expressed as the number of mosquitoes positive for viral RNA in abdomen/number tested. Bars and error bars show the mean percentage of positive mosquitoes and the standard error, respectively. Bars with the same letter within the same virus isolate are not significantly different ($p < 0.05$), while bars without or different letter are significantly different ($p > 0.05$), based on Tukey's test for means comparison.

4. Materials and Methods

Collection sites: Four districts in Chachoengsao province situated in eastern Thailand, with the highest (Muang and Bang Pakong) and lowest (Krong Kruang and Ratchasan) dengue (DF/DHF/DSS) incidence (per 100,000 population) reported during the past 20 years were selected. Within these districts the sub-districts (tambon) with the highest (Na Muang, Bang Pakong, and Tha Kam) and lowest (Krong Kruang, Gon Kheo, and Bang Kla) dengue incidence from 1999 to 2002 were selected as mosquito collection sites (Figure 4). Sub-districts were classified as high and low dengue risk areas based on annual incidence of DHF/DSS (cases/100,000 habitants), and the number of houses and inhabitants per square kilometer. The selected locations covered a wide geographical distribution of *A. aegypti*, including urban and rural environments. The shortest distance between study areas (districts) was ca. 20 km, and between sub-districts was ca. more than 30 km.

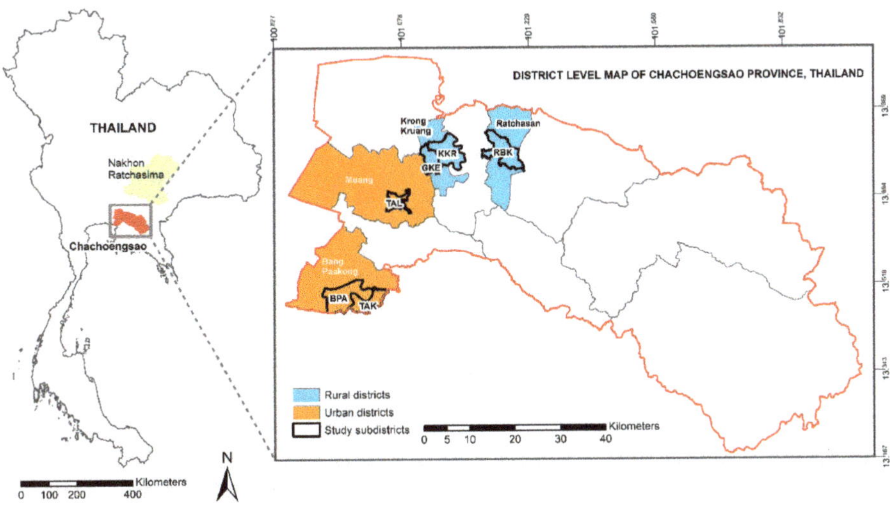

Figure 4. Geographic localization of the collection sites of *Aedes aegypti* used depicting the habitat type corresponding to each district.

Mosquito collections: *A. aegypti* were collected from two sites in each sub-district, (Table 2). Samples from high dengue incidence areas were collected from houses with recent report of dengue cases and from low incidence areas in house without dengue cases for the last two years. The mosquitoes were collected as fourth stage larvae and/or pupae, at least 200 per collection site, from water containers found outside and inside the house within a radius of ca. 2 km, at collection sites separated by at least ca.5 km. Field collected mosquitoes (F0 generation) were reared to adults under room conditions (26 °C ± 2 °C, RH 75% ± 10%). After morphological identification, based on the illustrated key to the mosquitoes of Thailand [33], adults were kept in cages and females were blood-engorged on mice to produce eggs. The onward generations of mosquitoes were reared in plastic pans (33 cm × 25 cm × 11 cm) containing 3 L of aged tap water with a density of 180–210 larvae per container. The larvae were fed on a diet of mouse food powder and maintained at 26 °C ± 2 °C, RH 75% ± 10%. Adults were fed on 3% sucrose soaked cotton and females were allowed to mate and then blood-feed on mice to enable egg production. The F0 mosquitoes were stored at −80 °C for future genetic analysis. Oral infection experiments were performed with females from the first (F1) or second (F2) generations generated from the F0 proved to be free of dengue virus infection. The mosquito strains used for oral infection experiments and the dengue background of the district where the mosquitoes were collected are given in Table 2 and Figure 4.

Virus strains: The four virus isolates were obtained from the sera of patients from Nakorn Ratchasima Provincial Hospital, Northeastern Thailand, diagnosed with DF and DSS in 2002. Clinical diagnosis confirmation and serological assays (ELISA IgG and IgM) were performed at the hospital, while virus isolation and serotype determination were performed by the staff of the Arbovirus Section, National Institute of Health, Medical Science Center, Nonthaburi. The clinical diagnosis and clinical severity gradings of each isolate was classified using the World Health Organization criteria (WHO,

1986) [34]. The serotype was determined as dengue virus type 2 by reverse transcription-polymerase chain reaction (RT-PCR). Relevant patient information is summarized in Table 3.

All isolates were passaged four times in *Aedes albopictus* clone C6/36 cell line from the patient serum before using for experiments. To ensure that the virus specimens were not altered significantly from their wild type character as found in the host patient, none of the isolate has been purified by any methodology before inoculation into mosquito cells.

Plaque assay: A seed virus was prepared by inoculation into a monolayer culture of *Ae. albopictus* clone C6/36 cell line and incubated at 28 °C for 8 days in Eagle's medium supplemented with 2% heat-inactivated fetal bovine serum (FBS) and 0.2 mM each of nonessential amino acids plus antibiotics (20 µg/mL Gentamicin, 5 µg/mL Amphotericin B, 200 U/mL). The infected culture fluid was harvested 8 days after inoculation, aliquoted, and stored at −80 °C until used. The virus titer of the isolates was measured by the focus formation test by using BHK-21 cells on 96-well plates [35]. The virus isolates were then used for preparation of the infectious meal, which was a blood virus sucrose solution (BVS).

Oral Infection: Only 4 to 5 day-old female mosquitoes were used to minimize age factors effects. About 60 females were placed in cylindrical pint cardboard cages covered at one end with fine non-wettable nylon mesh. These females were deprived of food for 36–40 h prior to the infectious meal, then were allowed to feed on the infectious meal (BVS) consisting of equal volumes of isolate virus suspension, washed rabbit erythrocytes, and 10% sucrose solution (as a source of energy). Drops of the infectious meal were placed on the mesh covering the cardboard cage containing mosquitoes as previously described [11,25]. Feeding time was limited to 1 h, mosquitoes were cold anesthetized and fully engorged mosquitoes were collected with an aspirator at 30 min intervals, transferred into clean carton cups and maintained for up to 14 days at 30 ± 1 °C as a virus extrinsic incubation period (EIP). After the EIP completed survived mosquitoes were frozen at −80 °C, then transfer individually into 1.5 mL tube and kept at −80 °C until further use for viral RNA detection. Mosquito strains were orally challenged on different days; however, each strain was challenged simultaneously with the four virus isolates on the same day. All feeding suspensions contained the same virus titer (1×10^2 PFU/mL) with an aliquot of the same virus pool. The titers of the post feeding virus suspensions did not change significantly.

Detection of virus infection: After has completed the EIP (14 days) mosquito head and abdomen were severed and homogenized in Trizol LS reagent for total RNA extraction, and then assayed for dengue virus genome by using RT-PCR method. The detection of viral RNA in the homogenized abdomen was interpreted as indicating that the mosquito midgut had become infected. Detection of RNA viral in the homogenized head indicated disseminated from the infected midgut to a secondary target organs.

RNA extraction and RT-PCR: Total RNA was extracted from individual mosquito head and abdomen severed using Trizol LS (Invitrogen) according to the manufacturer's recommendations. Briefly, the head and abdomen of mosquito were separately homogenized in 300 µL of Trizol reagent then 99.9 µL of chloroform was added. The resulting mixed suspension was centrifuged at 14,000 rpm/min for 10 min at 4 °C. RNA was precipitated from the aqueous phase (ca. 145 µL) by mixing with 133.2 µL of isopropyl alcohol. Isopropyl alcohol-RNA precipitate was recovered by centrifugation and the RNA pellet was washed once with 90 µL of 75% ethanol followed by a second wash with absolute ethanol and then air dried. The RNA pellet was resuspended in 10 µL of RNase-free water and used as a template RNA in reverse transcriptase PCR (RT-PCR). Synthetic oligonucleotide primer pairs were designed based on published sequence data for dengue virus serotype specific primer, sense (D2-S) and complementary (D2-C) genome nucleotide (nt) regions 1203 to 1222 and 1432 to 1413 respectively [36,37].

Dengue virus RNA was assayed by using one step RT-PCR kit (Quiagen) following the manufacturer's recommendations with slight modifications. Briefly, 2 µL of RNA, 8 µL of reaction mix, and RNase free water was reverse transcribed and amplified following the thermal cycle protocol recommended in the kit instruction manual, with some modifications, as follows: one cycle at 53 °C for 30 min, one cycle at 94 °C for 2 min, followed by 45 cycles at 94 °C for minute, 53 °C for 1 min, and 68 °C for 2 min, followed by one cycle at 68 °C for 7 min. The reaction mixture solution contained a mixture of 5 µL of 5× buffer (contains 12.5 mM $MgCl_2$), 1 µL of dNTP mix (containing 10 mM of each dNTP), 0.5 µL (25 pmol) of corresponding primers, D2-S (5′-GTTCGTCTGCAAACACTCCA-3′) and D2-C (5′-GTGTTATTTTGATTTCCTTG-3′), and 1 µL of enzyme mix (an optimized combination of Omnscript Reverse Transciptase, Sensiscript Reverse Transcriptase, and HotStarttTap DNA Polymerase). Nine microliters of PCR product was subjected to agarose gel electrophoresis, and amplified DNA fragments were visualized with ethidium bromide staining.

Statistical analysis: One-way analysis of variance was used to test the variation among disseminated infection rates, vector competence measures, and virus infectivity rates. Tukey's HSD was used for pairwise comparisons. Analogous non-parametric test were used when departures from normality or equality of variance were encountered. Statistical Package for the Social Sciences (SPSS) 11.5 software package was used to generate all the results and graphs in this study.

5. Conclusions

In this study, we analyzed 32 recently collected vector–virus pairs (eight geographically different mosquito strains versus four isolate virus strains). Among vector–virus pairs infection rates were different; indeed, many of those were significantly different. This leads us to conclude that the observed variations in susceptibility among mosquito strains is probably due to the combination of certain genetic elements of the vector and the virus as has been previously observed in similar researches. Furthermore, these results suggest that the efficacy of dengue virus circulation in a given locale may vary according to the interaction of virus strains with the origin of the vector mosquitoes. This study contributes to the knowledge of the role of the association between mosquito genetic determinants of susceptibility and genetic variation of dengue virus in the occurrence of different disease severity.

Author Contributions: Conceptualization, methodology, formal analysis, investigation, data curation, and writing—original draft preparation, R.E.M.-V.; writing—review and editing, R.E.M.-V., P.K., D.M., and I.F.C.; funding acquisition, R.E.M.-V. and P.K. All authors have read and agreed to the published version of the manuscript.

Funding: This research was funded by the TRF/BIOTEC Special Programme for Biodiversity Research and Training (BRT R245004), the UNDP/World Bank/World Health Organization/Special Programme for Tropical Diseases Research and Training (TDR/RCS/A00786), and the Mahidol University Research Grant (SCBI-47-T-217).

Acknowledgments: The authors are grateful to Pathon Swanpanyalert and Surapee Anantapreecha for providing viral isolates; Sutee Yoksan for his assistance in the virus seed preparation and titration; Noppawan Phumala Morales and Somboon Srimarat for their help during the mosquito collection. We also thank the Chachoengsao provincial and local (districts and subdistricts) public health officials for field assistance and the many homeowners who generously let us into their homes.

Conflicts of Interest: The authors declare no conflict of interest. The funders had no role in the design of the study; in the collection, analyses, or interpretation of data; in the writing of the manuscript, or in the decision to publish the results.

References

1. Bhatt, S.; Gething, P.W.; Brady, O.J.; Messina, J.P.; Farlow, A.W.; Moyes, C.L.; Drake, J.M.; Brownstein, J.S.; Hoen, A.G.; Sankoh, O.; et al. The global distribution and burden of dengue. *Nature* **2013**, *496*, 504–507. [CrossRef] [PubMed]

2. Messina, J.P.; Brady, O.J.; Golding, N.; Kraemer, M.U.G.; Wint, G.R.W.; Ray, S.E.; Pigott, D.M.; Shearer, F.M.; Johnson, K.; Earl, L.; et al. The current and future global distribution and population at risk of dengue. *Nat. Microbiol.* **2019**, *4*, 1508–1515. [CrossRef] [PubMed]

3. Gubler, D.J.; Clarck, G.G. Dengue/dengue haemorrhagic fever, the emergence of a global health problem. *Emerg. Infect. Dis.* **1995**, *1*, 55–57. [CrossRef] [PubMed]

4. Monath, T.P. Yellow fever and dengue: The interactions of virus, vector and host in the reemergence of epidemic disease. *Semin. Virol.* **1994**, *5*, 133–145. [CrossRef]

5. Nisalak, A.; Hoke, C.H.; Nimmannitya, S.; Endy, T.P.; Vaughn, D.W.; Burke, D.S.; Scott, R.M.; Thisayakorn, U.; Kalayanarooj, S.; Innis, B.L. Serotype-specific dengue virus circulation and dengue disease in Bangkok, Thailand from 1973 to 1999. *Am. J. Trop. Med. Hyg.* **2003**, *68*, 191–202. [CrossRef]

6. Kouri, G.P.; Guzman, M.G.; Bravo, J.R. Why dengue haemorrhagic fever in Cuba? 2: An integral analysis. *Trans. R. Soc. Trop. Med. Hyg.* **1987**, *81*, 821–823. [CrossRef]

7. Leitmeyer, K.C.; Vaughn, D.W.; Watts, D.M.; Salas, R.; Villalobos, I.; Ramos, C.; Rico-Hesse, R. Dengue virus structural differences that correlate with pathogenesis. *J. Virol.* **1999**, *73*, 4738–4747. [CrossRef]

8. Pandey, B.D.; Igarashi, A. Severity-related molecular differences among nineteen strains of dengue of dengue type 2. *Microbiol. Immunol.* **2000**, *44*, 179–188. [CrossRef]

9. Aitken, T.H.G.; Downs, W.G.; Shope, R.E. *Aedes aegypti* strain fitness for yellow fever virus transmission. *Am. J. Trop. Med. Hyg.* **1977**, *26*, 985–990. [CrossRef]

10. Rosen, L.; Roseboom, L.E.; Gubler, D.J.; Lein, J.C.; Chaniotis, B.N. Comparative susceptibility of mosquito species and strain to oral and parenteral infection with dengue and Japanese encephalitis viruses. *Am. J. Trop. Med. Hyg.* **1985**, *34*, 603–615. [CrossRef]

11. Gubler, D.J.; Nalim, S.; Tan, R.; Saipan, H.; Saroso, J.S. Variation in susceptibility to oral infection with dengue viruses among geographic strains of *Aedes aegypti*. *Am. J. Trop. Med. Hyg.* **1979**, *28*, 1045–1052. [CrossRef] [PubMed]

12. Tabacnick, W.J.; Powell, J.R. A world-wide survey of genetic variation in the yellow fever mosquito, *Aedes aegypti*. *Genet. Res.* **1979**, *34*, 215–229. [CrossRef] [PubMed]

13. Apostol, B.L.; Black, W.C.I.V.; Reiter, P.; Miller, B.R. Population genetics with RAPD-PCR markers: The breeding structure of *Aedes aegypti* in Puerto Rico. *Heredity* **1996**, *76*, 325–334. [CrossRef] [PubMed]

14. Bosio, C.F.; Beaty, B.J.; Black, W.C. Quantitative genetics of vector competence for dengue-2 virus in *Aedes aegypti*. *Am. J. Trop. Med. Hyg.* **1998**, *59*, 965–970. [CrossRef] [PubMed]

15. Gorrochotegui-Escalante, N.; Munoz, M.L.; Fernandez-Salas, I.; Beaty, B.J.; Black, W.C., IV. Genetic isolation by distance among *Aedes aegypti* populations along the northeastern coast of Mexico. *Am. J. Trop. Med. Hyg.* **2000**, *60*, 292–299. [CrossRef]

16. Bosio, C.F.; Fulton, R.E.; Salasek, M.J.; Beaty, B.J.; Black, W.C. Quantitative trait loci that control vector competence for dengue-2 virus in the mosquito *Aedes aegypti*. *Genetics* **2000**, *156*, 687–698.

17. Tabanick, W.J. Evolutionary genetics and arthropod-borne disease. The yellow fever mosquito. *Am. Entomol.* **1991**, *37*, 14–23. [CrossRef]

18. Ballinger-Crabtree, M.E. Use of genetic polymorphisms detected by RAPD-PCR for differentiation and identification of Aedes aegypti subspecies and populations. *Am. J. Trop. Med. Hyg.* **1992**, *47*, 893–901. [CrossRef]

19. Hardy, J.L. Susceptibility and resistance of vector mosquitoes. In *Arboviruses: Epidemiology and Ecology*; Monath, T.P., Ed.; CRC Press: Boca Raton, FL, USA, 1988; Volume I, pp. 87–126.

20. Tardieux, I.; Poupel, O.; Lapchin, L.; Rodhain, F. Variation among strains of *Aedes aegypti* in susceptibility to oral infection with dengue virus type 2. *Am. J. Trop. Med. Hyg.* **1990**, *43*, 308–313. [CrossRef]

21. Vaughn, D.W.; Green, S.; Kalayanarooj, S.; Innis, B.L.; Nimmannitya, S.; Suntayakorn, S.; Endy, T.P.; Raengsakulrach, B.; Rothman, A.L.; Ennis, F.A.; et al. Dengue viremia titer, antibody response pattern, and virus serotype correlate with disease severity. *J. Infect. Dis.* **2000**, *181*, 2–9. [CrossRef]

22. Wu, S.J.; Grouard-Vogel, G.; Sun, W.; Mascola, J.R.; Brachtel, E.; Putvatana, R.; Louder, M.K.; Filgeira, L.; Marovich, M.A.; Wong, H.K.; et al. Human skin Langerhans cells are targets of dengue virus infection. *Nat. Med.* **2000**, *6*, 816–820. [CrossRef] [PubMed]

23. Morens, D.M.; Marchette, N.J.; Chu, M.C.; Halstead, S.B. Growth of dengue type 2 virus isolates in human peripheral blood leukocytes correlates with severe and mild dengue disease. *Am. J. Trop. Med. Hyg.* **1991**, *45*, 644–651. [CrossRef] [PubMed]

24. Armstrong, P.M.; Rico-Hesse, R. Differential susceptibility of Aedes aegypti to infection by the American and Southeast Asian genotypes of Dengue type 2 virus. *Vector Borne Zoonotic Dis.* **2001**, *1*, 159–168. [CrossRef] [PubMed]

25. Morales, R.E.; Morita, K.; Eshita, Y.; Tsuda, Y.; Fukuma, T.; Takagi, M. Infection and dissemination of two dengue type-2 viruses isolated from patients exhibiting different disease severity in orally infected Aedes aegypti from different geographic origin. *Med. Entomol. Zool.* **2002**, *53*, 21–27. [CrossRef]

26. Armstrong, P.M.; Rico-Hesse, R. Efficiency of dengue serotype 2 virus strains to infect and disseminate in Aedes aegypti. *Am. J. Trop. Med. Hyg.* **2003**, *68*, 539–544. [CrossRef]

27. Tran, T.K.; Vazeille-Falcoz, M.; Mousson, L.; Hoang, T.H.; Rodhain, F.; Huong, N.T.; Failloux, A.-B. *Aedes aegypti* in Ho Chi Minh city (Viet Nam): Susceptibility to dengue type 2 virus and genetic differentiation. *Trans. R. Soc. Trop. Med. Hyg.* **1999**, *93*, 581–586.

28. Vazeille-Falcoz, M.; Mousson, L.; Rodhain, F.; Failloux, A.B. Variation in oral susceptibility to dengue type 2 virus of population of *Aedes aegypti* from the islands of Tahiti and Moorea, French Polynesia. *Am. J. Trop. Med. Hyg.* **1999**, *60*, 292–299. [CrossRef]

29. Fouque, F.; Vazeille, M.; Mousson, L.; Gaborit, P.; Carinci, R.; Issaly, J.; Rodhain, F.; Failloux, A.B. *Aedes aegypti* in French Guiana: Susceptibility to a dengue virus. *Trop. Med. Int. Health* **2001**, *6*, 76–82. [CrossRef]

30. Bennet, K.E.; Olson, K.E.; Munos, M.L.; Fernandez-Salas, I.; Farfan-Ale, J.; Higgs, S.; Black, W., IV; Beaty, B.J. Variation in vector competence for dengue 2 virus among 24 collections of *Aedes aegypti* from Mexico and the United States. *Am. J. Trop. Med. Hyg.* **2002**, *67*, 85–92. [CrossRef]

31. Franz, A.W.; Kantor, A.M.; Passarelli, A.L.; Clem, R.J. Tissue Barriers to Arbovirus Infection in Mosquitoes. *Viruses* **2015**, *7*, 3741–3767. [CrossRef]

32. Pompon, J.; Manuel, M.; Ng, G.K.; Wong, B.; Shan, C.; Manokaran, G.; Soto-Acosta, R.; Bradrick, S.S.; Ooi, E.E.; Missé, D.; et al. Dengue Subgenomic Flaviviral RNA Disrupts Immunity in Mosquito Salivary Glands to Increase Virus Transmission. *PLoS Pathog.* **2017**, *13*, e1006535. [CrossRef] [PubMed]

33. Rattanarithikul, R.; Harbach, R.E.; Harrison, B.A.; Panthusiri, P.; Jones, J.W.; Coleman, R.E. Illustrated keys to the mosquitoes of Thailand. II. Genera Culex and Lutzia. *Southeast Asian J. Trop. Med. Public Health* **2005**, *36*, 1–97. [PubMed]

34. World Health Organization. Dengue Guidelines for Diagnosis, Treatment, Prevention and Control: New Edition. 2009. Available online: https://apps.who.int/iris/handle/10665/44188 (accessed on 19 October 2020).

35. Mangada, M.N.; Igarashi, A. Molecular and in vitro analysis of eight dengue type 2 viruses isolated from patients exhibiting different disease severities. *Virology* **1998**, *244*, 458–466. [CrossRef] [PubMed]

36. Deubel, V.; Kinney, R.M.; Trent, D.W. Nucleotide sequence and deduced aminoacid sequence of the structural proteins of Dengue type 2 virus: Jamaica genotype. *Virology* **1986**, *165*, 365–377. [CrossRef]

37. Woodring, J.L.; Higgs, S.; Beaty, B.J. Natural Cycles of Vector-Borne Pathogens. In *The Biology of Diseases Vectors*; University Press of Colorado: Boulder, CO, USA, 1996; pp. 51–72.

Publisher's Note: MDPI stays neutral with regard to jurisdictional claims in published maps and institutional affiliations.

Article

Trends of the Dengue Serotype-4 Circulation with Epidemiological, Phylogenetic, and Entomological Insights in Lao PDR between 2015 and 2019

Elodie Calvez [1,*], Virginie Pommelet [2], Somphavanh Somlor [1], Julien Pompon [3,4], Souksakhone Viengphouthong [1], Phaithong Bounmany [1], Thep Aksone Chindavong [1], Thonglakhone Xaybounsou [1], Phoyphaylinh Prasayasith [1], Sitsana Keosenhom [1], Paul T. Brey [5], Olivier Telle [6,7], Marc Choisy [8,9], Sébastien Marcombe [5] and Marc Grandadam [1,10]

[1] Arbovirus and Emerging Viral Diseases Laboratory, Institut Pasteur du Lao PDR, Vientiane 01030, Laos; s.somlor@pasteur.la (S.S.); s.viengphouthong@pasteur.la (S.V.); p.bounmany@pasteur.la (P.B.); t.chindavong@pasteur.la (T.A.C.); t.xaybounsou@pasteur.la (T.X.); p.prasayasith@pasteur.la (P.P.); s.keosenhom@pasteur.la (S.K.); marc.grandadam@def.gouv.fr (M.G.)

[2] Epidemiology Unit, Institut Pasteur du Lao PDR, Vientiane 01030, Laos; v.pommelet@pasteur.la

[3] Department of Emerging Infectious Diseases, Duke-NUS Medical School, Singapore 169857, Singapore; julien.pompon@ird.fr

[4] MIVEGEC, University of Montpellier, CNRS, IRD, 34394 Montpellier, France

[5] Medical Entomology and Vector Borne Disease Unit, Institut Pasteur du Lao PDR, Vientiane 01030, Laos; p.brey@pasteur.la (P.T.B.); s.marcombe@pasteur.la (S.M.)

[6] Centre de Sciences Humaines (CHS), Centre National de la Recherche Scientifique (CNRS), Delhi 110001, India; telle.olivier@gmail.com

[7] Center for Policy Research (CPR), Delhi 110001, India

[8] Nuffield Department of Medicine, University of Oxford, Oxford OX3 7LF, UK; mchoisy@oucru.org

[9] Oxford University Clinical Research Unit, Ho Chi Minh City 700000, Vietnam

[10] Institut de Recherche Biomédicale des Armées, 91220 Brétigny-sur-Orge, France

* Correspondence: e.calvez@pasteur.la

Received: 30 July 2020; Accepted: 30 August 2020; Published: 3 September 2020

Abstract: Dengue outbreaks have regularly been recorded in Lao People's Democratic Republic (PDR) since the first detection of the disease in 1979. In 2012, an integrated arbovirus surveillance network was set up in Lao PDR and an entomological surveillance has been implemented since 2016 in Vientiane Capital. Here, we report a study combining epidemiological, phylogenetic, and entomological analyzes during the largest DENV-4 epidemic ever recorded in Lao PDR (2015–2019). Strikingly, from 2015 to 2019, we reported the DENV-4 emergence and spread at the country level after two large epidemics predominated by DENV-3 and DENV-1, respectively, in 2012–2013 and 2015. Our data revealed a significant difference in the median age of the patient infected by DENV-4 compared to the other serotypes. Phylogenetic analysis demonstrated the circulation of DENV-4 Genotype I at the country level since at least 2013. The entomological surveillance showed a predominance of *Aedes aegypti* compared to *Aedes albopictus* and high abundance of these vectors in dry and rainy seasons between 2016 and 2019, in Vientiane Capital. Overall, these results emphasized the importance of an integrated approach to evaluate factors, which could impact the circulation and the epidemiological profile of dengue viruses, especially in endemic countries like Lao PDR.

Keywords: dengue; DENV-4; epidemic; Lao PDR; phylogeny; *Aedes* vectors

---

1. Introduction

In the context of globalization of trade and travel, the arboviruses' epidemiology profiles have changed and their expansion is in constant progression [1]. Dengue fever is the most prevalent human arboviral disease in the world. Recent World Health Organization (WHO) statistics revealed an increase of the number of dengue cases reported from 505,430 cases in 2000 to 4.2 million in 2019, among which 70% of the burden is supported by Asia, and a modelling study estimated that 390 million people are infected by dengue virus per year [2–4]. Even if these data should be interpreted cautiously, due to changes in declaration systems and the increased number of contributing countries, they do reflect an alarming evolution of dengue virus (DENV) epidemiology. In 2017, WHO estimated that, every year, more than 500,000 people, including a high proportion of children, experience severe clinical presentations that require hospitalization. The number of fatal dengue cases has increased from 960 in 2000 to 4032 in 2015 [4].

Dengue fever is due to an infection of one of the four DENV serotypes transmitted to humans through the bite of an infected female mosquito of the *Aedes* genus [5]. In Lao People's Democratic Republic (Lao PDR), the main vector in urban areas is *Aedes aegypti* [6], whereas *Aedes albopictus* is considered a secondary vector, specifically in suburban, rural, and forested areas [5,7,8].

DENV are single-stranded, positive-sense RNA viruses belonging to the *Flavivirus* genus, *Flaviviridae* family. Antigenic studies identified four distinct serotypes (DENV-1 to 4) [9]. Genetic studies confirmed the segregation of DENV strains into four main groups matching with the serotypes, but also into genotype subdivisions that are often characteristic of the geographical origin of the viral strains [10–14]. Previous genetic studies have suggested that the four DENV serotypes emerged from sylvatic cycles in Asia [11,15,16]. Dengue outbreaks caused by the four serotypes have regularly been recorded in Asia over the last decades [17]. DENV-4 was identified for the first time in the Philippines and in Thailand in 1953 [17]. Since then, DENV-4 has been reported yearly in the Indochinese peninsula, but with a global burden significantly lower compared to the three other DENV serotypes [17]. From a genetic point of view, DENV-4 isolates segregate in four different genotypes: Genotype I (Asia) is frequently reported in South East Asia [18]; Genotype II (Asia, South Pacific, and South America); Genotype III (Thailand); and Genotype IV (sylvatic strains from Malaysia) [11,16,19–21].

Lao PDR is a land-locked country located in the middle of the Indochinese peninsula, surrounded by China (North), Cambodia (South), Thailand and Myanmar (West), and Vietnam (East). Lao PDR was a least-developed country of 7 million people in 2018, among which nearly 40% are under 20 years old (WHO and www.lsb.gov.la). The country is divided into 18 provinces. Since the first report of dengue hemorrhagic fever (DHF) cases in Lao PDR in 1979, DENV outbreaks have been regularly declared by the country and dengue fever has become a major national public health problem [22–24]. All four dengue serotypes now circulate in rural and urban areas in Lao PDR [22,23,25–27]. In 2008, a DENV-1 epidemic was described only in the North-West part of the country [28]. Then, in 2013, an extensive dengue outbreak with a predominance of DENV-3 caused 48,772 cases and 95 deaths at the country scale [29].

Interestingly, DENV-4 has been rarely detected in Lao PDR. Previous studies reported a lower prevalence of this serotype compared to the others in 1990 in Vientiane Capital and in other provinces in 2008–2009 [22,29,30]. Only sporadic cases have been detected in villages located around Vientiane Capital in 2006–2007 [30]. In 2012, an integrated arbovirus surveillance system, involving virologists and medical entomologists, was set up in Vientiane Capital by the Institut Pasteur du Laos and was gradually extended to different provinces in Lao PDR for the detection and monitoring of arbovirus epidemics [19,20,31]. This surveillance system detected DENV-4 transmission in 2013 in Vientiane Capital's suburban areas and subsequently its spread into the different districts of the city and then throughout the country.

This study aims to describe the DENV-4 switch from an endemic-sporadic circulation to a country-wide epidemic in Lao PDR in the context of multiple DENV serotypes co-circulation. A retrospective epidemiological study was conducted in Vientiane Capital to characterize the impact of the disease burden on the population. Furthermore, an entomological study was implemented in Vientiane Capital to estimate the population dynamics of the vectors at different time points between 2016 and 2019.

2. Results

2.1. Dengue Surveillance Activity between 2012 and 2019.

From 2012 to 2019, 15,152 samples were collected through the arbovirus surveillance system in 15 of the 18 provinces of Lao PDR (Table S1; Figure S1).

Of the 15,152 samples tested, 8771 (57.9%) were confirmed for a DENV infection by means of real-time RT-PCR and/or NS1 antigen detection. Among the confirmed cases, 8315 (54.9%) were found positives for DENV by RT-PCR and 448 (5%) cases were confirmed by NS1 antigen detection in plasma or in urine samples. The DENV serotypes were determined by real-time RT-PCR or sequencing after a viral culture for 3616 cases (43.5%).

Overall, 11,858 samples (78.4%) were collected in Vientiane Capital, notwithstanding the progressive extension of the surveillance system to other Lao provinces from 2015 and onwards. In 2012 and 2013, DENV serotype identification was mainly performed for isolates collected in Vientiane Capital, which represented, respectively, 80% and 87% of the samples tested. In 2014, only 134 suspected cases were investigated. The dramatically low number of samples tested by the arbovirus surveillance network was consistent with the poor syndromic activity recorded at the country level by the National Center for Laboratory and Epidemiology in charge of the centralization of the mandatory clinical declaration system (Figure 1). Furthermore, since 2012, the laboratory-based survey seemed to follow the same dengue epidemiological profile as the national syndromic surveillance system (Figure 1). A total of 14 (10.4%) were confirmed by DENV RT-PCR in 2014 and the DENV serotype could be determined for 11 samples (78.6%) (Figure 2).

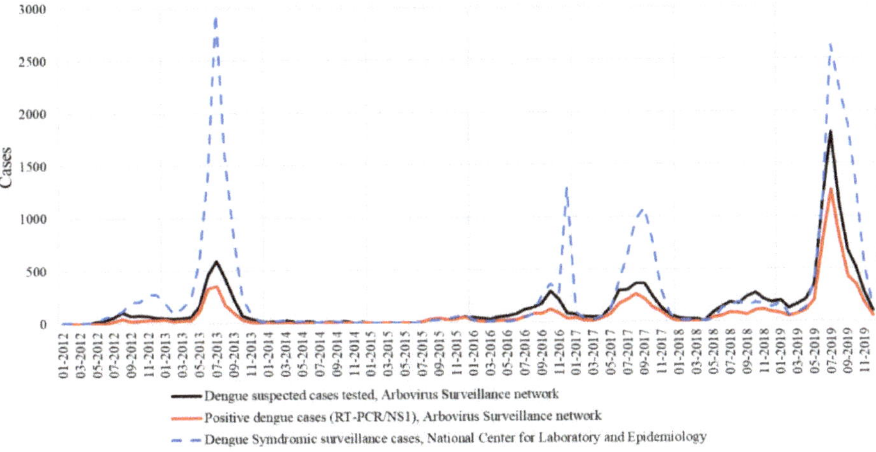

Figure 1. National syndromic surveillance system versus laboratory-based survey in Lao PDR.

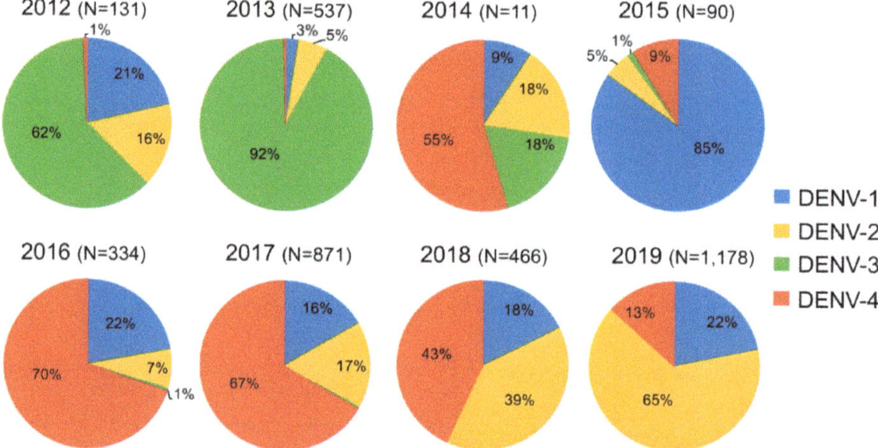

Figure 2. Dengue serotype distribution between 2012 and 2019 from samples collected by the Institut Pasteur du Laos arbovirus surveillance network. In parentheses, number of samples tested per year.

Among the cases confirmed by RT-PCR in Vientiane Capital, 85.7% and 62.4% were respectively serotyped in 2016 and in 2018 (Table S1). The proportion of samples serotyped varied from year to year (from 28.2% in 2019 to 76.6% in 2014), but the proportion of samples from Vientiane Capital remained constant over time (between 50% and 65%) (Table S1).

In 2012 and 2013, all four DENV serotypes were detected, but DENV-3 serotype was predominant (62% in 2012 and 92% in 2013) (Figure 2). This serotype gradually decreased and vanished in December 2013 [29]. Since then, only 2 DENV-3 sporadic cases were collected in Saravane in 2016 and 3 cases were detected in Vientiane Capital in 2017. In 2015, DENV-1 was at the origin of 85% of the confirmed dengue cases at the country level and this serotype continued to circulate at a low level in Lao PDR until 2019.

Between 2012 and 2015, DENV-4 samples represented less than 9% of the total cases (Figure 2) and were only recorded in the sub-urban districts of Vientiane Capital and in Vientiane Province, which are mainly rural areas. In 2014, clusters of DENV-4 cases were identified in Vientiane Capital. At the end of 2015, the dengue epidemiological profile changed in Vientiane capital with the identification of grouped cases of DENV-4 in urban districts of Vientiane Capital followed by a rapid increase of the proportion of DENV-4 cases in Vientiane Capital from 9% in 2015 to 70% and 67%, respectively, in 2016 and 2017. The outbreak peaked between June and August 2017 (Figure 3). The same trend was observed at the country level (Figure 3). In 2018, the proportion of DENV-4 samples decreased to 43% and DENV-2 became the predominant serotype at the end of the year. In 2019, at the country level, DENV-4 still represented 13% of the cases serotyped (13.2% in Vientiane Capital) with a co-circulation of DENV-1 and DENV-2 (Figure 2).

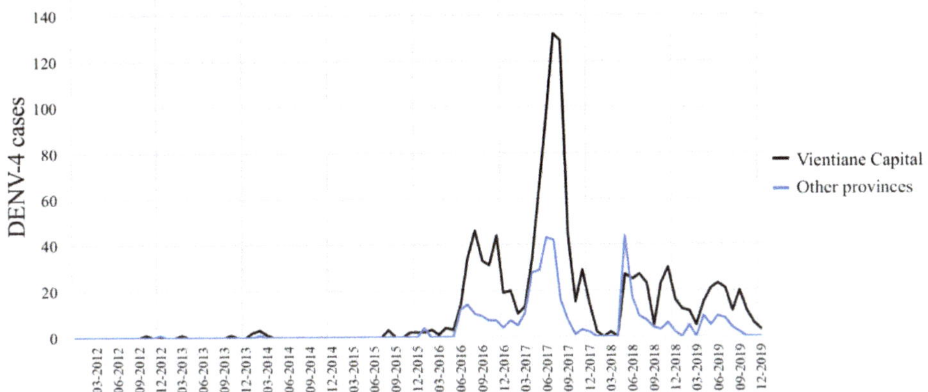

Figure 3. Dengue serotype 4 distribution between 2012 and 2019 per month from samples collected by the Institut Pasteur du Laos arbovirus surveillance network in Vientiane Capital and in other Lao provinces.

2.2. Epidemiological Analysis of the Patients Infected by DENV-4

Demographic and clinical characteristics of the patients reported by the surveillance network are presented in Table 1. Patients infected with DENV-4 serotype were significantly older (mean age 27.4 vs. 24.5, $p < 0.0001$) than patients infected with other DENV serotypes (Table 1 and Figure 4). Sex ratio, days to diagnosis, and severity were no different for DENV-4 when compared to the other serotypes.

Table 1. Demographic and clinical characteristics of the 15,152 patients investigated by the Institut Pasteurs du Laos arbovirus surveillance system between 2012 and 2019.

	All Samples (*n* = 15,152)	Dengue Positive [a] (*n* = 8771)	Sample Serotyped (DENV-1 to DENV-4) (*n* = 3616)	DENV-4 (*n* = 1187)
Mean age (SD)	23.5 (15.7) *n* = 15,062	23.9 (14.4) *n* = 8718	24.5 (14.5) *n* = 3589	27.4 (14.6) *n* = 1181
Sex				
Female (%)	6197 (40.9)	3775 (43.0)	1526 (42.3)	620 (52.2)
Male (%)	5945 (39.2)	3243 (39.3)	1403 (38.8)	557 (46.9)
Unknown (%) [†]	3010 (19.9)	1550 (17.7)	682 (18.9)	10 (0.8)
Mean number of days of fever (SD)	3.8 (2.4)	3.6 (1.8)	3.4 (1.8)	3.6 (1.7)
Clinical diagnosis [b]				
DF (%)	11,910 (78.6)	7036 (80.2)	2868 (79.4)	1154 (97.2)
DHF (%)	213 (1.4)	168 (1.9)	49 (1.4)	13 (1.1)
DSS (%)	27 (0.2)	23 (0.3)	17 (0.5)	9 (0.8)
Unknown (%) [†]	3002 (19.8)	1541 (17.6)	677 (18.8)	11 (0.9)

Note: [a] Dengue-positive: RT-PCR and or NS1-positive. [b] Data concerning clinical diagnosis were collected from 2015 onwards. DF: Dengue fever, DHF: Dengue hemorrhagic fever, DSS: Dengue shock syndrome. [†] No clinical data.

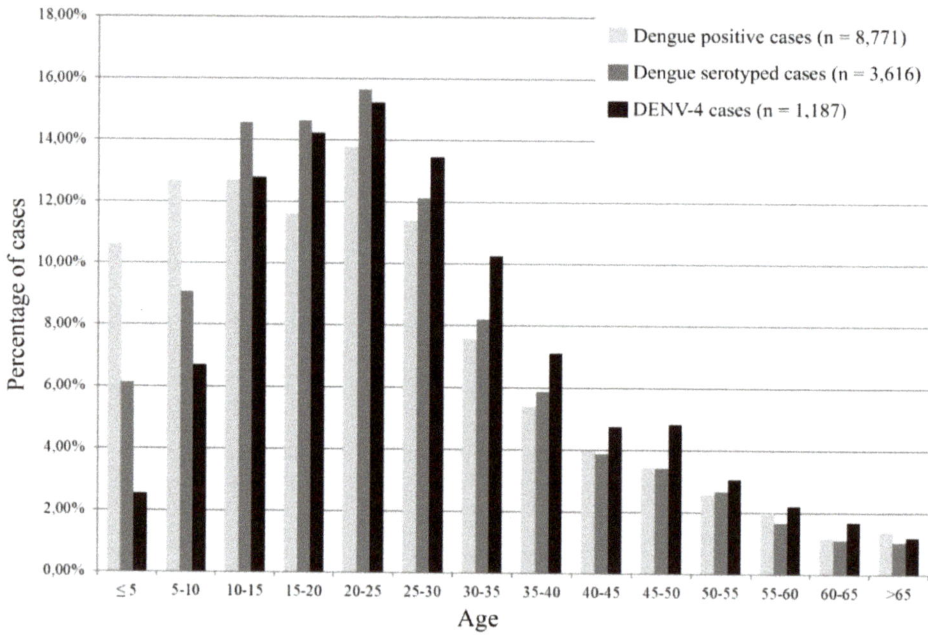

Figure 4. Age distribution of patients investigated by the Institut Pasteur du Laos arbovirus surveillance network.

Among the 52 DENV fatal cases investigated by the arbovirus surveillance network between 2012 and 2019, 9 were infected by DENV-4. These nine cases were identified between 2016 and 2019. In Vientiane Capital, four cases were detected in September 2016, August 2017, January 2019, and December 2019. In Attapeu province, 2 DENV-4 fatal cases were identified in July and September 2017. Two fatal cases were also recorded in June 2018 in Xayaboury and Champassak provinces and 1 case in Saravane province in April 2019 (Figure S1).

2.3. Phylogenetic Analysis of DENV-4

A panel of 45 samples, i.e., plasma ($n = 29$) or cell supernatant ($n = 16$), from DENV-4 cases collected between 2013 and 2019, were investigated for the phylogenetic analysis (Table 2). This panel included samples from different Lao provinces, cases imported from foreign Asian countries, patients with different degrees of dengue severity, and fatal cases. The sequence analysis based on a fragment spanning 1485 nucleotides of the envelop gene revealed that DENV-4 isolates grouped in two different genotypes (i.e., Genotype I and Genotype II) (Figure 5).

Table 2. References of Lao DENV-4 isolates.

Sample Identification	Years of Collection	Isolation Source	Genbank Number
LaoPDR-Vientiane, 2013-2769	2013	Cell supernatant	MT122852
LaoPDR-Vientiane (ex Thailand), 2014-2864	2014	Cell supernatant	MT122853
LaoPDR-Vientiane (ex Malaysia), 2014-2867	2014	Cell supernatant	MT122854
LaoPDR-Vientiane, 2015–3131	2015	Cell supernatant	MT122855
LaoPDR-Vientiane, 2015–3401	2015	Cell supernatant	MT122856

Table 2. *Cont.*

Sample Identification	Years of Collection	Isolation Source	Genbank Number
LaoPDR-Vientiane, 2015–3480	2015	Cell supernatant	MT122857
LaoPDR-Vientiane, 2016–3599	2016	Cell supernatant	MT122858
LaoPDR-Vientiane, 2016–3869	2016	Cell supernatant	MT122859
LaoPDR-Vientiane, 2016–3932	2016	Cell supernatant	MT122860
LaoPDR-Vientiane, 2016–4108	2016	Cell supernatant	MT122861
LaoPDR-Vientiane, 2016–4132	2016	Cell supernatant	MT122862
LaoPDR-Vientiane, 2016–4291	2016	Cell supernatant	MT122863
LaoPDR-Vientiane, 2016–4441 (Fatal case)	2016	Plasma	MT122864
LaoPDR-Saravane, 2017–5506	2017	Plasma	MT122865
LaoPDR-Saravane, 2017–5534	2017	Cell supernatant	MT122866
LaoPDR-Attapeu, 2017–5797	2017	Cell supernatant	MT122867
LaoPDR-Vientiane, 2017–5842	2017	Cell supernatant	MT122868
LaoPDR-Vientiane, 2017–5871	2017	Plasma	MT122869
LaoPDR-Attapeu, 2017–5876 (Fatal case)	2017	Plasma	MT122870
LaoPDR-Vientiane, 2017–5988	2017	Plasma	MT122871
LaoPDR-Vientiane, 2017–6237	2017	Plasma	MT122872
LaoPDR-Attapeu, 2017–6243	2017	Plasma	MT122873
LaoPDR-Vientiane, 2017–6358 (Fatal case)	2017	Plasma	MT122874
LaoPDR-Attapeu, 2017–6583 (Fatal case)	2017	Plasma	MT122875
LaoPDR-Vientiane, 2017–6509	2017	Plasma	MT122876
LaoPDR-Bolikhamxay, 2017–7305	2017	Plasma	MT122877
LaoPDR-Vientiane, 2017–7321	2017	Plasma	MT122878
LaoPDR-Attapeu, 2018–7509	2018	Plasma	MT122879
LaoPDR-Vientiane, 2018–7590	2018	Plasma	MT122880
LaoPDR-Xayaboury, 2018–7626 (Fatal case)	2018	Plasma	MT122881
LaoPDR-Savannakhet, 2018–7842	2018	Plasma	MT122882
LaoPDR-Saravane, 2018–7983	2018	Plasma	MT122883
LaoPDR-Vientiane-2018–8364	2018	Plasma	MT122884
LaoPDR-Vientiane, 2018–9161 (Fatal case)	2018	Plasma	MT122885
LaoPDR-Champassak, 2018-DS18-698-16 (Fatal case)	2018	Cell supernatant	MT122886
LaoPDR-Attapeu, 2019–9310	2019	Plasma	MT122887
LaoPDR-Saravane, 2019–9566 (Fatal case)	2019	Plasma	MT122888
LaoPDR-Vientiane, 2019–9831	2019	Plasma	MT122889
LaoPDR-Vientiane, 2019–10299	2019	Plasma	MT122890
LaoPDR-Saravane, 2019–10310	2019	Plasma	MT122891
LaoPDR-Saravane, 2019–10439	2019	Plasma	MT122892
LaoPDR-VientianeProvince, 2019–11393	2019	Plasma	MT122893
LaoPDR-Vientiane, 2019–15448	2019	Plasma	MT122894
LaoPDR-Vientiane, 2019–15460	2019	Plasma	MT122895
LaoPDR-Vientiane, 2019–15690	2019	Plasma	MT122896

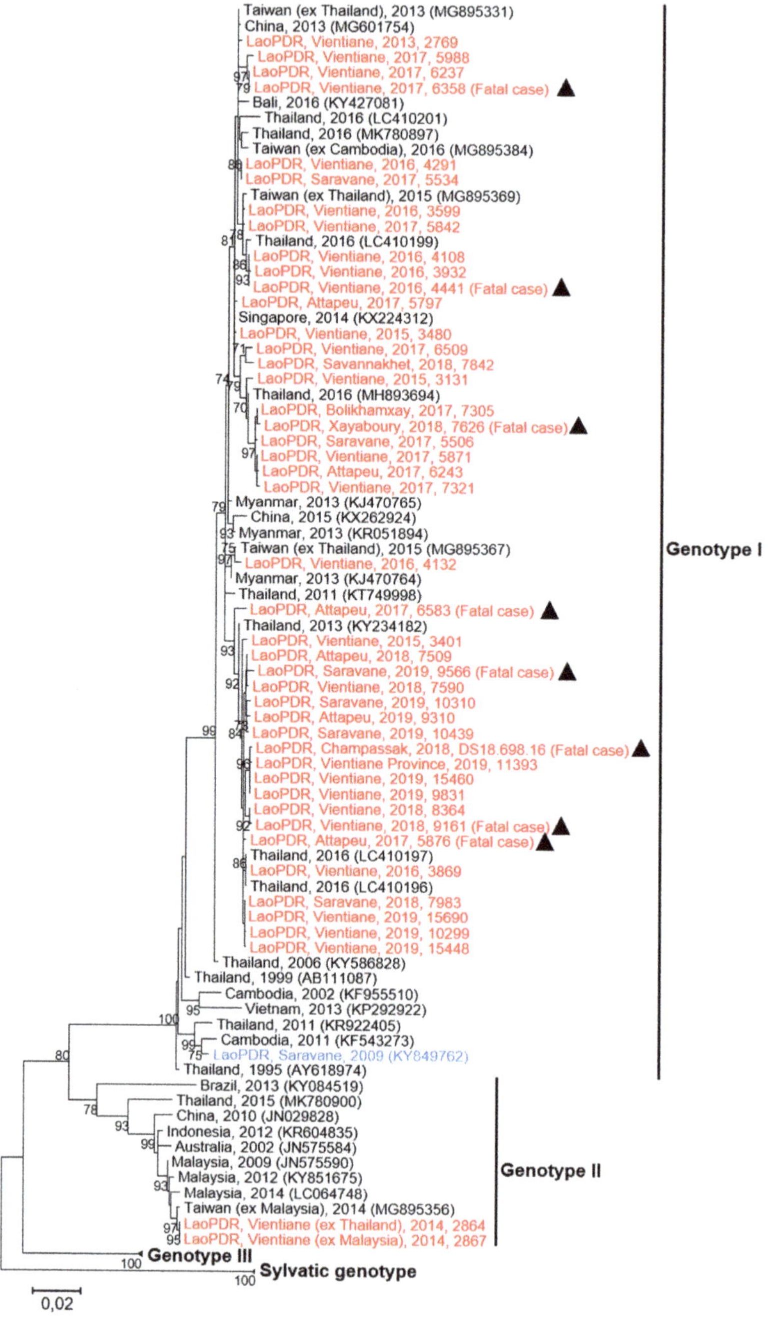

Figure 5. Maximum likelihood phylogenetic tree of DENV-4 sequences from Lao PDR. The tree was constructed on the envelope protein gene (1485 nt). Only the bootstrap values > 70 are shown. Scale bar indicates the nucleotide substitution per site. The Lao strains sequenced in this study are indicated in red and the triangles highlight the strains from fatal cases. The Lao strain previously described is indicated in blue. Black triangles identify isolates obtained from fatal cases.

Genotype II was identified in samples from European tourists traveling in Southeast Asia (Thailand and Malaysia) in 2014 and who developed a dengue-like syndrome during their stay in Vientiane. These isolates clustered with DENV-4 strains identified in Malaysia in 2009 and 2014 (>99% identity) (Figure 5) [32].

All autochthonous DENV-4 sequences belonged to the Genotype I. Among the Genotype I, the Lao strains isolated between 2013 and 2019 displayed a maximum nucleotide distance of 2.7% from each other. Interestingly, our series of isolates showed 3.1% to 4.1% of divergence with a Lao strain (KY849762) isolated in 2009 in Saravane province (Figure 5) [30]. When compared to DENV-4 reference strains collected in other Asian countries from 2011 to 2016, less than 2.5% of nucleotide divergence was found with the Lao DENV-4 isolates (Figure 5). In this study, eight DENV-4 isolates from fatal cases could be analyzed. These isolates belonged to the Genotype I but segregated in different phylogenetic clusters formed by the different Lao isolates characterized since 2013 (Figure 5). Globally, the phylogenetic analysis revealed the active circulation of DENV-4 Genotype I since at least 2013 in Lao PDR. Furthermore, several clusters were detected and could emphasize several introduction events in the country from neighboring countries such as Thailand (Figure 5).

2.4. Entomological Survey on the Aedes sp. Vectors in Vientiane Capital

The results of the abundance of the adult dengue vectors collected between 2016 and 2019 are presented in Figure 6. A total of 1276 specimens were collected and *Aedes aegypti* was the most abundant species, representing more than 86% of the total. *Aedes albopictus* represented 14% of the total. The percentage of *Ae. albopictus* varied between 3.2% in 2018 to 57.8% in 2016 when the study only started in May. In 2017 and 2019, the percentages were 12.6% and 9.9%, respectively. Every year, the vectors were more abundant during the rainy season between May and August, but the mosquito populations remained throughout the dry season albeit at a low level. Of note, no collections were made during the month of April; and in 2016, the collection started in May only.

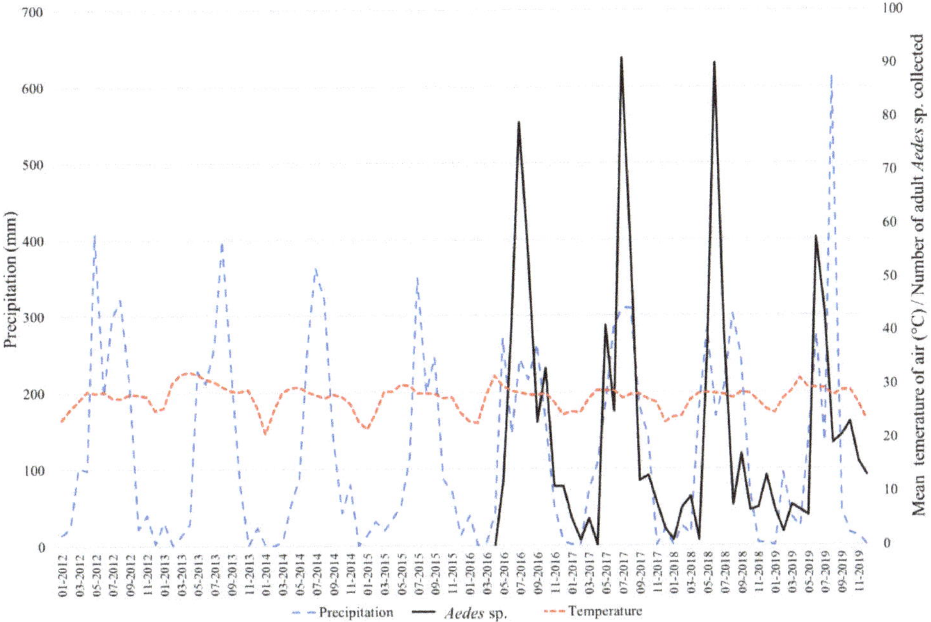

Figure 6. Monthly rainfall (mm; in blue) and mean temperature in the air (°C; in red) recorded between 2012 and 2019 in Vientiane Capital (17°58′12″ N, 102°34′14″ E). Number of *Aedes aegypti* and *Aedes albopictus* adult mosquitoes collected monthly between 2016 and 2019 in Vientiane Capital (in black).

3. Discussion

Since the first report on dengue in Lao PDR, limited data have been published on DENV serotypes/genotypes circulating at the country level [22]. Nevertheless, recent studies suggest a complex and dynamic dengue virus circulation countrywide, but without reference to DENV-4 [22,28–30]. This study is a first step in filling this gap by providing genetic and epidemiologic information on the first DENV-4 epidemic recorded in Lao PDR. The progressive increase of DENV-4 burden occurred in a context of the co-circulation of different DENV serotypes previously described and maintained over the eight years of surveillance (i.e., 2012–2019) [29–31]. Since 2009, DENV-4 circulated sporadically in the country [29,30]. Sporadic cases or small clusters of grouped cases were observed in Vientiane Capital between February 2014 and August 2015. Subsequently, the proportion of DENV-4 cases gradually increased among the samples tested by the arbovirus surveillance network in Vientiane Capital and became predominant in 2016 at the country level until October 2018. Since then, DENV-2 has overcome DENV-4 in Vientiane Capital and at the country level [33]. However, DENV-4 still circulated at a significant level in 2019 (13%; Figure 2). These data reveal a circulation of DENV-4 during at least six years (2014–2019) in Lao PDR and could be correlated with an insufficient immunity protection of the Lao population against this serotype as suggested by the impact of DENV-4 in the different age groups.

This study highlighted the significant difference in the age of the patients infected with DENV. The syndromic patients infected with DENV-4 were significantly older than the patients infected with other serotypes. The Lao population is young (~40% under 20 years old) (WHO and www.lsb.gov.la) and this data emphasize the risk of dengue emergence in Lao PDR due to a high percentage of the population immunologically naive for DENV. In this study, the mean age (27.4) of the patients infected by DENV-4 appeared to be consistent with studies from other DENV-endemic countries [34,35]. However, the fact that syndromic DENV-4 patients were older could suggest the recent emergence of this serotype and/or a low DENV-4 circulation in Lao PDR at least since 2009 [30]. These findings demonstrate that a large proportion of the Lao population is susceptible to DENV-4 and highlights the need to study the immune status of populations living in dengue-endemic countries in order to predict and prevent future DENV outbreaks as previously done in Lao PDR [22] and in Asia in general [36–38]. Furthermore, this study highlights the need to perform serotype-specific studies of the immune status of the population.

Here, we describe the pre-epidemic circulation of DENV-4 in Lao PDR and the specific features of this serotype during the outbreak in Vientiane Capital and at the country level in the context of DENV serotypes co-circulation. Clusters of DENV-4 infections were identified sporadically several years before the outbreak started. Indeed, serotype-specific surveillance networks are able to finely monitor the profile of DENV circulating and detect early warning signals. Strengthening the prediction capacity of DENV outbreaks by combining syndromic surveillance systems with a laboratory surveillance system has been demonstrated in Lao PDR ([30]; present study). Even if a systematic exhaustive investigation of suspected cases was clearly unrealistic, an appropriate sample size allowed for the drawing of an accurate picture of DENV serotypes' history in a country over the last eight years. In this way, the actual disease burden of dengue could be calculated, including at the serotype level. The amplitude over time and the burden by serotype and by age groups could be used as proxies to estimate the level of herd immunity in the Lao population. Interestingly, the local lab-based surveillance system in Vientiane city followed the trends of the syndromic surveillance data. Based on this approach, mathematical models could be developed to determine parameters including costs of long-term sustainable, combined surveillance systems.

The disease severity of dengue fever was not associated with the predominance of DENV-4 in Lao PDR. Indeed, no significant difference in the disease severity was found between the DENV serotypes in the confirmed samples collected by the arbovirus surveillance network. This finding is in agreement with previous observations, such as during a DENV-4 outbreak in Brazil, which failed to show any link between this serotype and dengue severe symptoms [20]. Furthermore, in Lao PDR, the fatal cases due to a DENV-4 infection were recorded during the entire outbreak and no link could be established with the number of positive DENV-4 cases. Indeed, in our study, the same number of fatal cases was recorded in 2017 and 2019, respectively, during the peak of the outbreak and at the end of the outbreak. Moreover, dengue fatal cases were observed during the active DENV transmission period, as well as when dengue cases were recorded at a low level.

Previous phylogenetic studies showed the recent circulation of two DENV-4 genotypes (Genotype I and II) in Southeast Asia [18,32]. Both genotypes were detected in Lao PDR, but in two different and independent types of patients, i.e., Lao native autochthonous cases and imported cases by foreigners. All autochthonous cases recorded at the country level investigated here belonged to DENV-4 Genotype I. The first evidence of the presence of DENV-4 Genotype I in Lao PDR came from a strain that was first identified in Saravane Province isolated in 2009 [30]. Our serial investigation suggests that this genotype has spread and circulated at least since 2013 at the country level. The genetic difference (2.7% of nucleotide divergence) between the former strain and the recent isolates could be the result of several introductions of DENV-4 Genotype I in Lao PDR since 2009 from different Asian countries. The composition of the different clusters and the circulation of the Lao isolates over several years are a signature of successful consecutive emergences of those independent introduction events. DENV-4 Genotype II was only detected in two imported cases from Thailand/Malaysia in 2014 (Figure 5). Interestingly, the predominant circulation of the DENV-4 Genotype I in Lao PDR contrasted with some results obtained in South East Asia. Indeed, in the Philippines, a genotype turnover from Genotype I to Genotype II was observed before 2013, and this seemed to match with a high DENV-4 circulation in the country between 2012 and 2016 [39]. The results found in the Philippines were consistent with those from Malaysia in 2001 that showed the circulation of the DENV-4 Genotype II [40]. However, both introductions of DENV-4 Genotype II were recorded in February 2014, the dry season in Lao PDR (October–April) and a decrease of vector density. This context, associated with the abnormal lack of rainfalls until August 2015, could explain the non-emergence of DENV-4 Genotype II in Vientiane Capital. However, these dramatically unfavorable conditions for DENV transmission did not hamper the maintenance of a sufficient inoculum of a mixt population of DENV-4 Genotype I isolates at the origin of the rapid spread-off since September 2015. Hence, the genotype selection could impact the epidemiology of DENV emergence and expansion of the virus. Indeed, as previously described in India, China, or the WHO Pacific region, a DENV genotype switch could increase the risk of dengue outbreak [41–43]. This selection of DENV genotype could be influenced by several factors, such as serotype-specific herd immunity, viremia titer in humans, and the ability of local mosquito populations, such as *Aedes aegypti*, to transmit the virus under the influence of environmental and climatic conditions [44–49]. These factors need to be investigated for DENV serotypes and genotypes, which circulated in a specific geographical context to improve the knowledge on DENV epidemiology and to prevent DENV outbreaks [50].

The entomological results demonstrate the presence of both *Ae. aegypti* and *Ae. albopictus* in Vientiane Capital with an increase of vector density during the rainfall season as previously described in Asia. The dynamic of DENV transmission in human population follows the trend of the dynamic of *Aedes* vectors [51]. Moreover, predominance of *Ae. aegypti* was highlighted since 2016 in Vientiane Capital. Vector competence of *Ae. aegypti* from Lao PDR was investigated for DENV-1 and showed that 50% of the females orally exposed to the virus could transmit DENV-1 [52]. These data indicate a high efficiency of transmission of this serotype by *Ae. aegypti* from Lao PDR compared to other *Ae. aegypti* populations from the Pacific region (3–37% transmission efficiency), from French Guiana (<10% transmission efficiency), and from Florida (33% transmission efficiency) in the USA [53–55]. DENV transmission ability appeared to be a specific interaction between the mosquito population and a virus strain [49,56]. Vazeille et al. [54] suggested a possible competition between serotypes at the midgut level in co-infected mosquitoes leading to a drastically different transmission potential and, in this case, favoring the competitive displacement of DENV-1 by DENV-4. Thus, the ability of Lao vectors for DENV-4 must be investigated to determine how this factor could influence the onset of new dengue epidemics in Lao PDR.

DENV emergence and spread are complex mechanisms, especially in endemic countries like Lao PDR. This study shows the need to combine results obtained by the DENV laboratory-based surveillance system and epidemiologic studies, combined with entomological data, along with more research oriented approaches, such as virus genetic, viral kinetic, or virus/vector interaction findings, to determine the impact of these factors on DENV transmission in each specific geographical context.

4. Materials and Methods

4.1. Human Samples Collection

In 2012, an arbovirus surveillance system was set up and coordinated by the Institut Pasteur du Laos. It was first implemented in Vientiane Capital and then progressively extended to hospitals in 18 provinces, with Vientiane Capital's hospital network being the most active. From 2012 to 2019, human samples were collected through this surveillance hospital network. Suspected dengue fever cases were defined according to the WHO's definition (fever onset $\geq$ 38 °C for less than 7 days with at least one of the following accompanying symptoms: headache; myalgia; arthralgia; retro-orbital pain; digestive troubles or hemorrhaging) [29]. After obtaining informed consent, clinicians filled a standardized clinical report form (CRF). Cases were classified according to the WHO criteria (1997), defining the severity as Dengue fever (DF), Dengue hemorrhagic fever (DHF), or Dengue shock syndrome (DSS). Venous blood samples (5 mL) were taken from patients. Samples were stored at 4 °C during transportation to the Institut Pasteur du Laos for analysis and serotypes and genotypes investigation.

4.2. Ethical Statement

Ethics approval was obtained from the Lao National Ethics Committee for Health Research (N°2018.116). Oral and written informed consent were obtained from all participants, or a parent or legal guardian.

4.3. Dengue Virus Screening

Samples collected by the arbovirus surveillance system were screened by RT-PCR as previously described [29] with a pan-dengue real-time RT-PCR [57] and the DENV serotypes were determined with specific real-time RT-PCR [58].

4.4. Gene E Sequencing Analysis

Viral genomic RNA was extracted from 45 human plasma using a Nucleospin DX or Nucleospin 96 core kit purification kit (Macherey-Nagel) according to the manufacturer's instructions (Table 2).

Gene E sequencing (1485nt) was performed using primers FGT1, FGT2, and F3 (Table 3). First Stand cDNA were generated using a Maxima H Minus First Stand cDNA Synthesis kit (Thermo Scientific, Waltham, MA, USA) and the PCR was performed using a Phusion Flash High-Fidelity PCR Master Kit (New England Biolabs® Inc, Waltham, MA, USA). Amplified fragments were purified using ExoSAP-IT™ PCR Product Cleanup Reagent (Thermo Fisher Scientific, Waltham, MA, USA). Purified fragments were sequenced using a BigDye Terminator v3.1 Cycle sequencing kit (Applied Biosystem, Waltham, MA, USA) on a Genetic Analyzer 3500xL (Applied Biosystem, Waltham, MA, USA).

Sequences were analyzed using Chromas software (www.technelysium.com.au) and aligned with the multiple sequence alignment software Clustal W integrated in BioEdit version 7.0.5.3 software (Manchester, United Kingdom) [59,60]. For the phylogenetic analysis, a maximum likelihood tree was constructed using MEGA version 7 (www.megasoftware.net), with a kimura-2 parameter model with a bootstrap of 1000 replication [61] as previously described for dengue virus phylogenetic analysis [29,43,62–64]).

Table 3. List and positions of primers used for DENV-4 complete envelope gene RT-PCR and sequencing.

Fragment	Forward	Genome Position	Reverse	Genome Position
FGT1	5′CAT-TCA-GGA-ATG-GGA-TTG-GA3′	732–751	5′ACA-GTC-CAC-AAT-GGA-GAY-AC3′	1362–1381
FGT2	5′AGG-AGG-AGT-TGT-GAC-ATG3′	1268–1285	5′TTG-GGC-GCA-TCA-TCA-CAT3′	1981–1998
F3	5′GAG-ATG-GCA-GAA-ACW-CAG-C3′	1869–1887	5′TTA-GAT-CAA-CCA-CGA-GGC-T3′	2593–2611

The genome positions are given according to the dengue virus serotype 4 reference genome (GenBank: NC_002640).

4.5. Epidemiological Analysis

Demographic and clinical data were extracted from the surveillance network's de-identified database. Continuous variables were summarized using mean and standard deviation (SD), and categorical variables were summarized using frequencies and percentages. Continuous data were compared using parametric Student test and rates using the Chi-square and Fisher exact test. Statistical significance was set at the 5% level. Analyses were conducted using STATA software (version 14.0; StatCorp LP, College Station, TX, USA).

4.6. Mosquito Surveillance

Five mosquito sentinel sites in five different villages of Vientiane Capital were chosen for the mosquito surveillance and the GPS locations are presented in Table 4. The vectors abundance study took place during both dry and rainy seasons between 2016 and 2019. BG sentinel traps® (Biogents, Regensburg, Germany) were used to follow the dynamics of mosquito population abundance in the city (4 traps per locations, except at Kao-Gnot with two traps). After every weekly collection, the adult mosquitoes were brought back to the laboratory for morphological identification.

Table 4. Mosquito collection locations in Vientiane Capital.

District	Village	Latitude	Longitude
Sittattanak	Donkoy	17.562677	102.390768
Sittattanak	Kao-Gnot	17.962684	102.615035
Xaysettha	Sengsavang	17.995816	102.664895
Xaithany	Sivilay	18.003705	102.380003
Sittattanak	Saphanthong Tai	17.949470	102.628487

Supplementary Materials: The following are available online at http://www.mdpi.com/2076-0817/9/9/728/s1, Figure S1: Map of Lao PDR, Table S1: Geographic origins of the DENV samples serotyped by the Institut Pasteur du Laos arbovirus surveillance system.

Author Contributions: Conceptualization: E.C., V.P., S.M., and M.G.; methodology, E.C., V.P., S.M., and M.G.; validation, E.C., V.P., S.M., and M.G.; formal Analysis, E.C., V.P., and S.M.; investigation, E.C., V.P., S.V., P.B., T.A.C., S.S., and S.M.; resources, S.S., S.V., T.A.C., T.X., P.P., P.B., S.K., and S.M.; writing—original draft preparation, E.C., V.P., S.S., S.M., and M.G.; writing—review and editing, E.C., V.P., J.P., P.T.B., M.C., S.M., and M.G.; visualization, E.C., V.P., and S.M.; supervision, M.G. and E.C.; project administration, M.G.; funding acquisition, M.G., P.T.B., O.T., and M.C. All authors have read and agreed to the published version of the manuscript.

Funding: This work was funded by the Agence Française de Développement grant n° CZZ 2146 01 (Ecomore2 project) and by UNITEDengue and Global Partnership Programme, Canada (ASEAN-GPP Grant Phase 3—Laboratory Capacity Development for diagnostics of Emerging Dangerous Pathogens).

Acknowledgments: We thank Lee Ching Ng, Director, Environmental Health Institute (EHI), National Environment Agency, Chanditha Hapuarachchi, and Carmen Koo for their sequencing technical assistance. We also thank the Institut Pasteur du Laos staff for the mosquito collections and identification (Phoutmany Thammavong, Phonesavanh Luangamath, Kaithong Lakeomany, Somphat Nilaxay, and Vaekey Vungkyly).

Conflicts of Interest: The authors declare no conflict of interest.

References

1. Cao-Lormeau, V.-M. Tropical Islands as New Hubs for Emerging Arboviruses. *Emerg. Infect. Dis.* **2016**, *22*, 913–915. [CrossRef] [PubMed]
2. Bhatt, S.; Gething, P.W.; Brady, O.J.; Messina, J.P.; Farlow, A.W.; Moyes, C.L.; Drake, J.M.; Brownstein, J.S.; Hoen, A.G.; Sankoh, O.; et al. The global distribution and burden of dengue. *Nature* **2013**, *496*, 504–507. [CrossRef] [PubMed]
3. Brady, O.J.; Gething, P.W.; Bhatt, S.; Messina, J.P.; Brownstein, J.S.; Hoen, A.G.; Moyes, C.L.; Farlow, A.W.; Scott, T.W.; Hay, S.I. Refining the global spatial limits of dengue virus transmission by evidence-based consensus. *PLoS Negl. Trop. Dis.* **2012**, *6*, e1760. [CrossRef] [PubMed]
4. World Health Organization. Dengue and Severe Dengue Update 23 June 2020. 2020. Available online: https://www.who.int/news-room/fact-sheets/detail/dengue-and-severe-dengue (accessed on 14 July 2020).
5. Gubler, D. Epidemic dengue/dengue hemorrhagic fever as a public health, social and economic problem in the 21st century. *Trends Microbiol.* **2002**, *10*, 100–103. [CrossRef]
6. Marcombe, S.; Fustec, B.; Cattel, J.; Chonephetsarath, S.; Thammavong, P.; Phommavanh, N.; David, J.-P.; Corbel, V.; Sutherland, I.W.; Hertz, J.C.; et al. Distribution of insecticide resistance and mechanisms involved in the arbovirus vector Aedes aegypti in Laos and implication for vector control. Barker CM, editor. *PLoS Negl. Trop. Dis.* **2019**, *13*, e0007852. [CrossRef]
7. Hawley, W.A. The biology of Aedes albopictus. *J. Am. Mosq. Control. Assoc. Suppl.* **1988**, *1*, 1–39.
8. Tangena, J.-A.; Marcombe, S.; Thammavong, P.; Chonephetsarath, S.; Somphong, B.; Sayteng, K.; Grandadam, M.; Sutherland, I.W.; Lindsay, S.W.; Brey, P.T. Bionomics and insecticide resistance of the arboviral vector Aedes albopictus in northern Lao PDR. *PLoS ONE* **2018**, *13*, e0206387. [CrossRef]
9. Katzelnick, L.C.; Fonville, J.M.; Gromowski, G.D.; Bustos-Arriaga, J.; Green, A.; James, S.; Lau, L.; Montoya, M.; Wang, C.; VanBlargan, L.A.; et al. Dengue viruses cluster antigenically but not as discrete serotypes. *Science* **2015**, *349*, 1338–1343. [CrossRef]
10. Henchal, E.A.; Putnak, J.R. The dengue viruses. *Clin. Microbiol. Rev.* **1990**, *3*, 376–396. [CrossRef]
11. Weaver, S.; Vasilakis, N. Molecular evolution of dengue viruses: Contributions of phylogenetics to understanding the history and epidemiology of the preeminent arboviral disease. *Infect. Genet. Evol.* **2009**, *9*, 523–540. [CrossRef]
12. Gubler, D. Dengue and dengue hemorrhagic fever. *Clin. Microbiol. Rev.* **1998**, *11*, 480–496. [CrossRef] [PubMed]
13. Kuno, G.; Chang, G.-J.J.; Tsuchiya, K.R.; Karabatsos, N.; Cropp, C.B. Phylogeny of the genus Flavivirus. *J. Virol.* **1998**, *72*, 73–83. [CrossRef] [PubMed]
14. Rico-Hesse, R. Microevolution and virulence of dengue viruses. *Adv. Exp. Med. Biol.* **2003**, *59*, 315–341.
15. Rico-Hesse, R. Molecular evolution and distribution of dengue viruses type 1 and 2 in nature. *Virology* **1990**, *174*, 479–493. [CrossRef]
16. Wang, E.; Ni, H.; Xu, R.; Barrett, A.D.T.; Watowich, S.J.; Gubler, D.J.; Weaver, S. Evolutionary relationships of endemic/epidemic and sylvatic dengue viruses. *J. Virol.* **2000**, *74*, 3227–3234. [CrossRef]
17. Messina, J.; Brady, O.J.; Scott, T.W.; Zou, C.; Pigott, D.M.; Duda, K.A.; Bhatt, S.; Katzelnick, L.; Howes, R.E.; Battle, K.E.; et al. Global spread of dengue virus types: Mapping the 70 year history. *Trends Microbiol.* **2014**, *22*, 138–146. [CrossRef]
18. Hamel, R.; Surasombatpattana, P.; Wichit, S.; Dauvé, A.; Donato, C.; Pompon, J.F.; Vijaykrishna, D.; Liegeois, F.; Vargas, R.M.; Luplertlop, N.; et al. Phylogenetic analysis revealed the co-circulation of four dengue virus serotypes in Southern Thailand. *PLoS ONE* **2019**, *14*, e0221179. [CrossRef]
19. Lanciotti, R.S.; Gubler, D.J.; Trent, D.W. Molecular evolution and phylogeny of dengue-4 viruses. *J. Gen. Virol.* **1997**, *78*, 2279–2284. [CrossRef]
20. Heringer, M.; Souza, T.M.A.; Lima, M.D.R.Q.; Nunes, P.C.G.; Faria, N.R.D.C.; De Bruycker-Nogueira, F.; Chouin-Carneiro, T.; Nogueira, R.M.R.; Dos Santos, F.B. Dengue type 4 in Rio de Janeiro, Brazil: Case characterization following its introduction in an endemic region. *BMC Infect. Dis.* **2017**, *17*, 410. [CrossRef]
21. Klungthong, C.; Zhang, C.; Mammen, M.P.; Ubol, S.; Holmes, E.C. The molecular epidemiology of dengue virus serotype 4 in Bangkok, Thailand. *Virology* **2004**, *329*, 168–179. [CrossRef]

22. Fukunaga, T.; Phommasack, B.; Bounlu, K.; Saito, M.; Tadano, M.; Makino, Y.; Kanemura, K.; Arakaki, S.; Shinjo, M.; Insisiengmay, S. Epidemiological situation of dengue infection in Lao P.D.R. *Trop. Med.* **1994**, *35*, 219–227.

23. Khampapongpane, B.; Lewis, H.C.; Ketmayoon, P.; Phonekeo, D.; Somoulay, V.; Khamsing, A.; Phengxay, M.; Sisouk, T.; Sisouk, P.; Bryant, J.E. National dengue surveillance in the Lao People's Democratic Republic, 2006–2012: Epidemiological and laboratory findings. *West. Pac. Surveill Response J.* **2014**, *5*, 7–13.

24. Soukaloun, D. Dengue infection in Lao PDR. *Southeast Asian J. Trop. Med. Public Health.* **2014**, *45*, 113–119. [PubMed]

25. Bounlu, K.; Tadano, M.; Makino, Y.; Arakaki, S.; Kanemura, K.; Fukunaga, T. A seroepidemiological study of dengue and Japanese encephalitis virus infections in Vientiane, Lao PDR. *Jpn. J. Trop. Med. Hyg.* **1992**, *20*, 149–156. [CrossRef]

26. Makino, Y.; Saito, M.; Phommasack, B.; Vongxay, P.; Kanemura, K.; Pothawan, T.; Bounsou; Insisiengmay, S.; Sompaw; Fukunaga, T. Arbovirus Infections in Pilot Areas in Laos. *Trop. Med.* **1995**, *36*, 131–139.

27. Guo, X.; Zhao, Q.; Wu, C.; Zuo, S.; Zhang, X.; Jia, N.; Liu, J.; Zhou, H.-N.; Zhang, J.-S. First isolation of dengue virus from Lao PDR in a Chinese traveler. *Virol. J.* **2013**, *10*, 70. [CrossRef]

28. Dubot-Pérès, A.; Vongphrachanh, P.; Denny, J.; Phetsouvanh, R.; Linthavong, S.; Sengkeopraseuth, B.; Khasing, A.; Xaythideth, V.; Moore, C.E.; Vongsouvath, M.; et al. An Epidemic of Dengue-1 in a Remote Village in Rural Laos. *PLoS Negl. Trop. Dis.* **2013**, *7*, e2360. [CrossRef]

29. Lao, M.; Caro, V.; Thiberge, J.-M.; Bounmany, P.; Vongpayloth, K.; Buchy, P.; Duong, V.; Vanhlasy, C.; Hospied, J.-M.; Thongsna, M.; et al. Co-Circulation of Dengue Virus Type 3 Genotypes in Vientiane Capital, Lao PDR. *PLoS ONE* **2014**, *9*, e115569. [CrossRef]

30. Castonguay-Vanier, J.; Klitting, R.; Sengvilaipaseuth, O.; Piorkowski, G.; Baronti, C.; Sibounheuang, B.; Vongsouvath, M.; Chanthongthip, A.; Thongpaseuth, S.; Mayxay, M.; et al. Molecular epidemiology of dengue viruses in three provinces of Lao PDR, 2006–2010. *PLoS Neglected Trop. Dis.* **2018**, *12*, e0006203. [CrossRef]

31. Somlor, S.; Vongpayloth, K.; Diancourt, L.; Buchy, P.; Duong, V.; Phonekeo, D.; Ketmayoon, P.; Vongphrachanh, P.; Brey, P.T.; Caro, V.; et al. Chikungunya virus emergence in the Lao PDR, 2012–2013. *PLoS ONE* **2017**, *12*, e0189879. [CrossRef]

32. Yang, C.-F.; Chang, S.-F.; Hsu, T.-C.; Su, C.-L.; Wang, T.-C.; Lin, S.-H.; Yang, S.-L.; Lin, C.-C.; Shu, P.-Y. Molecular characterization and phylogenetic analysis of dengue viruses imported into Taiwan during 2011–2016. *PLoS Neglected Trop. Dis.* **2018**, *12*, e0006773. [CrossRef] [PubMed]

33. Calvez, E.; Somlor, S.; Viengphouthong, S.; Balière, C.; Bounmany, P.; Keosenhom, S.; Caro, V.; Grandadam, M. Rapid genotyping protocol to improve dengue virus serotype 2 survey in Lao PDR. *PLoS ONE* **2020**, *15*, e0237384. [CrossRef] [PubMed]

34. Lestari, C.S.W.; Yohan, B.; Yunita, A.; Meutiawati, F.; Hayati, R.F.; Trimarsanto, H.; Sasmono, R. Phylogenetic and evolutionary analyses of dengue viruses isolated in Jakarta, Indonesia. *Virus Genes* **2017**, *53*, 778–788. [CrossRef] [PubMed]

35. Shrivastava, S.; Tiraki, D.; Diwan, A.; Lalwani, S.K.; Modak, M.; Mishra, A.C.; Arankalle, V.A. Co-circulation of all the four dengue virus serotypes and detection of a novel clade of DENV-4 (genotype I) virus in Pune, India during 2016 season. *PLoS ONE* **2018**, *13*, e0192672. [CrossRef] [PubMed]

36. Vongpunsawad, S.; Intharasongkroh, D.; Thongmee, T.; Poovorawan, Y. Seroprevalence of antibodies to dengue and chikungunya viruses in Thailand. *PLoS ONE* **2017**, *12*, e0180560. [CrossRef] [PubMed]

37. Tsai, J.-J.; Liu, C.-K.; Tsai, W.-Y.; Liu, L.-T.; Tyson, J.; Tsai, C.-Y.; Lin, P.-C.; Wang, W.-K. Seroprevalence of dengue virus in two districts of Kaohsiung City after the largest dengue outbreak in Taiwan since World War II. *PLoS Neglected Trop. Dis.* **2018**, *12*, e0006879. [CrossRef] [PubMed]

38. Fox-Lewis, A.; Hopkins, J.; Sar, P.; Sao, S.; Pheaktra, N.; Day, N.P.J.; Blacksell, S.D.; Turner, P. Seroprevalence of Dengue Virus and Rickettsial Infections in Cambodian Children. *Am. J. Trop. Med. Hyg.* **2019**, *100*, 635–638. [CrossRef]

39. Luz, M.A.D.V.; Nabeshima, T.; Moi, M.L.; Dimamay, M.T.A.; Pangilinan, L.-A.S.; Dimamay, M.P.S.; Matias, R.R.; Mapua, C.A.; Buerano, C.C.; De Guzman, F.; et al. An Epidemic of Dengue Virus Serotype-4 during the 2015–2017: The Emergence of a Novel Genotype IIa of DENV-4 in the Philippines. *Jpn. J. Infect. Dis.* **2019**, *72*, 413–419. [CrossRef]

40. Abubakar, S.; Wong, P.-F.; Chan, Y.F. Emergence of dengue virus type 4 genotype IIA in Malaysia. *J. Gen. Virol.* **2002**, *83*, 2437–2442. [CrossRef]

41. Ahamed, S.F.; Rosario, V.; Britto, C.; Dias, M.; Nayak, K.; Chandele, A.; Murali-Krishna, K.; Shet, A. Emergence of new genotypes and lineages of dengue viruses during the 2012–15 epidemics in southern India. *Int. J. Infect. Dis.* **2019**, *84*, S34–S43. [CrossRef]

42. Jiang, L.; Jing, Q.L.; Liu, Y.; Cao, Y.M.; Su, W.Z.; Biao, D.; Yang, Z.C. Molecular characterization and genotype shift of dengue virus strains between 2001 and 2014 in Guangzhou. *Epidemiol. Infect.* **2016**, *145*, 760–765. [CrossRef] [PubMed]

43. Dupont-Rouzeyrol, M.; Aubry, M.; O'Connor, O.; Roche, C.; Gourinat, A.-C.; Guigon, A.; Pyke, A.; Grangeon, J.-P.; Nilles, E.J.; Chanteau, S.; et al. Epidemiological and molecular features of dengue virus type-1 in New Caledonia, South Pacific, 2001–2013. *Virol. J.* **2014**, *11*, 61. [CrossRef] [PubMed]

44. Hang, V.T.T.; Holmes, E.C.; Veasna, D.; Quy, N.T.; Hien, T.T.; Quail, M.; Churcher, C.; Parkhill, J.; Cardosa, J.; Farrar, J.; et al. Emergence of the Asian 1 Genotype of Dengue Virus Serotype 2 in Viet Nam: In Vivo Fitness Advantage and Lineage Replacement in South-East Asia. *PLoS Negl. Trop. Dis.* **2010**, *4*, e757. [CrossRef]

45. Quiner, C.A.; Parameswaran, P.; Ciota, A.T.; Ehrbar, D.J.; Dodson, B.L.; Schlesinger, S.; Kramer, L.D.; Harris, E. Increased Replicative Fitness of a Dengue Virus 2 Clade in Native Mosquitoes: Potential Contribution to a Clade Replacement Event in Nicaragua. *J. Virol.* **2014**, *88*, 13125–13134. [CrossRef]

46. Lambrechts, L.; Fansiri, T.; Pongsiri, A.; Thaisomboonsuk, B.; Klungthong, C.; Richardson, J.H.; Ponlawat, A.; Jarman, R.G.; Scott, T.W. Dengue-1 Virus Clade Replacement in Thailand Associated with Enhanced Mosquito Transmission. *J. Virol.* **2011**, *86*, 1853–1861. [CrossRef]

47. Hanley, K.A.; Nelson, J.T.; Schirtzinger, E.E.; Whitehead, S.S.; Hanson, C.T. Superior infectivity for mosquito vectors contributes to competitive displacement among strains of dengue virus. *BMC Ecol.* **2008**, *8*, 1. [CrossRef]

48. Fansiri, T.; Fontaine, A.; Diancourt, L.; Caro, V.; Thaisomboonsuk, B.; Richardson, J.H.; Jarman, R.G.; Ponlawat, A.; Lambrechts, L. Genetic Mapping of Specific Interactions between Aedes aegypti Mosquitoes and Dengue Viruses. *PLoS Genet.* **2013**, *9*, e1003621. [CrossRef]

49. Lambrechts, L. Quantitative genetics of Aedes aegypti vector competence for dengue viruses: Towards a new paradigm? *Trends Parasitol.* **2011**, *27*, 111–114. [CrossRef]

50. Cologna, R.; Armstrong, P.M.; Rico-Hesse, R. Selection for Virulent Dengue Viruses Occurs in Humans and Mosquitoes. *J. Virol.* **2005**, *79*, 853–859. [CrossRef]

51. Wai, K.T.; Arunachalam, N.; Tana, S.; Espino, F.; Kittayapong, P.; Abeyewickreme, W.; Hapangama, D.; Tyagi, B.K.; Htun, P.T.; Koyadun, S.; et al. Estimating dengue vector abundance in the wet and dry season: Implications for targeted vector control in urban and peri-urban Asia. *Pathog. Glob. Health* **2012**, *106*, 436–445. [CrossRef] [PubMed]

52. Miot, E.F.; Calvez, E.; Aubry, F.; Dabo, S.; Grandadam, M.; Marcombe, S.; Oke, C.; Logan, J.G.; Brey, P.T.; Lambrechts, L. Risk of arbovirus emergence via bridge vectors: Case study of the sylvatic mosquito Aedes malayensis in the Nakai district, Laos. *Sci. Rep.* **2020**, *10*, 7750–7759. [CrossRef] [PubMed]

53. Calvez, E.; Guillaumot, L.; Girault, D.; Richard, V.; O'Connor, O.; Paoaafaite, T.; Teurlai, M.; Pocquet, N.; Cao-Lormeau, V.-M.; Dupont-Rouzeyrol, M. Dengue-1 virus and vector competence of Aedes aegypti (Diptera: Culicidae) populations from New Caledonia. *Parasites Vectors* **2017**, *10*, 381. [CrossRef] [PubMed]

54. Vazeille, M.; Gaborit, P.; Mousson, L.; Girod, R.; Failloux, A.-B. Competitive advantage of a dengue 4 virus when co-infecting the mosquito Aedes aegypti with a dengue 1 virus. *BMC Infect. Dis.* **2016**, *16*, 318. [CrossRef] [PubMed]

55. Richards, S.L.; Anderson, S.L.; Alto, B.W. Vector competence of Aedes aegypti and Aedes albopictus (Diptera: Culicidae) for dengue virus in the Florida Keys. *J. Med. Entomol.* **2012**, *49*, 942–946. [CrossRef]

56. Lambrechts, L.; Chevillon, C.; Albright, R.G.; Thaisomboonsuk, B.K.; Richardson, J.H.; Jarman, R.G.; Scott, T.W. Genetic specificity and potential for local adaptation between dengue viruses and mosquito vectors. *BMC Evol. Boil.* **2009**, *9*, 160. [CrossRef]

57. Warrilow, D.; Northill, J.A.; Pyke, A.; Smith, G.A. Single rapid TaqMan fluorogenic probe based PCR assay that detects all four dengue serotypes. *J. Med. Virol.* **2002**, *66*, 524–528. [CrossRef]

58. Ito, M.; Takasaki, T.; Yamada, K.-I.; Nerome, R.; Tajima, S.; Kurane, I. Development and Evaluation of Fluorogenic TaqMan Reverse Transcriptase PCR Assays for Detection of Dengue Virus Types 1 to 4. *J. Clin. Microbiol.* **2004**, *42*, 5935–5937. [CrossRef]

59. Hall, T.A. Bioedit: A user-friendly biological sequence alignment editor and analysis program for Windows 95/98/NT. *Nucl. Acids Symp. Ser.* **1999**, *41*, 95–98.

60. Thompson, J.D.; Higgins, D.G.; Gibson, T.J. CLUSTAL W: Improving the sensitivity of progressive multiple sequence alignment through sequence weighting, position-specific gap penalties and weight matrix choice. *Nucleic Acids Res.* **1994**, *22*, 4673–4680. [CrossRef]

61. Kumar, S.; Stecher, G.; Tamura, K. MEGA7: Molecular Evolutionary Genetics Analysis Version 7.0 for Bigger Datasets. *Mol. Biol. Evol.* **2016**, *33*, 1870–1874. [CrossRef]

62. Afreen, N.; Naqvi, I.H.; Broor, S.; Ahmed, A.E.; Kazim, S.N.; Dohare, R.; Kumar, M.; Parveen, S. Evolutionary Analysis of Dengue Serotype 2 Viruses Using Phylogenetic and Bayesian Methods from New Delhi, India. *PLoS Neglected Trop. Dis.* **2016**, *10*, e0004511. [CrossRef] [PubMed]

63. Lee, K.-S.; Lai, Y.-L.; Lo, S.; Barkham, T.; Aw, P.; Ooi, P.-L.; Tai, J.-C.; Hibberd, M.L.; Johansson, P.; Khoo, S.-P.; et al. Dengue Virus Surveillance for Early Warning, Singapore. *Emerg. Infect. Dis.* **2010**, *16*, 847–849. [CrossRef] [PubMed]

64. Ali, A.; Ali, I. The Complete Genome Phylogeny of Geographically Distinct Dengue Virus Serotype 2 Isolates (1944–2013) Supports Further Groupings within the Cosmopolitan Genotype. *PLoS ONE* **2015**, *10*, e0138900. [CrossRef] [PubMed]

MDPI
St. Alban-Anlage 66
4052 Basel
Switzerland
Tel. +41 61 683 77 34
Fax +41 61 302 89 18
www.mdpi.com

Pathogens Editorial Office
E-mail: pathogens@mdpi.com
www.mdpi.com/journal/pathogens